Expression of Eukaryotic
Viral and Cellular Genes

ORGANIZING COMMITTEE

Nils Oker-Blom
Chairman

David Baltimore   Leevi Kääriäinen   Ralf Pettersson
Lennart Philipson   Hans Söderlund

Anja Nykänen
Editorial Secretary

0372573 - 3

For

Expression of Eukaryotic Viral and Cellular Genes

Eighth Sigrid Jusélius Foundation Symposium: Helsinki, Finland.
June 1980

Edited by

RALF F. PETTERSSON

Department of Virology, University of Helsinki, Helsinki, Finland

LEEVI KÄÄRIÄINEN

Department of Virology, University of Helsinki, Helsinki, Finland

HANS SÖDERLUND

*Department of Biochemistry, University of Helsinki,
Helsinki, Finland*

NILS OKER-BLOM

Department of Virology, University of Helsinki, Helsinki, Finland

1981

ACADEMIC PRESS

A Subsidiary of Harcourt Brace Jovanovich, Publishers
London · New York · Toronto · Sydney · San Francisco

ACADEMIC PRESS INC. (LONDON) LTD
24/28 Oval Road
London NW1 7DX

United States Edition published by
ACADEMIC PRESS INC.
111 Fifth Avenue
New York, New York 10003

British Library Cataloguing in Publication Data

Expression of eukaryotic viral and cellular genes.
1. Gene expression – Congresses
2. Eukaryotic cells – Congresses
I. Pettersson, R. F.
574.87'322   QH450
ISBN 0-12-553120-6

Filmset in 'Monophoto' Times New Roman by Eta Services (Typesetters) Ltd., Beccles, Suffolk
Printed in Great Britain by Thomson Litho Ltd., East Kilbride, Scotland

LIST OF PARTICIPANTS

Full addresses of contributors appear on the first page of each chapter

AMINOFF, C. G., Helsinki, Finland
BALTIMORE, D., Cambridge, U.S.A.
VON BONSDORFF, B., Helsinki, Finland
VON BONSDORFF, C.-H., Helsinki, Finland
BOURNE, A., London, U.K.
BROWNLEE, G. G., Cambridge, U.K.
CANTELL, K., Helsinki, Finland
CARBON, J., Santa Barbara, U.S.A.
CHAMBON, P., Strasbourg, France
CHANG, A. C. Y., Stanford, U.S.A.
DE LA CHAPELLE, A., Helsinki, Finland
CLARKE, L., Santa Barbara, U.S.A.
COHEN, S. N., Stanford, U.S.A.
DARNELL, J. E., New York, U.S.A.
EK, J., Pori, Finland
ENARI, T. M., Espoo, Finland
FARRAND, R. A., London, U.K.
FIERS, W., Gent, Belgium
GAHMBERG, C. G., Helsinki, Finland
GAROFF, H., Heidelberg, West Germany
GILBERT, W., Cambridge, U.S.A.
GRAF, T., Heidelberg, West Germany
HALKKA, O., Helsinki, Finland
HALONEN, P., Turku, Finland
HOVI, T., Helsinki, Finland
HUANG, A. S., Boston, U.S.A.
JÄNNE, J., Helsinki, Finland
JÄNNE, O., Oulu, Finland
JÄRNEFELT, J., Helsinki, Finland
JÄRVI, O., Turku, Finland
KÄÄRIÄINEN, L., Helsinki, Finland
KAESBERG, P., Madison, U.S.A.
KAFATOS, F. C., Cambridge, U.S.A.
KALKKINEN, N., Helsinki, Finland
KERÄNEN, S., Helsinki, Finland
KNOWLES, J., Espoo, Finland
KOURILSKY, P., Paris, France

KRUG, R. M., New York, U.S.A.
KULONEN, E., Turku, Finland
LEHTOVAARA P., Helsinki, Finland
LEINIKKI, P., Tampere, Finland
MÄENPÄÄ, P., Kuopio, Finland
MÄKELÄ, O., Helsinki, Finland
O'MALLEY, B. W., Houston, U.S.A.
MÄNTYJÄRVI, R., Kuopio, Finland
MIKOLA, J., Jyväskylä, Finland
VAN MONTAGU, M., Gent, Belgium
NATHANS, D., Baltimore, U.S.A.
NEVALAINEN, H., Espoo, Finland
OKER-BLOM, N., Helsinki, Finland
PALVA, I., Helsinki, Finland
PENTTINEN, K., Helsinki, Finland
PERSSON, H., Uppsala, Sweden
PETTERSSON, R. F., Helsinki, Finland
PETTERSSON, U., Uppsala, Sweden
PFEIFER, S., Helsinki, Finland
PHILIPSON, L., Uppsala, Sweden
PYHTILÄ, R., Oulu, Finland
RAINA, A., Kuopio, Finland
RANKI, M., Helsinki, Finland
SALKINOJA-SALONEN, M., Helsinki, Finland
SALMI, A., Turku, Finland
SARVAS, M., Helsinki, Finland
SCHIMKE, R. T., Stanford, U.S.A.
SHATKIN, A. J., Nutley, U.S.A.
SÖDERLUND, H., Helsinki, Finland
SORSA, M., Helsinki, Finland
SORSA, V., Helsinki, Finland
STEHELIN, D., Lille, France
STEITZ, J. A., New Haven, U.S.A.
TUOHIMAA, P., Tampere, Finland
VAHERI, A., Helsinki, Finland
VARMUS, H. E., San Francisco, U.S.A.
WEISSMANN, C., Zürich, Switzerland
WIKGREN, B.-J., Åbo, Finland

PREFACE

Fifty years have now elapsed since Fritz Arthur Jusélius, a prosperous businessman in the city of Pori on the south-west coast of Finland, created the Sigrid Jusélius Foundation in memory of his beloved daughter Sigrid who died of tuberculosis after a measles infection in 1898 at the age of only eleven years. The aim of the Foundation is to support, irrespective of race or language, international medical research in the fight against diseases which are particularly harmful to mankind, with an emphasis on microbiology. It is understandable that microbiology was central to the ideas of F. A. Jusélius, partly because at the turn of the century infectious diseases played an important role in medicine and partly because his daughter had succumbed to one. Unfortunately, owing to the very rigorous stipulations concerning the growth of the Foundation's capital, the distribution of grants could not start until 1948. Since then, though, most of the basic research in medicine in Finland has been supported by the Sigrid Jusélius Foundation.

In his will F. A. Jusélius stressed the internationality of the Foundation. In accordance with this wish, the Foundation also provides grants for foreign scientists to work in hospitals and laboratories in Finland and, from 1965 on, it has supported many international symposia. Most of the symposia so far have been concerned with basic biological problems like "Control of Cellular Growth in Adult Organisms", "Regulatory Functions of Biological Membranes", "Cell Interactions and Receptor Antibodies in Immune Response", "Biology of Fibroblasts" and, most recently, "Cell Interactions in Differentiation". Two symposia have been devoted to more clinical problems; one on "Amyloidosis" and the latest concerning "Population Genetic Studies on Isolates".

The principle of all the symposia has been to choose a specific problem and to try to elucidate this problem from as many angles as possible. The same is also true of this symposium. Under the common heading "Expression of Eukaryotic Viral and Cellular Genes" this meeting combines recent advances in molecular virology, molecular genetics and recombinant DNA research. Viruses are excellent models for studying eukaryotic gene structure and expression, and much of our current knowledge in certain areas comes from experiments carried out with animal viruses, e.g. the presence and synthesis of poly A tracts at the 3' end of messenger RNAs, the structure and synthesis of the 5' end present in eukaryotic messengers, synthesis and processing of heterogenous nuclear RNA, ribosome-binding to messenger RNAs, the splicing phenomena, the complete determination of the nucleotide sequences of certain DNA viruses, etc.

Recombinant DNA research has again expanded tremendously during the last two years. Virology in the future will be intimately connected to recombinant DNA research, and here again it is enough just to mention a few fields like production of vaccines against, for instance, influenza and hepatitis B, diagnostic procedures, production of interferon, etc.

In some fields of science the progress is so fast that publishing monographs does not seem warranted. This also holds true for certain aspects of this symposium. In spite of this, it was felt that since our approach has been to focus on one problem complex from as many angles as possible, the compilation of these papers could give new impulse to the discussion. Thanks to the collaboration of the participants of the symposium allowing fast publication, their papers retain their original freshness and it is therefore hoped that this book will be of value for those working in this interesting and thrilling field.

Helsinki, June 1980   Nils Oker-Blom
                                                     The Rector,
                                                     Professor,
                                                     University of Helsinki

CONTENTS

I

Structure and Expression of DNA Virus Genes

Molecular Analysis of Adenovirus Transformation

M. PERRICAUDET,[1] G. WESTIN,[2] J. M. LE MOULLEC,[1]
L. VISSER,[3] J. ZABIELSKI,[2] P. ALESTRÖM,[2]
G. AKUSJÄRVI,[2] A. VIRTANEN[2] and U. PETTERSSON[2]

[1] Institut Pasteur,
Unite de Genie Genetique,
28, rue du Dr Roux,
75724 Paris Cedex 15,
France

[2] Department of Microbiology,
The Biomedical Center,
Box 581,
S-751 23 Uppsala, Sweden

[3] State University of Utrecht,
Laboratory for Physiological Chemistry,
Vondellaan 24a,
3521 GG Utrecht,
The Netherlands

INTRODUCTION

The adenovirus system is often used as a model to study eukaryotic gene expression. The adenovirus group has in addition, attracted a great deal of attention due to the oncogenic potential of its members. In 1962, Trentin and his colleagues discovered that human adenovirus type 12 (ad12) causes tumors in newborn hamsters. Since then a number of adenovirus serotypes have been shown to be oncogenic for rodents and the human adenoviruses are subdivided into highly, weakly, and non-oncogenic serotypes. Almost all human adenoviruses, including the non-oncogenic serotypes, are, however, able to transform rat or hamster cells in tissue culture. Like many other DNA tumor viruses, the adenoviruses preferentially transform cells which are non-permissive for viral replication. The ultimate consequence of viral transformation is the integration of viral DNA sequences in the genome of the transformed cell. The sequences present in cells transformed by adenovirus type 2 (ad2) have been studied in great detail, using nucleic acid hybridization methods. Sharp *et al.* (1974) demonstrated for the first time that rat cells transformed by ad2 do not contain a complete copy of the viral genome. Subsequently it has been shown that the presence of subgenomic fragments of viral DNA in cells transformed by adenovirus types 2 and 5 is a rule rather than an exception (Gallimore *et al.*, 1974). Graham and coworkers (1974) have identified the transforming genes of adenovirus type 5; by the use of specific restriction enzyme fragments they were able to show that genes mandatory for transformation are located at the left-hand end of the adenovirus genome. Subsequently van der Eb *et al.* (1977) showed that the smallest fragment required to achieve transformation comprises the leftmost 4.5% of the adenovirus type 5 genome. However, cells transformed by this small fragment have a different phenotype than cells transformed by virus or large fragments. The minimum fragment which has been used to obtain complete transformation includes the leftmost 15% of the viral genome (van der Eb *et al.*, 1979). Promotor mapping studies have identified two major early promotors in the leftmost early region of ad2, also known as region E1 (Fig. 1). In this way region E1 is subdivided into the two transcription units, E1A and E1B (Wilson *et al.*, 1979). In addition, the mRNA for the quasi-late polypeptide IX is transcribed from a separate promotor in region E1 (Aleström *et al.*, 1980).

Studies on the organization of the adenovirus genome have progressed very rapidly. Many genes have been mapped by hybridization coupled with *in vitro* translation (Halbert *et al.*, 1979; Lewis *et al.*, 1979) and large regions of the genome have been analyzed by DNA sequencing methods (for review see Tooze, 1980). The sequenced regions include the left-hand end of the ad5 genome, carrying all genes required for transformation.

In this chapter we describe a molecular analysis of mRNAs from the transforming region of the adenovirus genome and a preliminary study of integrated viral sequences in two adenovirus-transformed cell lines.

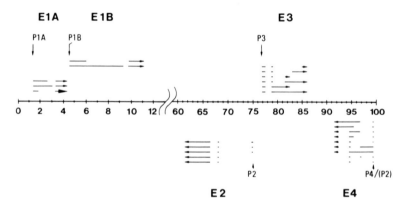

FIG 1. Schematic drawing, indicating regions which are transcribed early after adenovirus infection. Region E1 contains two promoters (P1A, P1B) which subdivide it into transcription units E1A and E1B.

THE mRNAs FROM REGION E1A

Region E1A comprises the leftmost 4.5% of the ad2 genome. Electron microscopy and S1 nuclease mapping have identified three distinct mRNAs in this region, a 13S, a 12S and a 9S mRNA (Berk and Sharp, 1978; Spector et al., 1978; Chow et al., 1979). The 9S mRNA appears to be synthesized exclusively late after infection and is presumably not involved in transformation. All the E1A mRNAs are believed to be transcribed from a single promotor, located about 475 nucleotides from the left-hand end of the ad5 genome (Baker and Ziff, 1980). They are spliced and carry identical 5′ and 3′ ends, but have intervening sequences of different sizes. In order to examine the mRNAs from the transforming region of subgroup C human adenoviruses we have used molecular cloning techniques. Double-stranded cDNA copies were synthesized with early mRNA as a template and inserted into the Pst I cleavage site of the pBR322 plasmid by dG/dC tailing. Recombinant plasmids corresponding to the 12S and the 13S mRNAs from region E1A have been identified and part of their sequences have been determined. By comparing the established sequences with the sequence of region E1A of the closely related ad5 DNA (van Ormondt et al., 1978) it was possible to deduce the structure of two spliced mRNAs (Fig. 2). The splice in the 12S mRNA deletes sequences between nucleotides 975 and 1228 and the small splice in the 13S mRNA deletes sequences between nucleotides 1113 and 1228. The two mRNAs thus use a common acceptor site for splicing but different donor sites. In both mRNAs the splice results in a frame shift which allows the DNA sequence to be utilized in the most efficient way and a very AT-rich region containing numerous termination codons is excluded from both mRNAs. From the structure of the two mRNAs it is possible to predict the amino acid sequence of the corresponding protein products. The polypeptide specified by the 13S mRNA will be 288 amino acids long (32K) whereas the polypeptide corresponding to the 12S mRNA will be 242 amino acids long (26K), assuming that the first AUG following the cap is used for initiation of translation. Both polypeptides have unusual amino acid compositions. They are particularly rich in proline and glutamic acid, and a region with alternating glutamic acid and proline residues is present in both polypeptides. Due to the structure of their mRNAs, the 26K and 32K polypeptides will be completely overlapping, their only difference being a deletion of 46 amino acids from the shorter polypeptide.

The polypeptides encoded by region E1A have also been analyzed by several investigators using mRNA selection coupled with in vitro translation (Halbert et al., 1979; van der Eb et al., 1979). In this way it has been possible to show that four polypeptides in the molecular weight range

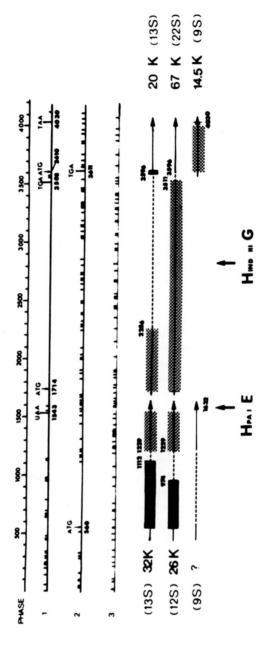

FIG. 2. Schematic drawing indicating the mRNA sequences which are transcribed from the transforming region of the adenovirus type 5 genome. The positions of splices as well as reading frames are indicated.

42–53 000 are encoded by this region. The discrepancy between the molecular weight estimated by SDS polyacrylamide gel electrophoresis and that predicted from the DNA sequence may be due to the unusual amino acid composition of the polypeptides. The difference between the number of mRNAs and the number of polypeptide chains encoded by this region is puzzling. Studies using tryptic fingerprint analysis of the proteins have shown that the four polypeptides in the 42–53K size range all are related and that the 12S and the 13S mRNAs from region E1A each gives rise to two polypeptide species, presumably through post-translation modification. However, we cannot yet exclude the possibility of having additional mRNAs in this region although neither electron microscopy nor S1 nuclease mapping give support for this.

THE mRNAs FROM REGION E1B

The mRNAs of region E1B were analyzed in a similar way. Bacterial clones containing sequences from region E1B were identified by hybridization, using the Hind III-C fragment (map position 8.0–17.0) as a probe. Two different clones were identified, one containing, and one lacking a cleavage site for endonuclease Bgl II (located at map position 9.0 in the ad2 genome). Results from S1 nuclease mapping studies and electron micros-copy, although not completely consistent, predict the presence of two differently spliced mRNAs from region E1B (Chow et al., 1979; Berk and Sharp, 1978). These mRNAs have common 5' and 3' ends and differ by the size of their intervening sequences. They also have a common promoter which has been mapped around nucleotide 1675 by sequencing the 5' ends of the E1B mRNAs (Baker and Ziff, 1979). By sequence analysis of selected regions of the clones it was possible to correlate our results with the established DNA sequence for region E1B (Maat and van Ormondt, 1979; Maat et al., 1979) and in this way predict the structure of a 13S and a 22S mRNA from region E1B (Fig. 2). The 22S mRNA is spliced between nucleotides 3512 and 3595. It is interesting to note that the coding potential of the 22S mRNA will not be altered by the splice since the termination codon occurs before the splice and the polypeptide will thus be translated in an uninterrupted reading frame. Thus, this mRNA resembles the mRNA for the small t-antigen of SV40 which also is read in a contiguous sequence from a spliced mRNA (Reddy et al., 1979; Fiers et al., 1978). The 13S mRNA is spliced in a different fashion; the acceptor site is shared with the 22S mRNA but the donor site is located at position 2257 in the DNA sequence. From the sequence we can predict that the 13S mRNA will be translated before as well as after the splice although in different reading frames. Only five codons will, however, be read after the splice and it is interesting to note that four of these will specify proline residues. The protein products of the 13S and the 22S mRNAs are predicted to have molecular weights of 20K and 67K respectively if we assume that translation is initiated after the first AUG in the sequence. Polypeptides with similar molecular weights have been identified by hybridization selection of mRNAs from region E1B followed by *in vitro* translation (Halbert *et al.*, 1979; van der Eb *et al.*, 1979).

It should be emphasized that, although in the case of the 20K polypeptide it has been shown by amino acid analysis that it is initiated at the first AUG after the cap, this may not be the case for the larger polypeptide.

THE GENE FOR POLYPEPTIDE IX

The gene for polypeptide IX, a structural protein of the virion, has also been mapped in region E1B. This polypeptide is produced in small amounts before the onset of viral DNA replication but accumulates late after infection and thus it resembles the quasi-late proteins of bacteriophage T4 in its kinetics of appearance. We have sequenced the complete gene for polypeptide IX as well as a cDNA copy of the entire mRNA including the poly(A) addition site (Aleström *et al.*, 1980). The mRNA for polypeptide IX overlaps the spliced mRNAs in region E1B. A promotor-like sequence as well as the cap site is identified in the intervening sequence which is common to both the 13S and the 22S mRNAs from region E1B (Fig. 2). The coding part of the 9S mRNA overlaps completely with the 3' non-coding part of the spliced E1B mRNAs and it is interesting that the 20K polypeptide which is encoded by the 13S mRNA is terminated by the sequence (A)UGA. The AUG part of the sequence is used for initiation of polypeptide IX synthesis and consequencely polypeptide IX is translated from a different reading frame than the C-terminal part of the 20K polypeptide. By comparing the DNA sequences of the gene with the cDNA sequence of the polypeptide IX mRNA it was surprising to find that the two sequences are colinear. The polypeptide IX mRNA, unlike most other eukaryotic mRNAs, thus matures without splicing.

POSSIBLE ROLE OF DIFFERENT PROTEINS IN ADENOVIRUS TRANSFORMATION

Van der Eb and his colleagues (1977) have shown that fragment Hpa I-E which comprises as little as 4.5% of the ad5 genome is sufficient to immortalize cells (Fig. 3). The immortalized cells have an infinite life-span but have few of the other phenotypic properties which are characteristic for cell transformation, The T-antigen staining pattern is also very weak and atypical. In contrast, cells transformed by the Hind III-G fragment which comprises the leftmost 8 % of the viral genome have similar phenotypic properties as cells transformed by the complete virus. This may suggest that the 20K polypeptide, encoded by region E1B, plays a critical role in transformation since the major difference in sequence content between fragments Hpa I-E (0–4.5 map units) and Hind III-G (0–8 map units) is that most sequences encoding the 20K polypeptide are present in the larger fragment (Fig. 2) whereas almost half of the sequences encoding the 67K polypeptide will be missing. The acceptor site for splicing of the 13S mRNA from region E1B is absent as well as the codons for five amino acids, but presumably a suitable acceptor site is acquired during integration. The role of the E1A region in transformation is less obvious. Apparently cells become immortalized by one or more of its gene products since fragment Hpa I-E is sufficient to render cells an infinite life-span. It has recently been shown by Jones and Shenk (1979) and by Berk *et al.* (1979) that mutations in the E1A region result in a defective expression of mRNA sequences from the other early regions. Thus, it appears that a product of region E1A has a pleiotropic effect, either controlling transcription or the subsequent processing of early transcripts. The major role of region E1A in transformation may consequently be to provide the product required for expression of the E1B region. It remains however, to explain how region E1A alone can immortalize cells. Conceivably the factor in question activates a cellular gene which alters the growth potential of the cells.

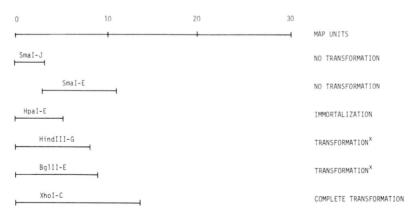

FIG. 3. Transforming potential of different fragments of adenovirus DNA (data from van der Eb *et al.*, 1979). [x] Atypical T-antigen staining pattern.

THE ORGANIZATION OF THE TRANSFORMING REGION OF ADENOVIRUS TYPE 12—A HIGHLY ONCOGENIC ADENOVIRUS SEROTYPE

Members of the subgroup A adenoviruses differ from other adenovirus serotypes by their oncogenic potential. In an attempt to study the molecular nature of this difference we have examined cloned mRNA sequences from adenovirus type 12. Double-stranded cDNA copies of early ad12 mRNA were inserted into the Pst I cleavage site of the pBR322 plasmid. The recombinant plasmids were propagated in *E. coli* and 10 000 bacterial clones have been screened for sequences from the left-hand end of the ad12 genome by hybridization using the Eco RI-C fragment of ad12 as a probe (map position 0–16.5). Thirty clones were found to give a positive hybridization signal. Plasmid DNA was extracted from all positive clones and analyzed by restriction enzyme cleavage, using endonucleases Pst I, Hinf I, and Ava II. The results showed that the clones could be divided into two groups, each representing one spliced mRNA from region E1A. No clones corresponding to mRNAs from region E1B have been detected among the 10 000 clones so far examined. In order to study the structure of the corresponding mRNAs, plasmid DNA from two clones each representing a different spliced mRNA was subjected to partial sequence analysis. By comparing our sequence information with the established sequence for the left-hand end of the ad12 genome (Fujinaga *et al.*, 1979) it was possible to deduce the structure of two mRNAs and their corresponding polypeptides. The results show that the clones represent two mRNAs which have identical 5' and 3' ends and differ by the size of their intervening sequences. The longer mRNA lacks sequences between nucleotide 1070 and 1143 and the shorter mRNA lacks sequences between nucleotides 977 and 1143. As is the case for the mRNAs from the E1A region of ad5 and ad2, both splicing events will result in a frame shift and the polypeptides which can be translated from these two mRNAs will be 266 and 235 amino acids long. Their amino acid compositions resemble those of the corresponding proteins from ad5. It is interesting to note that, although the sequences of the transforming regions of the highly and non-oncogenic serotypes have diverged considerably at the nucleotide level, the proteins have very similar primary structures.

When examining the splice points of the two ad12 mRNAs a number of interesting observations can be made. As is the case for the ad5 mRNAs, the two ad12 mRNAs have a common acceptor but different donor sites for splicing. The sequences around the acceptor sites in ad5 and ad12 are identical for nine nucleotides and have clearly been conserved during adenovirus evolution (Fig. 4). This is also the case for the donor sites which are used to splice the long mRNAs since at these sites eleven nucleotides are

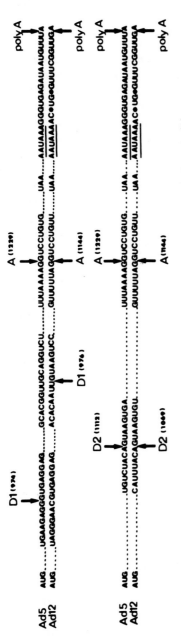

FIG. 4. A comparison between donor and acceptor sites in region E1A of ad5 and ad12.

Fig 4.

identical in ad5 and ad12. In contrast, the donor sites giving rise to the short mRNAs differ in structure as well as in location. The sequence at nucleotide 974 in ad5 which corresponds to a donor site is largely conserved in ad12 since eight nucleotides are shared between the two serotypes (Fig. 4). However, at the exact position of the splice the ad5 sequence reads GGGU whereas in ad12 the sequence reads ACGU. The nucleotide preceding the GU (a dinucleotide which is present at the 5′ side of most introns (Breathnach *et al.*, 1978)) has obviously been changed from a purine to a pyrimidine which apparently makes this donor site non-functional. Instead the donor site has been moved to nucleotide 1069 in the ad12 sequence where the sequence reads UGUAAGUC (Fig. 4). Only a few nucleotides at this position are shared between the two serotypes. It is interesting that this splice point still has an exceptional structure since a pyrimidine precedes the GU dinucleotide at the 5′ side of the intron. When comparing the ad5 and ad12 sequences with the established DNA sequence for ad7 (Dijkema *et al.*, 1980), a weakly oncogenic serotype, it is apparent that the donor site at nucleotide 1069 in the ad12 sequence is conserved in ad7 and we would like to propose that it also is functional in this serotype. This may suggest that in adenovirus evolution the highly and weakly oncogenic serotypes have evolved together from the common ancestor.

Clearly the structure of the E1A mRNAs does not explain the difference between oncogenic and non-oncogenic adenoviruses. We are currently extending our analysis to the mRNAs of region E1B. Available sequence information suggests that no large polypeptides can be encoded by this region in ad12 (Fujinaga *et al.*, 1979). If the established nucleotide sequence is confirmed there must be a fundamental difference in the protein products of region E1B from ad5 and ad12 or in the pattern of splicing giving rise to the mRNAs. Further analysis of the protein products from the E1B region are clearly required to resolve this issue.

INTEGRATED SEQUENCES IN CELLS TRANSFORMED BY SUBGROUP C ADENOVIRUSES

Hybridization studies have so far only permitted relatively approximate estimates of the amount of viral sequences present in transformed cells. Thanks to the rapid development of molecular cloning techniques it is now possible to isolate and directly study single-copy genes from mammalian cells. In the present study we have isolated and characterized integrated viral sequences from the cell line C31 transformed by ad5 (Visser *et al.*, 1979) and the cell line F19 transformed by ad2 (Gallimore *et al.*, 1974). DNA samples from the two cell lines were cleaved with the Eco RI restriction endonuclease and fractionated by sucrose gradient centrifugation. Fragments in the size-range from 8 to 20 kilobases (kb) were pooled and cloned in the lambda vector Charon 4A.

Recombinant phages carrying adenovirus sequences were identified by the plaque hybridization method of Benton and Davis (1977). A number of recombinants have been isolated from both cell lines, and DNA from three clones—F19/1A, F19/2B and C31/11—has been analyzed by hybridization and electron microscopy.

Analysis of DNA from Recombinant Phages by Hybridization

DNA from the two recombinant phages F19/1A and F19/2B was labeled *in vitro* by nick translation and hybridized to Sma I and Xba I fragments of ad2 DNA which had been transferred to nitrocellulose by the method of Southern (1975). The two DNA preparations give identical results, showing hybridization exclusively to fragments Sma I-E and Xba I-E. Fragment Sma I-E is located between map positions 2.85 and 10.7 and fragment Xba I-E between map positions 0 and 3.85. Consequently the viral DNA sequences in the two recombinants must originate between map positions 2.85–3.85 and the recombinants thus contain a surprisingly small segment of viral DNA. DNA from recombinant C31/11 was labeled in a similar way and hybridized to blotted Sma I and Xho I fragments of ad2 DNA. The result showed hybridization to fragments Sma I-J (0–2.85), Sma I-E (2.85–10.7), Sma I-F (11.3–18.6) and Xho I-C (0–15.5). Thus we conclude that this recombinant most likely contains a contiguous sequence of viral DNA, starting close to the left-hand end of the DNA and ending somewhere between position 11.3 and 15.5.

Analysis of DNA from the Recombinant Phages by Electron Microscopy

In order to further investigate the structure of the integrated viral DNA an electron microscopic study was carried out. Heteroduplex molecules were first formed between DNA from each recombinant phage and DNA from the Charon 4A vector. Typical heteroduplex molecules were observed with a characteristic substitution loop in the center of the molecule. The substitution loop in each case had one arm with a length of 14 kb which is the expected length of fragments C plus D of the Charon 4A DNA. The other arm had a length of 16.56 ± 1.38 kb when the heteroduplex was formed between DNA from Charon 4A and clone F19/1A and a length of 17.17 ± 0.99 when the heteroduplex was formed between DNA from Charon 4A and clone F19/28. In the longer arm of the substitution loop a characteristic structure consisting of a short stem and a loop was frequently observed.

14 M. PERRICAUDET *et al.*

This is presumably due to the presence of an inverted repeat structure in the inserted DNA. In order to compare the inserts in recombinants F19/1A and F19/2B heteroduplex molecules were also formed between DNA from the two recombinants. An analysis of the resulting heteroduplex molecules in the electron microscope revealed perfectly matched duplex molecules as well as molecules containing small loops located 17–21 kb from one end of the molecule. Similar loops were also observed when DNA from either recombinant phage was denatured and annealed with itself.

In order to determine the location of the viral sequences in recombinants 1A and 2B heterotriplex molecules were constructed. For this purpose DNA from either of the two recombinants was mixed with Charon 4A DNA as well as fragment Sma I-E (map positions 2.85 to 10.7). The mixture was denatured and allowed to renature and the resulting molecules were examined in the electron microscope. Molecules in which single strands of fragments Sma I-E had hybridized to one arm of the substitution loop in the heteroduplex molecule were occasionally observed. The area of homology between fragments Sma I-E and the recombinant is located within the longer arm of the substitution loop as expected. It is clearly separate from the "loop and stem structure" which was frequently observed in this arm of heterduplexes between recombinant DNA molecules and Charon 4A DNA. The length of the homology region was estimated to be 100 nucleotides and it is located approximately 140 nucleotides from one end of the Sma I-E fragment. Many heterotriplex molecules contained two hybridized Sma I fragments and in one case three Sma I fragments were hybridized to the long arm of the substitution loop of the heterotriplex molecule. The distance between the homology regions was 300–400 nucleotides (Fig. 5).

Examination of DNA from the C31/11 recombinant gave a different result. Heteroduplex molecules formed between DNA from the recombinant and adenovirus type 2 DNA resulted in heteroduplex molecules with an area of homology which was 4.4 kb long. When hybridizing the adenovirus, single strands will form circular structures due to the presence of inverted repeats. In many cases this structure was visible and it was consequently possible to determine the distance between the end of the adenovirus single strand and the start of the homology region. From the measurements it

FIG. 5. A schematic drawing of a heteroduplex molecule between fragment Sma I-E and cloned DNA from the ad2 F19 transformed cell line. A stem and loop structure is indicated as well as the hybridization between fragment Sma I-E and three small inserts of viral DNA. The junctions with the vector arms (〜) are also indicated.

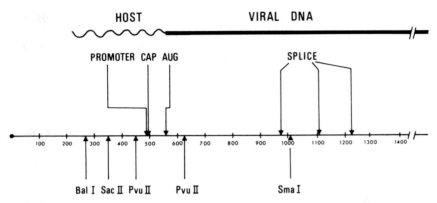

FIG. 6. A schematic drawing illustrating the junction between host and integrated viral sequences in one clone from the ad5 C31 cell line.

appears that the integrated viral DNA starts 578 ± 69 nucleotides from one and ends 4400 further away. By combining the data obtained by electron microscopy and hybridization it is possible to conclude that the inserted fragment starts 1.5% from the left-hand terminus of the adenovirus genome and ends around map position 14.2. The cap site for the region E1A mRNAs has previously been mapped at nucleotide 499 (Baker and Ziff, 1979). The results therefore suggest that the promotor region as well as the cap site for the region E1A mRNAs are missing from the integrated segment (Fig. 6). The absence of the initiator AUG as well as the cap site has been confirmed by sequence analysis.

Clearly none of the clones so far analyzed appear to contain sequences which could give rise to the transformed phenotype when expressed. Presumably additional insertions have occurred in these cells, which include the entire transforming region and the cap site. We are currently analyzing additional clones from these and other adenovirus-transformed cell lines. From our results it is apparent that the pattern of integration in adenovirus-transformed cell lines is more complicated than had been expected. Apparently sequences which have little or no relevance for transformation are integrated and the structure of the integrated sequences from the F19 cell line suggests that substantial rearrangements of the integrated sequences occur during or after integration.

SUMMARY

Double-stranded cDNA copies of early mRNA from adenovirus-infected cells have been cloned, using the *E. coli* plasmid pBR322 as a vector.

Recombinant plasmids containing sequences from the transforming region of the adenovirus genome have been identified and used for a structural analysis. The structure of four different spliced mRNAs from the transforming region can be deduced from the results. We also report a preliminary analysis of integrated viral sequences from two adenovirus-transformed cell lines.

REFERENCES

Aleström, P., Akusjärvi, G., Perricaudet, M., Mathews, M. B., Klessig, D. and Pettersson, U. (1980). *Cell* **19**, 671–681.
Baker, C. and Ziff, E. (1979). *Cold Spring Harb. Symp. quant. Biol.* **44**, 471–476.
Benton, W. D. and Davis, R. J. (1977). *Science, N.Y.* **196**, 180–182
Berk, A. J. and Sharp, P. A. (1978). *Cell* **14**, 695–711.
Berk, A. J., Lee, F., Harrison, T., Williams, J. and Sharp, P. A. (1979). *Cell* **17**, 935–944.
Breathnach, R., Benoist, C., O'Hare, K., Gannon, G. and Chambon, P. (1978). *Proc. natn. Acad. Sci. U.S.A.* **75**, 4853–4857.
Chow, L. T., Broker, T. and Lewis, J. B. (1979). *J. molec. Biol.* **134**, 265–303.
Dijkema, R., Dekker, B. M. M. and van Ormondt, H. (1980). *Gene* **9**, 141–156.
Fiers, W., Contreras, R., Haegeman, G., Rogiers, R., van der Voorde, A., van Hewverswyn, H., van Herreweghe, J., Volchaert, G. and Ysebaert, M. (1978). *Nature, Lond.* **27**, 113–120.
Fujinaga, K., Sawada, T., Vemizu, Y., Yamashita, T., Shinojo, H., Shiroki, K., Sugisaki, H., Sugimoto, K. and Takanagin, M. (1979). *Cold Spring Harb. Symp. quant. Biol.* **44**, 519–532.
Gallimore, P. H., Sharp, P. A. and Sambrook, J. (1974). *J. molec. Biol.* **89**, 49–72.
Graham, F. L., Abrahams, P. J., Mulder, C., Heinjneker, H. L., Warnaar, S. O., De vries, F. A. J., Fiers, W. and van der Eb, A. J. (1974). *Cold Spring Harb. Symp. quant. Biol.* **39**, 637–650.
Halbert, D. W., Spector, D. J. and Raskas, H. J. (1979). *J. Virol.* **31**, 621–629.
Jones, N. and Shenk, T. (1979). *Proc. natn. Acad. Sci. U.S.A.* **76**, 3665–3669.
Lewis, J. B., Esche, H., Smart, J. E., Stillman, B., Hartee, M. L. and Mathews, M. B. (1979). *Cold Spring Harb. Symp. quant. Biol.* **44**, 493–508.
Maat, J. and van Ormondt, H. (1979). *Gene* **6**, 75–90.
Maat, J., van Beveren, C. P. and van Ormondt, H. (1979). *Gene* **6**, 27–38.
Reddy, V. B., Ghosh, P. K., Lebowitz, P., Piatak, M. and Weissman, S. M. (1979). *J. Virol.* **30**, 279–296.
Sharp, P. A., Pettersson, U. and Sambrook, J. (1974). *J. molec. Biol.* **86**, 709–726.
Southern, E. (1975). *J. molec. Biol.* **98**, 503–518.
Spector, D. J., McGrogan, M., and Raskas, M. J. (1978). *J. molec. biol.* **196**, 395–414.
Tooze, J. (1980). *In* "Molecular Biology of Tumor Viruses", part 2, Appendices D-1. Cold Spring Harbor Laboratory, New York.
Trentin, J. J., Yabe, Y. and Taylor, G. (1962). *Science, N.Y.* **137**, 835–841.
van der Eb, A. J., Mulder, C., Graham, F. L. and Houweling, A. (1977). *Gene* **2**, 115–132.

van der Eb, A. J., van Ormondt, H., Schrier, P. I., Lupker, J. H., Jochemsen, H., van den Elsen, P. J., De Leys, R. J., Maat, J., van Beveren, C. P., Dijkema, R. and De Waard, A. (1979). *Cold Spring Harb. Symp. quant. Biol.* **44**, 383–399.

van Ormondt, H., Maat, J., De Waard, A. and van der Eb, A. J. (1978). *Gene* **4**, 309–328.

Visser, L., van Maarschalkerweerd, M. W., Rozijn, T. H., Wassenaar, A. D. C., Reemst, A. M. C. B. and Sussenbach, J. S. (1979). *Cold Spring Harb. Symp. quant. Biol.* **44**, 541–550.

Wilson, M. C., Fraser, N. W. and Darnell, J. E. (1979). *Virology* **94**, 175–184.

Adenovirus mRNA Production is Controlled both by Differential Transcription and Differential Stability of mRNA

J. E. DARNELL, Jr., M. WILSON, J. R. NEVINS and E. ZIFF

The Rockefeller University, New York, New York 10021, U.S.A.

INTRODUCTION

Almost twenty years ago large nuclear RNA molecules were discovered in cultured mammalian cells. Some of these large molecules were shown to be precursors to rRNA. Because some of the nuclear RNA had characteristics of cytoplasmic mRNA it also appeared possible that they might be precursors to mRNA. These early findings made the definition of transcriptional units—regions of DNA including the stop and start sites for RNA synthesis—a necessary first step in determining the level or levels at which gene control lies in eukaryotic cells (Darnell et al., 1977; Evans et al., 1977; Goldberg et al., 1977; Weber et al., 1977; Darnell, 1979). It was necessary to prove or reject the large nuclear RNA transcripts as mRNA precursors. If both short and long transcriptional units had existed simultaneously in the DNA encoding a messenger RNA, then the long transcription units might not have produced true messenger-RNA precursors. The first large nuclear transcripts that were proved to be mRNA precursors were those formed late in adenovirus infection from the right-hand two-thirds of the virus genome. No short primary transcripts arose from this region (Darnell et al., 1977; Evans et al., 1977; Goldberg et al., 1977; Goldberg and Schwartz, 1977; Weber et al., 1977) and therefore mRNAs must be derived from the large primary transcript that covered more than 25 000 bases of the DNA strand copied in the rightward direction. The characterization of additional early adenovirus transcription units and cell transcription units followed. The definition of adenovirus transcription units involved both "nascent chain" analysis of pulse-labeled RNA (Weber et al., 1977) and UV transcription analysis (Goldberg et al., 1977; reviewed in Darnell, 1979). In addition to the adenovirus transcription units, cell mRNA had a target size for UV inactivation that was approximately three to five times larger than the mRNA (Goldberg *et al.*, 1977).

Together with the data on transcription unit size and the discoveries of capping (Shatkin, 1976), polyadenylation (reviewed in Bavermann, 1976; Edmonds and Winters, 1976), methylation (Shatkin, 1976) and finally splicing (Berget *et al.*, 1977; Chow *et al.*, 1977; Klessig, 1977), a common pathway for formation of most mRNA molecules could be suggested (Evans *et al.*, 1977). The likelihood of extensive processing during mRNA formation has renewed the old conjectures (Darnell, 1968; Scherrer and Marcaus, 1968; Georgier, 1969) that gene regulation might include differential processing, but now the hypotheses can be stated in more specific guise.

In this chapter we will describe direct measurements on control of mRNA formation that take advantage of the transcriptional map of adenovirus type 2 (ad2). The experiments demonstrate various levels at which mRNA concentrations from a variety of ad2 transcriptional units are controlled. Both positive and negative control at the initiation of transcription appear to occur during the regulation of several different ad2 transcriptional units. At the other end of the spectrum of possible control points for mRNA concentrations is stabilization of mRNA in the cytoplasm. For at least one early transcriptional unit, and perhaps for others, an increase in one or two different spliced mRNAs appears to depend on lengthening the half-life of one of them. These early results illustrate that a variety of methods are used for gene control in the levels of ad2 mRNA concentration. Thus, as was the case in studying steps in mRNA biosynthesis, HeLa cells infected with ad2 remain an attractive system to continue studies aimed at discovering the proteins and cell structures that presumably participate in controlling mRNA concentrations.

TRANSCRIPTION UNIT MAP

Philipson and colleagues and Sharp, Flint and their colleagues (reviewed in Flint, 1977) used the mRNA from cells both early and late after infection to determine which regions of the adenovirus DNA, employed as labeled DNA, would hybridize to mRNA. Separated strands of various regions were prepared to allow determination of the direction of transcription as well. From these studies a "messenger RNA map" was produced. Pulse-labeled nuclear RNA obtained from isolated nuclei (Weber *et al.*, 1977) or extracted from pulse-labeled cells (Evans *et al.*, 1977; Goldberg *et al.*, 1977) was analyzed to establish which of the various ad2 mRNAs derive from independent transcriptional units. The apparent origin of transcription could also be located by determining which restriction fragments of the ad2 genome hybridized short RNA molecules after a pulse. The analysis of such pulse-labeled RNA coupled with the analysis of RNA synthesis after UV irradiation (Goldberg et al., 1977) (which damages promoter distal transcription but leaves promoter proximal RNA synthesis unimpaired) revealed the apparent start sites for several separate early promoters, two intermediate promoters and the major late promoters. Gelinas and Roberts (Gelinas and Roberts, 1977) then identified a similar if not identical capped oligonucleotide on all the late mRNAs. The capped 5' end of the mRNA was then shown by electron microscopy (Berget et al., 1977; Chow et al., 1977) to be homologous to the approximate region where the RNA initiation site had been mapped (Darnell et al., 1977; Goldberg et al., 1977; Weber et al., 1977). The existence of adenovirus mRNA with sequence homology to non-contiguous regions of the adenovirus genome, coupled with the evidence that the transcription unit contained all the sequences present in the mRNA was the basis for concluding that RNA·RNA splicing occurs (Darnell, 1979).

The time seemed ripe therefore to try to establish definitely the sequence of the start site for RNA synthesis late in ad2 infection. The DNA sequence of over 1000 bases was determined around 16, the site at which synthesis had been judged to begin by the earlier UV and pulse-label studies. In addition, the exact sequence of the 5'-eleven bases found on all mRNAs was determined (Ziff and Evans, 1978). Only one region of DNA (16.45) between ∼15 and 18 on the genome was found capable of encoding the 5' end of the late mRNAs. Analysis of nuclear RNA from cells or isolated nuclei showed no transcribed oligonucleotides "upstream" from the cap site, although all the oligonucleotides predicted from this sequence "downstream" could be found (Baker and Ziff, 1979). Thus it appeared that the cap site was the start site for late ad2 RNA synthesis.

This general approach of determining the DNA sequences around a presumed promoter site and determining the sequence of the capped oligonucleotide of mRNAs derived from that promoter had identified the apparent start site for a series of early transcriptional units in addition to the large late transcriptional unit. A common hexa- or heptanucleotide sequence, TATAAA, occurs ∼25 to 30 nucleotides "upstream" from all of the proposed ad2 cap sites. This same sequence appears in the same place before a number of cap sites in cellular genes as well. Thus it appears that one portion of a common signal for RNA polymerase II exists ∼35 to 30 nucleotides before the RNA initiation site (Baker and Ziff, 1979).

Cap addition is then very quick after RNA chain initiation. Chains less than 100 nucleotides in length from the late ad2 promoter have already received a cap (Babich, Nevins and Darnell, unpublished). Finally, it should be noted that Roeder and his colleagues (Weil et al., 1979) have described an *in vitro* transcription initiation system that utilized ad2 DNA, polymerase II, and a crude cell extract from infected and uninfected cells. In this system initiation occurs at the same late promoter site as that which functions inside the cell and the transcripts are capped. Knowing the apparent start site for RNA synthesis has enabled us to determine whether transcriptional control in the nucleus can be strictly demonstrated during ad2 infection.

CONTROL OF PROTEIN IN TRANSCRIPTIONAL UNIT

Two early promoters were established at positions 1.4 and 4.5 and an intermediate promoter at 9.1 by identification of short UV-resistant promoter proximal RNA segments (Wilson *et al.*, 1978). These presumed starting regions were confirmed by sequence analysis of caps in mRNA and corresponding sites in DNA (Baker and Ziff, 1979). The promoters at 1.4 (termed 1A) and 4.5 (termed 1B) produce mRNAs both early and late in infection and can be scored as active sites of transcription by hybridization of UV-resistant short (<500 bases) nuclear RNA to fragments from 0–3, 3–4.4, 4.4–8, and 8–11. The third promoter at ~ 9 produces the mRNA for the capsid protein IX beginning about 6 hours after infection (Spector *et al.*, 1978; infection produced by high multiplicities of ~ 3000 particles per cell). Short UV-resistant promoter proximal RNA for the protein IX was not detected at earlier times. This is the most sensitive assay presently available for control at the site of initiation (Wilson *et al.*, 1978). Only a measurement of the synthesis rate of the exact capped oligonucleotide complementary to the 9–11 region would improve this conclusion. It therefore appears that initiating events for RNA synthesis are limited at ~ 9.1 early in infection even though a neighboring promoter at 4.5 is active. While it is not known in our experiments where high multiplicity has been used whether both 1B and protein IX promoters are used in the same DNA molecule this must be the case when single particles infect cells, since protein IX is formed without any DNA replication (Flint, 1977).

CONTROL AT 16.45: THE MAJOR LATE PROMOTER

The beginning of labeling of late mRNAs from the major late promoter at 16.45 occurs after 10 to 12 hours of infection. The rate of RNA synthesis from the rightward reading strand from the region of 16 was compared to that from the region between 3–11 which includes all of the transcriptional units described above (Fraser *et al.*, 1979). Compared to early in infection there was at least a 50-fold increase in rightward transcribed RNA complementary to the 11–18 region late in infection. The nature of the first RNA synthesized from the 16 region is of interest because there may be some changes in the products from this region during infection. Late in infection only about one in every four polymerases that starts RNA synthesis at 16.45 completes the transcriptional unit all the way to 98 (Darnell *et al.*, 1977; Fraser *et al.*, 1979). The other initiation events result in premature termination producing capped RNA chains that are from ∼100 to ∼2000 bases in length that correspond to the beginning of the transcriptional unit. No role for these premature stop sites has been indicated yet but it remains a possibility that as RNA synthesis increases at 16 a differential and controlled termination could occur.

SURVEY OF PROMOTER ACTIVITY DURING EARLY INFECTION

After the definition of the ad2 promoters, a survey of patterns of transcription unit function can be made by pulse-labeling cells for 2 to 5 min and hybridizing total nuclear RNA to DNA on filters containing the start sites for various early transcription units (Nevins *et al.*, 1979). First, the transcription of region 1A reaches a maximum, followed closely by regions 3 and 4. All these transcription units reach a maximum rate of transcription in 2 to 3 hours. Transcription of region 2 however only reaches a maximum after ∼6 hours of infection. In the initial studies on the transcription of the various early regions it appeared that transcription units in regions 2 and 4 and possibly region 3 underwent a substantial decline in transcriptional rate later in infection. This decline has now been confirmed with additional experiments (Nevins and Jensen-Winkler, 1980). The decrease in transcriptional rate from region 4 as well as possibly region 2 is blocked by cycloheximide treatment, indicating a need for protein synthesis in order to decrease transcription rate from these regions. The protein that mediates the suppression of transcription in region 4 is suggested by studies of ts125, a mutant virus that does not make viral DNA at 41°C (Ginsberg *et al.*, 1977) and produces a temperature-sensitive version of the ∼72 000 dalton protein that binds to single-stranded DNA. The mutant virus goes through the same program of control of early transcription as does the wild type when grown at 32°C, but fails to properly shut off region 4 synthesis when grown at elevated temperatures (Nevins and Jensen-Winkler, 1980). Thus a temperature-sensitive transcriptionally active protein would appear to have been uncovered.

SUMMARY OF TRANSCRIPTIONAL CONTROLS

A summary of transcriptional controls thus far established for the ad2 infectious cycle can be made: (1) RNA synthesis from region 1A probably precedes synthesis from other regions. Evidence from virus mutants (Berg *et al.*, 1979; Jones and Shenk, 1979) implicate the RNA from 1A or its protein product in assisting the synthesis of perhaps all the early transcription units. The function of region 1A however, can apparently be carried out by the cell if high multiplicities of infection are used or if a long time elapses between infection and assay for early transcription (Nevins and Shenk, unpublished observations). (2) Synthesis from regions 3 and 4 follows 1A quickly but region 2 has a definite lag in maximal activation. (3) Regions 2 and 4 undergo a suppression of synthesis after 4 to 5 hours declining to as little as 10% of their maximal rates. A new protein, perhaps the 72K protein, is needed for the suppression of at least region 4. (4) Transcription of the protein IX mRNA early in infection is undetectable. Transcription begins at ~6 to 8 hours and continues. Thus transcriptional control at the initiation step appears to be almost absolute for this transcriptional unit. (5) There is at least a 50–fold increase in transcription from the start site at the late transcriptional unit at 16 hours compared to 5 hours. This increase includes the prematurely terminated transcripts and is not simply a relief of premature termination.

GENE CONTROL AT OTHER LEVELS

Since the discovery of poly(A) on mRNA, and especially since the discovery of splicing, ideas about differential processing of primary transcripts have been discussed (Evans *et al.*, 1977). Can a cell exercise a choice in conducting any of the necessary post-transcriptional steps and thereby regulate what mRNAs appear in the cytoplasm? Can a cell choose variably one of two or more possible poly(A) sites or splice sites in a transcriptional unit containing more than one of either of these sites? While it is difficult to rule out the possibility of occasionally using such variable choices the evidence is against such controls in the formation of ad2 mRNA as a major factor in determining individual mRNA levels. For example, ad2 late transcription appears to give rise to one mRNA each time the late transcript is synthesized (Nevins *et al.*, 1978). The single mRNA is chosen in two steps, first, the 3' end, the poly(A) site, and then the splice site (Nevins *et al.*, 1978). At the moment no variations in the choice of poly(A) sites or splice sites have been proven. In the manufacture of cell mRNAs it is also true that while only about one in three to four transcripts has poly(A) added, those that do are largely conserved (Harpold, Wilson and Darnell, unpublished observations). The unconserved nuclear transcripts appear to differ qualitatively from the conserved transcripts (Harpold *et al.*, 1979). Much more work is required to support a firm opinion but it is at least possible that variable controlled processing occurs rarely if at all.

However, new evidence on ad2 mRNA formation shows quite clearly that all regulation of mRNA concentration is not at the level of transcription. Transcriptional unit 1B produces two mRNAs spliced differently to give products of ~ 1200 and ~ 2000 nucleotides, respectively (Berk and Sharp, 1975; Wilson *et al.*, 1979). The ratios of these two forms of 1B mRNA differs in various circumstances. In transformed rat cells where presumably the same 1B transcriptional unit operates, the longer mRNA predominates in a molar ratio of at least 4 or 5 to 1 (Wilson *et al.*, 1979). During early lytic infection the ratio of the two RNAs is about equal (Spector *et al.*, 1978; Wilson *et al.*, 1979). Late in infection a change occurs in the two mRNAs from 1B so that the relative amount of the shorter mRNA increases perhaps 5-fold (Spector *et al.*, 1978). This result could be due either to: (1) more frequent production of 14S mRNA than 20S mRNA; or (2) more stable 14S mRNA in the cytoplasm late in infection. Using the accumulation of labeled mRNA as the guide to mRNA half-lives, the turnover rate of the 14S 1B mRNA has been shown to be ~ 5 to 10 times longer late in infection compared to early (Wilson and Darnell, unpublished observations). A similar set of circumstances appears to be true for a small RNA from the 1A region. Thus differential stabilization, and not either differential transcription or RNA processing of the 14S mRNA, accounts for its relative increase in concentration late in infection.

CONCLUSION

These early results on different types of control in the progression of ad2 through its infectious cycle suggests that ad2 may offer a legitimate model for a developmental program. Genes "come on" and "stay on" and "come on" and "go off" due to transcriptional controls. Other mRNAs change levels because of differential mRNA stabilization. The value of the adenovirus system, aside from the necessary fact that such an array of controls exists, is that infected cells all progress through the same pathways uniformly. An additional most important advantage of studying ad2 is that infected cells synthesize adenovirus RNA faster than RNA synthesis from any other single transcriptional unit yet described. Thus it is not an unrealistic hope that knowing which of the possible regulatory points are utilized in controlling a transcriptional unit, the proteins or cell structures involved in this control may be studied directly.

REFERENCES

Baker, C. D. and Ziff, E. B. (1980). *Cold Spring Harb. Symp. quant. Biol.* **44**, 415–428.

Berget, S. M., Moore, C. and Sharp, P. A. (1977). *Proc. natn. Acad. Sci. U.S.A.* **71**, 3171–3175.

Berk, A. J. and Sharp, P. A. (1978). *Cell* **14**, 695–711.

Berk, A. J., Lee, F., Harrison, T., Williams, J. and Sharp, P. A. (1979). *Cell* **17**, 935–944.

Brawerman, G. (1976). *In* "Progress in Nucleic Acid Research and Molecular Biology" (W. E. Cohn, ed.) Vol. 17, pp. 117–148. Academic Press, New York and London.

Chow, L. T., Gelinas, R., Broker, T. R. and Roberts, R. J. (1977). *Cell* **12**, 1–8.

Darnell, J. E. (1968). *Bact Rev.* **32**, 262–290.

Darnell, J. E. (1979). *In* "Progress in Nucleic Acid Research and Molecular Biology" (W. E. Cohn, ed.) Vol. 22, pp. 327–353. Academic Press, New York and London.

Darnell, J. E., Evans, R., Fraser, N., Goldberg, S. and Nevins, J. (1977). *Cold Spring Harb. Symp. quant. Biol.* **42**, 515–522.

Edmonds, M. and Winters, M. A. (1976). *In* "Progress in Nucleic Acid Research and Molecular Biology" (W. E. Cohn, ed.) Vol. 17, pp. 149–179. Academic Press, New York and London.

Evans, R., Fraser, N., Ziff, E., Weber, J. and Darnell, J. E. (1979). *Nature, Lond.* **278**, 367–370.

Flint, J. (1977). *Cell* **10**, 153–166.

Fraser, N. W., Seghal, P. B. and Darnell, J. E., Jr. (1979). *Proc. natn. Acad. Sci. U.S.A.* **76**, 2571–2575.

Gelinas, R. E. and Roberts, R. J. (1977). *Cell* **11**, 533–544.

Georgiev, G. P. (1969). *J. theor. Biol.* **24**, 473.

Ginsberg, H. S., Lundholm, U. and Linne, T. (1977). *J. Virol.* **23**, 142–151.

Goldberg, S., Schwartz, H. and Darnell, J. E. (1977). *Proc. natn. Acad. Sci. U.S.A.* **74**, 4520–4524.

Harpold, M. M., Evans, R. M., Salditt-Georgieff, M. and Darnell, J. E. (1979). *Cell* **17**, 1025–1035.

Jones, N. and Shenk, T. (1979). *Proc. natn. Acad. Sci. U.S.A.* **76**, 3665–3669.

Klessig, D. F. (1977). *Cell* **12**, 9–21.

Nevins, J. R. and Darnell, J. E., Jr. (1978). *Cell* **15**, 1477–1493.

Nevins, J. R. and Jensen-Winkler, J. (1980). *Proc. natn. Acad. Sci. U.S.A.* **77**, 1893–1897.

Nevins, J. R., Ginsberg, J. R., Blanchard, J.-M., Wilson, M. C. and Darnell, J. E. (1979). *J. Virol.* **32**, 727–733.

Shatkin, A. J. (1976). *Cell* **9**, 645–653.

Scherrer, K. and Marcaus, L. (1968). *J. cell Physiol.* **72** (Suppl. 1), 181–182.

Spector, D. J., McGrogran, M. and Raskas, H. J. (1978). *J. mol. Biol.* **126**, 395–414.

Weber, J., Jelenik, W. and Darnell, J. E. (1977). *Cell* **10**, 611–616.

Weil, P. A., Luse, D. S., Segall, J. and Roeder, R. G. (1979). *Cell* **18**, 469–484.

Wilson, M. C., Sawicki, S., Salditt-Georgieff, M. and Darnell, J. E. (1978). *J. Virol.* **25**, 97–103.

Wilson, M. C., Fraser, N. W. and Darnell, J. E., Jr. (1979). *Virology* **94**, 175–184.

Ziff, E. B. and Evans, R. M. (1978). *Cell* **15**, 1463–1475.

An Early Adenovirus Glycoprotein and its Association with Transplantation Antigens

HÅKAN PERSSON,[1] SUNE KVIST,[2] PER A. PETERSON[2] and LENNART PHILIPSON[1]

[1] *Department of Microbiology, Biomedical Center and* [2] *Department of Cell Research, The Wallenberg Laboratory, University of Uppsala, S-751 23 Uppsala, Sweden*

INTRODUCTION

The importance of cellular immunity against virus infections was first realized from clinical observations of viral infections in immunodeficient patients and subsequently from experimental findings with virus infections of thymectomized or antilymphocyte-serum-treated animals (Wheelock and Toy, 1973). In recent years it has become clear that the cellular immune defence against virus infections displays genetic restriction. Cytolytic T-lymphocytes generated during a viral infection can only lyse infected target cells which share major histocompatibility antigens with the cells that originally stimulated the development of the cytolytic T-cells (Doherty *et al.*, 1976). This genetic restriction which has been mapped to the loci controlling the expression of the transplantation antigens suggests a mechanism for how cytolytic T-lymphocytes recognize the target cells. Either the virus product and the histocompatibility antigens are separate entities on the surface of the target cells and the T-cells express independent receptors for the two types of molecules (*dual recognition*) or the virus and the transplantation antigens form a molecular complex recognized by one receptor on the T-lymphocytes (*altered self*) (Zinkernagel and Doherty, 1977). Evidence for both hypotheses have accumulated. Cocapping experiments (Henning *et al.*, 1976), antiserum-mediated inhibition of T-cell-dependent cytolysis (Schrader and Edelman, 1976; Hale *et al.*, 1978) and selective incorporation of H2 antigens into enveloped viruses (Bubbers *et al.*, 1978) may argue for the altered self concept. However, the development of precursor T-cells to mature T-lymphocytes with cytolytic activity requires recognition of the major histocompatibility antigens and occurs independently of recognition of non-self antigens (Zinkernagel *et al.*, 1978) a result which is compatible with dual recognition. To obtain unambiguous information about the molecular nature of the target recognition sites for the cytotoxic T-lymphocytes detailed biochemical studies are needed. We have approached this problem by examining the adenovirus system as a model system and we have found that adenovirus-infected or -transformed cells may provide tools for this analysis.

The adenovirus genome contains five major transcription units (E1A, E1B, E2, E3, and E4) which are expressed early after infection (for a review, see Philipson, 1979). A complex spectrum of spliced mRNAs are transcribed from these regions (Berk and Sharp. 1978; Chow et al., 1979; Kitchingman and Westphal, 1980). Twelve viral-coded early polypeptides have been identified by cell-free translation of purified early viral mRNA (Lewis et al., 1976; Harter and Lewis, 1978; Persson et al., 1979a). A 19 000 dalton glycoprotein (E3/19K) encoded by early region 3 (Persson et al., 1980) has been purified to homogeneity and it contains around 25 % carbohydrates by weight (Persson et al., 1979b, 1980). The glycoprotein is associated with the cell membrane where it is complexed with the cell transplantation antigen (Persson et al., 1980; Kvist et al., 1978).

The structure of the E3 mRNAs were recently deduced by RNA/DNA heteroduplex analysis (Chow et al., 1979; Kitchingman and Westphal, 1980). Several mRNAs with common 5' ends were found early after infection. These mRNAs differ in splicing patterns and several of them do not have coterminal 3' ends.

Some ad2-transformed cell lines like a hamster cell line (Ad2 HE4) and a Hooded Lister rat cell line (A2T2C4) express the glycoprotein E19K but other cell lines, although transformed by adenovirus, fail to express E19K (Persson et al., 1979b). These findings are of importance in evaluating whether this protein constitutes the Tumor Specific Transplantation Antigen (TSTA) of adenovirus-transformed cells.

An experimental system involving mouse adenovirus and inbred animals has established the importance of cellular immunity in adenovirus infection (Uetake and Inada, 1977). Three different approaches were used to study cell-mediated immunity to mouse adenovirus-infected cells: (i) inhibition of capsid protein synthesis in infected cells after incubation with immune spleen cells; (ii) direct and indirect inhibition of migration of immune mouse peritoneal exudate cells, i.e. macrophage migration inhibition tests; and finally (iii) specific ^{51}Cr release from infected cells after incubation with immune spleen cells. These studies reveal that cytotoxic T-lymphocytes do react with adenovirus-infected cells and the reaction is obviously independent of synthesis of viral DNA or capsid antigens, suggesting that an early protein might be involved in recognition. A viral surface antigen (S-antigen) was probably responsible for the cell-mediated immune reaction.

The cell-mediated lysis of mouse adenovirus-infected cells required H2 compatibility at the H2D or H2K locus between the target and the effector cell. Similar findings have been reported concerning cell-mediated immunity against cells infected with lymphocytic choriomeningitis (Zinkernagel and Doherty, 1974), vaccinia (Koszinowski and Thomsen, 1975), ectromelia (Gardner et al., 1975) and Sendai viruses (Doherty and Zinkernagel, 1976). Obviously the adenoviruses may serve as a convenient model to identify the polypeptides associated with the histocompatibility antigens in the cytolytic T-cell reaction with virus-infected target cells. This chapter summarizes our results in purifying and mapping the glycoprotein E19K on the viral genome. The primary structure of the glycoprotein was also determined by aligning the protein N-terminal sequence with the known DNA sequence (Hérissé et al., 1980). The role of this protein in the induction and recognition of the cytotoxic T-cells has also been investigated. The E19K appears to form a ternary complex with the major histocompatibility antigen subunits in the plasma membrane of infected and transformed cells.

PURIFICATION AND CHARACTERIZATION OF THE HUMAN ADENOVIRUS TYPE 2 GLYCOPROTEIN

Our purification scheme for the ad2 early glycoprotein from HeLa cells is shown in Fig. 1. A cycloheximide block was introduced between 2–5 hours after virus infection to enhance the abundance of viral mRNA. The block was reversed and the cells labeled with ^{35}S methionine from 5–9 hours post infection in the presence of cytosine arabinoside to prevent viral DNA synthesis. Mock-infected cells were treated in the same way. Cell extracts were prepared and incubated in a high salt buffer and subsequently centrifuged at high speed to separate the soluble from the insoluble protein. Quantitative analysis showed that around 80% of the E19K polypeptide was recovered in the pellet. The E17.5K was also preferentially associated with the pellet. The pellet fraction was solubilized overnight in a high salt buffer containing 1% of the neutral detergent Triton X-100. Solubilized material was fractionated by lectin (Lens culinaris) affinity chromatography and the E19K as well as the E17.5K had a high affinity for the lectin column. The E19K polypeptide was enriched 20-fold after this purification step and in the final purification step, DEAE-Sephadex chromatography, the majority of E19K did not bind to the column. The flow through material was therefore collected and analyzed by SDS-PAGE. Three polypeptides, E40K, E19K and E17.5K were detected with 90% of the radioactivity in the E19K polypeptide (Fig. 2). Tryptic peptide analysis of these three proteins showed that they are related to each other. The E40K is probably a dimer of

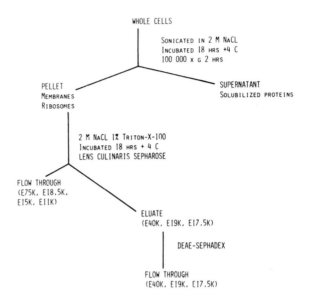

FIG. 1. Scheme for purification of an ad2 early glycoprotein. Ad2-infected HeLa cells were treated with cycloheximide (25 µg/ml) between 2–5 hours post infection, the cells were washed and labeled with ³⁵S methionine in the presence of cytosine arabinoside (25 µg/ml). At 9 hours post infection the cells were harvested, and subjected to the outlined purification scheme.

the E19K and the E17.5K may be a precursor to the mature glycoprotein. This purification scheme yielded a 200-fold enrichment of the E19K polypeptide with an estimated recovery of 12%. In order to study the kinetics of synthesis of the E19K, cell extracts were prepared from cells which had been pulse-labeled for 30 minutes with ³⁵S methionine at different times after ad2 infection. An antiserum against the highly purified E19K was used to immunoprecipitate the protein. The E19K polypeptide had already been detected at 2 hours and the rate of synthesis reached a maximum at 4 hours post infection. The E17.5K showed the same kinetics as the E19K polypeptide. Three hamster embryo cell lines transformed with UV-inactivated adenovirus type 2 (ad2HE$_1$, ad2HE$_4$, ad2HE$_5$) and two ad2-transformed rat cell lines (A2F19 and A2T2C4) were analyzed for the presence of this protein. Cell extracts were immunoprecipitated with the antiserum and one hamster cell line (ad2HE$_4$) and one rat cell line (A2T2C4) expressed this polypeptide. Immunoprecipitation with a serum prepared in syngeneic rats by injection of A2T2C4 cells also revealed the E19K polypeptide in the latter two cell lines.

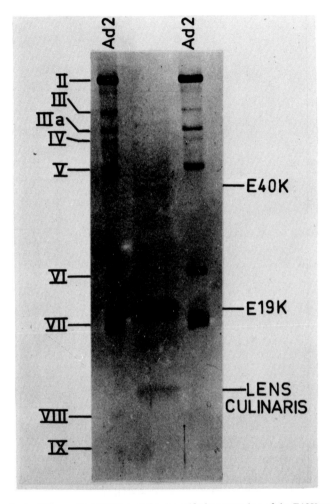

FIG. 2. Analysis of the purified E19K protein. A purified preparation of the E19K protein was analyzed by SDS-polyacrylamide gel (SDS-PAGE) electrophoresis with ad2 virus markers in adjacent slots. The gel was stained with Coomassie Brilliant Blue. *Lens culinaris* denotes the size of the lectin used for affinity chromatography in one of the purification steps.

MEMBRANE ASSOCIATION OF THE E19K GLYCOPROTEIN

Ad2-infected cells were radioactively labeled with ^{35}S methionine early after infection. Membranes were prepared from the cytoplasmic extracts by sedimentation in a discontinuous sucrose gradient. The cytosol fraction was also collected. Analysis by immunoprecipitation followed by SDS-PAGE

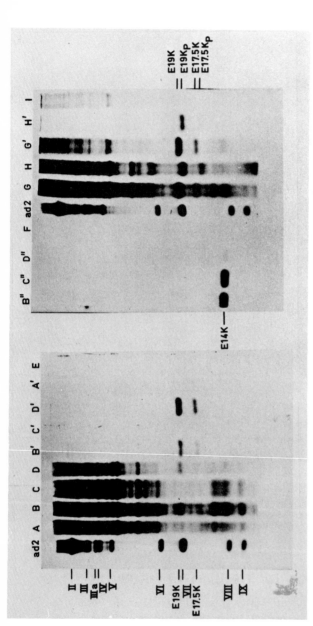

FIG. 3. Fractionation of cytoplasmic extracts prepared early after ad2 infection. Ad2-infected cells were labeled with ³⁵S methionine from 3 to 6 hours post infection. The cells were harvested and a cytoplasmic extract was prepared by homogenization. The cytoplasm was sedimented in a discontinuous sucrose gradient and the membranes collected at the interface between 80% and 20% sucrose layers. The top of the gradient represented the cytosol fraction. Mock-infected cells (A), cytoplasmic (B), cytosol (C) and membrane (D) fractions from ad2-infected cells. Immunoprecipitation of equal number of cells with anti-E19K serum of cytoplasmic (B′), cytosol (C′), membrane (D′) fractions of infected cells and a mock-infected cell extract (A′). Immunoprecipitation with an antiserum raised against a purified E14K protein (Persson *et al.*, 1978) from the cytoplasmic (B″), the cytosol (C″) and the membrane (D″) fractions of infected cells. Immunoprecipitation with a normal rabbit serum of the membrane (E) and cytosol (F) fractions.

The membranes were digested with proteinase K (400 μg/ml) for 15 min at 37°C and analyzed by SDS-PAGE before and after immunoprecipitation. (G) Membranes without and (H) membranes with protease treatment. Immunoprecipitation with anti-E19K serum of the untreated (G′) and protease-treated membrane (H′) fractions. Immunoprecipitation from untreated membranes with a normal rabbit serum (I). E19K$_P$ and E17.5K$_P$ denote polypeptides specifically immunoprecipitated with the anti-E19K serum from protease-treated membranes. The gels were analyzed by fluorography. Ad2 denotes here and in subsequent figures a ³⁵S-methionine-labeled ad2 virus marker.

revealed the presence of the E19K protein in the membrane fraction while it was absent in the cytosol fraction (Fig. 3). The E14K protein, previously purified and characterized as an ad2 early cytoplasmic protein (Persson *et al.*, 1978) was recovered in the cytosol.

Proteolytic digestion of the membrane fraction from infected cells revealed that the major part of the E19K and E17.5K polypeptides was protected. The size of both glycoproteins was reduced by around 500 daltons, suggesting that both proteins are inserted into the membrane with only a small portion of the polypeptide outside the membrane (Fig. 3).

The role of carbohydrates for membrane association of the E19K protein was investigated by fractionation of cells maintained in the presence of tunicamycin. A polypeptide immunologically and chemically related to the E19K glycoprotein with an apparent molecular weight of 14 500 ($E19K_0$) was observed in the presence of tunicamycin. This unglycosylated form of the E19K protein was associated with the membranes and no protein was found free in the cytosol. Proteolytic digestion of membranes from tunicamycin-treated cells showed that the molecular weight of the $E19K_0$ protein was also reduced by around 500 after proteolysis.

IN VITRO SYNTHESIS OF THE E19K PROTEIN

The E19K glycoprotein or $E19K_0$ could not be observed after *in vitro* translation in the reticulocyte cell-free system. Since cell-free systems fortified with rough microsomes from dog pancreas allow the synthesis of fully processed mature glycoproteins (Garoff *et al.*, 1978), early adenovirus RNA was translated in the reticulocyte system in the presence of microsomes. Under these conditions we observed synthesis of a 19 000 daltons polypeptide which was immunoprecipitated with the anti-E19K serum.

Proof that the E19K glycoprotein is coded by the viral genome was obtained by selection of early viral mRNA on the complementary strands of the ad2 genome. The selected RNA was translated *in vitro* in the presence of rough microsomes from dog pancreas and the translation productions were immunoprecipitated with anti-E19K serum. Early ad2 mRNA selected by hybridization to the viral r-strand directed the synthesis of the E19K glycoprotein under these conditions (Fig. 4A). The *in vitro*-synthesized glycoprotein was apparently inserted into the membrane vesicles during translation since all the E19K glycoprotein was associated with the microsomes after translation. Hybrid-arrested cell-free translation (Paterson *et al.*, 1977) in the presence of rough microsomes followed by specific immunoprecipitation with the anti-E19K serum located the gene for the glycoprotein within the Eco RI-D fragment of the ad2 genome (75.9–84.0

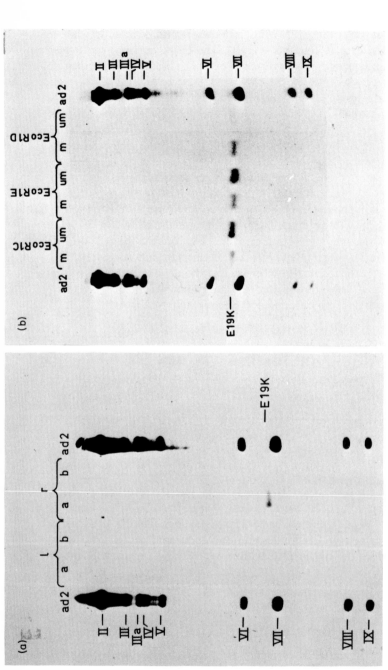

FIG. 4. Location of the gene for the ad2-coded E19K glycoprotein. (A) Ad2-specific early RNA was isolated by hybridization to the complementary strands of the viral genome. Hybridized RNA was translated in the reticulocyte system in the presence of rough microsomes from dog pancreas and the translation mixtures were immunoprecipitated with normal rabbit serum and the anti-E19K serum. The immunoprecipitates were analyzed by SDS-PAGE. The strands transcribed leftwards (l) and rightwards (r) are indicated at the top. (a) Anti-E19K serum; (b) normal rabbit serum.

(B) Ad2 early cytoplasmic RNA was hybridized in solution to the indicated restriction enzyme fragments. Each sample was divided into two aliquots; one was melted (m) before translation and the other was translated directly (um) in the reticulocyte system in the presence of rough microsomes from dog pancreas. The translation mixtures were immunoprecipitated with anti-E19K serum. The immunoprecipitates were analyzed by SDS-PAGE.

map units) (Fig. 4B). This result is consistent with the failure of the $AD_2^+ND_1$ hybrid virus to synthesize the E19K glycoprotein and locates the gene for this protein within early region 3 of the ad2 genome.

CELL-FREE SYNTHESIS WITH PURIFIED mRNAs

Restriction enzyme fragments from the E3 region of the viral genome were next prepared and used for selection of viral mRNAs (Fig. 5). The selected mRNAs were translated *in vitro* with and without microsomes and the products were analyzed by SDS-PAGE. The Eco RI-D and Hind III-H fragments selected mRNAs for two early viral polypeptides with molecular weights at 16 000 and 14 000 daltons (E3/16K and E3/14K) (Fig. 5C). The Hind III-L fragment selected the mRNA from the E3/16K protein while the mRNA for the E3/14K protein was predominantly selected by the Eco RI-E fragment. This fragment also selected the mRNA for a 14 500 daltons protein (E3/14.5K) as well as small amounts of the E3/16K protein. The E3/16K protein is around 1500 daltons larger than the unglycosylated $E3/19K_0$ protein synthesized in the presence of tunicamycin (Fig. 5C, slot b). Restriction enzyme fragments derived from early regions 2 and 4 (Eco RI $-$ B $+$ C, Fig. 5C) selected mRNAs encoding a different spectrum of early proteins (E2/75K, E4/26K, E4/18.5K, E4/12,5K and E4/11K). The cell-free system was also supplemented with rough microsomes before the start of incubation and the translation products were immunoprecipitated with an anti-E3/19K serum. Messenger RNA selected by hybridization to the Eco RI-D and Hind III-L fragments directed the synthesis of the mature glucoprotein under these conditions (Fig. 5C, slots c–e).

TRYPTIC FINGERPRINT ANALYSIS

A tryptic fingerprint analysis revealed that the mature E3/19K glycoprotein contained five ^{35}S-methionine-labeled tryptic peptides. The E3/16K protein contained the same tryptic peptides and one additional peptide. The latter peptide was more heavily labeled with ^{35}S methionine suggesting that it contains more than one methionine residue. The E3/14.5K and the E3/14K polypeptides contained ^{35}S-methionine-labeled tryptic peptides, mutually similar but different from any of the peptides found in the E3/16K protein.

FRACTIONATION OF mRNA CODING FOR E3/16K

Messenger RNA selected by hybridization was further purified by sucrose-gradient centrifugation. Labeled RNA selected on the Eco RI-D fragment

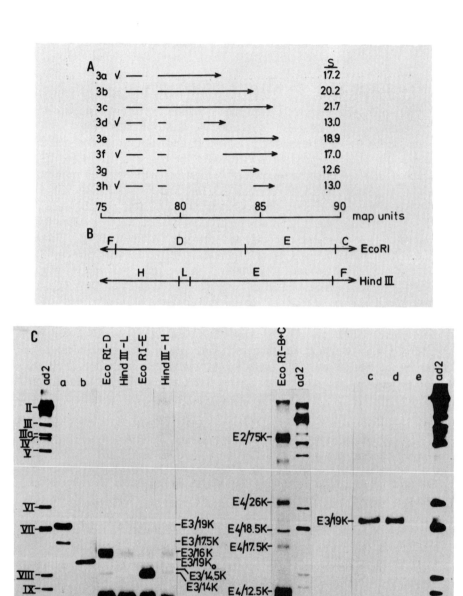

sedimented in two peaks (Fig. 6). The major peak had a sedimentation coefficient around 17S while the minor sedimented at around 13S. The 17S RNA directed the synthesis *in vitro* of the E3/16K protein while RNA from the 13S peak synthesized the E3/14K protein. The E3/16K protein was also to a lower extent synthesized from an mRNA sedimenting slightly faster than the 18S ribosomal RNA marker.

Purified E3/16K mRNAs were hybridized to ad2 DNA or restriction fragments thereof and the resulting heteroduplexes were visualized in the electron microscope. The 17S mRNA had its 5′ end located at coordinate 76.7 with a 3′ terminus located at position 82.8. The mRNA also contained an intron of around 430 nucleotides from coordinates 77.8 to 79.0 The 19–20S mRNA (fraction 10 in Fig. 6) was also analyzed by RNA/DNA heteroduplex formation. This mRNA was a mixture of the 3b and 3c species shown in Fig. 5A with a 5′ end at coordinate 76.7% with an intron between 77.8 and 79.0% and 3′ ends located at coordinates 84.6% (3b) or 85.8% (3c). Thus, the E3/16K protein is encoded by three different mRNAs having the same 5′-end structures. The larger mRNAs (species 3b and 3c in Fig. 5A) represent together about one third of the E3/16K mRNAs while the remaining two-thirds belong to the 3a species shown in Fig. 5A.

ALIGNMENT OF THE E3/16K PROTEIN WITH THE DNA SEQUENCE

Protein N-terminal radiosequence analysis was performed on the *in vitro*-synthesized E3/16K protein in order to align the protein in the DNA sequence of the Eco RI-D fragment (Hérissé *et al.*, 1980). The E3/16K protein was synthesized *in vitro* in the presence of different radiolabeled amino acids, one at a time, and the N-terminal sequence was determined by liquid-phase sequencer analysis. The protein contained methionine at positions 1 and 4, leucine at positions 6, 8, 9 and 11, isoleucine at position 5 and lysine at positions 19, 20 and 24 (Fig. 7).

FIG. 5. Adenovirus-coded early region 3 proteins. (A) Adenovirus type 2 early region 3 transcripts according to Chow *et al.* (1979) and Kitchingman and Westphal (1980). The calculated sedimentation coefficients are shown to the right and the major RNA species are indicated by check marks. (B) Maps of restriction enzyme fragments in early region 3. (C) Early viral RNA was hybridized to the indicated restriction enzyme fragments of ad2 DNA and the purified RNA was translated *in vitro*. The translation products were analyzed by SDS-PAGE followed by fluorography. Mature E3/19K glycoprotein (a) and E3/19K$_0$ protein synthesized in the presence of tunicamycin (b) after immunoprecipitation from cells labeled *in vivo*. The cell-free system was supplemented with rough microsomes and the translation products were immunoprecipitated with the anti-E3/19K serum (c,d) or a preimmune serum (e). The cell-free system was programmed with mRNA selected by hybridization to Eco RI-D (c,e) and Hind III-L (d).

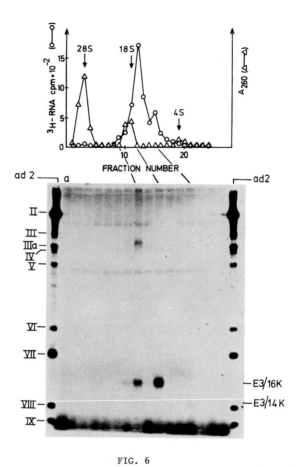

FIG. 6

FIG. 6. Size fractionation of early region 3 mRNAs. Early viral mRNA labeled with ^{3}H uridine and purified by hybridization to restriction enzyme fragments Eco RI-D was sedimented on a 15–30% sucrose gradient. RNA from each fraction was precipitated with ethanol and translated *in vitro*. The ^{35}S-methionine-labeled products were analyzed by SDS-PAGE followed by fluorography. (a) No mRNA added. Cytoplasmic RNAs prepared from mock-infected cells were used as size markers.

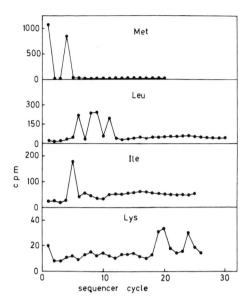

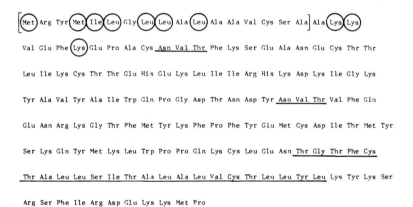

FIG. 7. Sequence analysis of the E3/16K protein. Radioactive E3/16K protein was prepared and the radioactivity in each cycle of sequence degradation was determined. The amino acids used for labeling are indicated. The predicted amino acid sequence for the E3/16K protein is shown at the bottom as deduced by alignment with the DNA sequence (Herissé *et al.*, 1980). Residues identified by sequence analysis are indicated by circles. A presumptive signal sequence in the protein is indicated in brackets. The two tripeptides (-Asn-Val-Thr-) are underlined as well as a predominant hydrophobic non-charged region close to the C-terminus of the protein.

The positions analyzed can be completely aligned with the DNA sequence and show that the protein initiates at coordinate 80.3 %. From here the DNA sequence contains an open reading frame for 159 amino acids (Hérissé *et al.*, 1980). The sequence contains a hydrophobic presumptive signal sequence of maximally 18 amino acids including the initiating methionine residue. Two presumptive glycosylation sites (-Asn-Val-Thr-) occur in the protein, and a region with uncharged predominantly hydrophobic amino acids is found close to the C-terminus of the protein.

INHIBITION OF T-LYMPHOCYTE-DEPENDENT LYSIS OF A2T2C4 CELLS

Spleen cells obtained from Hooded Lister rats primed *in vivo* with the virus-transformed A2T2C4 cells could lyse the transformed cells provided the primed spleen cells had been restimulated *in vitro* with the target cells prior to cytolysis. Lysis was observed at an effector-to-target cell ratio as low as 3:1. No cytolytic activity of the spleen cells was observed when the target cells were primary fibroblasts from the Hooded Lister rats or the adenovirus-transformed hamster cells, ad2 HE4, the latter synthesizing the E19K polypeptide. Cell-mediated cytolysis of the A2T2C4 cells was enhanced when B-cells were removed either by nylon-wool columns or by antirat IgG serum. Together these results suggest that T-lymphocytes were responsible for the cytolytic activity. To examine whether the T-lymphocyte-dependent lysis of the A2T2C4 cells recognized virus-genome-coded products in addition to the recognition of the syngeneic AgB antigens, the cytolytic action of T-lymphocytes was investigated in the presence of various concentrations of antiserum. Figure 8 shows that a rabbit anti-AgB antiserum efficiently blocked lysis but normal rabbit serum did not affect lysis. A syngeneic antiserum against A2T2C4 cells also inhibited lytic activity of the T-cells although this antiserum was only effective at high concentrations. As expected, normal rat serum did not display inhibitory capacity.

EVIDENCE FOR A TERNARY COMPLEX BETWEEN THE TRANSPLANTATION ANTIGENS AND THE E19K

Cellular glycoproteins were solubilized by neutral detergents followed by lectin affinity chromatography. When the glycoproteins from A2T2C4 cells were analyzed two polypeptide chains with molecular weights of 45 000 and 19 000 were immunoprecipitated with the syngeneic antiserum raised against the transformed A2T2C4 cells.

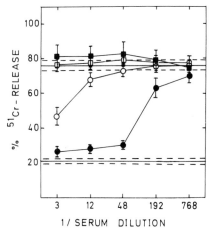

FIG. 8. Inhibition of T-cell-dependent cytolysis of A2T2C4 cells. Indicated dilutions of a rabbit antiserum against AgB antigens (●—●), a syngeneic rat antiserum against the A2T2C4 cells (○—○), normal rabbit serum (□—□) and normal rat serum (■—■) were incubated with ^{51}Cr-labeled A2T2C4 cells. After 30 min at 4°C educated rat spleen T-lymphocytes were added at an effector-to-target-cell ratio of 20:1. The incubations proceeded for 6 hours at 37°C and the ^{51}Cr released was measured after centrifugation. A release of 100% represents the amount of radioactivity recovered in the supernatant after Triton X-100 treatment of the target cells. The solid and the broken horizontal lines denote the mean and the standard deviation of the ^{51}Cr release in the presence (upper) and in the absence (lower) of the cytolytic T-lymphocytes alone.

It was next investigated whether the E3/19K was coprecipitated also with antisera of individual AgB antigen subunits. AgB alloantiserum precipitated the E3/19K together with the AgB antigen subunits from ^{3}H-tyrosine-labeled glycoproteins obtained from A2T2C4 cells, but only the AgB antigen chains were visualized from the labeled molecules derived from primary rat fibroblasts. The same results were obtained with the antiserum against β_2-microglobulin. These results are consistent with the idea that the two AgB antigen subunits formed a ternary complex with the viral E19K glyco-protein. It was also established that antisera to human HLA-antigens could coprecipitate the E3/19K glycoprotein in adenovirus-infected but not in mock-infected cells (Fig. 9) providing additional evidence for a ternary complex between the adenovirus glycoprotein and the transplantation antigens.

DISCUSSION

The adenovirus E3/16K protein is structurally related to the E3/19K glycoprotein. Addition of rough microsomes to the translation mixture

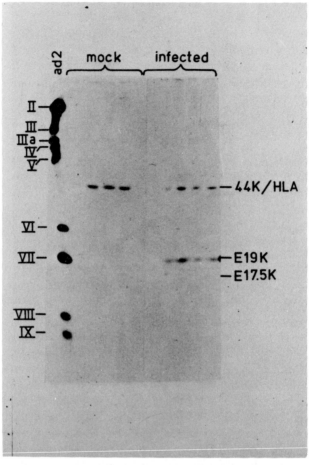

FIG. 9. Immunoprecipitation of the human transplantation antigens from uninfected and adenovirus-infected cell membranes. The cellular membrane fraction was solubilized with detergents and chromatographed over a lectin column. The fraction eluted with sugar was immunoprecipitated with an allogenic anti-HLA serum prepared against purified human transplantation antigens.

programmed with purified E3/16K mRNAs allowed the synthesis of the mature glycoprotein, establishing that the E3/16K protein is the precursor to the glycoprotein. The amino acid sequence of the E3/16K protein shows an 18-amino-acid-long hydrophobic sequence at its N-terminus. The following results suggest that this region constitutes a signal sequence:

(1) The E3/16K protein is around 1500 daltons larger than the tunicamycin product (E3/19K_0).

(2) In accordance with the predicted amino acid sequence for the E3/16K protein, showing methionine at positions 1 and 4, the tryptic fingerprint of this protein contains only one ^{35}S-methionine-labeled peptide not found in the E3/19K glycoprotein.

(3) Addition of rough microsomes before the start of cell-free synthesis followed by protease treatment showed that the E3/19K protein was inserted into the vesicles.

(4) Sequence analysis of the E3/19K glycoprotein labeled *in vivo* with ^{3}H lysine indicated a possible cleavage after position 17 in the E3/16K protein.

The signal sequence is contiguous with the sequence present in the mature protein and it is not encoded in a separate exon (as observed for some other proteins) as mouse immunoglobulin light chain (Bernard *et al.*, 1978) and conalbumin (Cochet *et al.*, 1979). There are two potential sites for carbohydrate attachment in the protein based on the presence of the sequence (-Asn-Val-Thr-) which is invariant for glycosylation. The C-terminus of the protein contains a 23-amino-acid-long region of uncharged predominantly hydrophobic amino acids, followed by a hydrophilic region of 15 amino acids. This resembles the structure for the HLA-antigens and glycophorin (Springer and Strominger, 1976; Jokinen *et al.*, 1979), two proteins which span the cell membrane with a hydrophilic C-terminus located on the cytoplasmic side of the membrane. The predicted structure for the E3/19K proteins suggests that this glycoprotein is associated with the membranes in a similar fashion. Both glycosylation sites are then located on the exterior side of the membrane.

The findings that the transplantation antigens of both rat and human cells form a complex with the E19K after solubilization of the membrane glycoproteins with neutral detergents provide suggestive evidence that the T-lymphocytes might recognize altered self rather than possess a dual recognition mechanism for the target cells. The identification of a close association between the E19K and the transplantation antigens may also explain the mechanism of cell-mediated immunity to virus-infected cells in the adenovirus system. Analysis of kinetics of polypeptide synthesis in ad2-infected cells suggests that the E19K protein has already been synthesized at 1–2 hours after the onset of productive infection (Persson *et al.*, 1979b). It is therefore tempting to speculate that this protein might have a specific function in membranes generating a favorable environment for adenovirus gene expression. The present study opens the field for a closer inspection of the role of virus-coded membrane proteins in the productive infection of the virus and also for the host defence mechanism involved in cell-mediated immunity. It is now possible to fragment the E3/19K glycoprotein and determine the regions of the polypeptide which interact with the transplantation antigens.

ACKNOWLEDGEMENTS

We would like to thank Drs M. Jansson, H. Jörnvall, O. Kämpe, C. Signäs, J. Zabielski and L. Östberg for their contributions in several experiments. This work was supported by grants from the Swedish Cancer Society and Knut and Alice Wallenberg's Foundation.

REFERENCES

Berk, A. J. and Sharp, P. A. (1978). *Cell* **14**, 695–711.
Bernard, O., Hozumi, N. and Tonegawa, S. (1978). *Cell* **15**, 1133–1144.
Bubbers, J. E., Chen, S. and Lilly, F. (1978). *J. exp. Med.* **147**, 340–351.
Chow, L. T., Broker, T. R. and Lewis, J. B. (1979). *J. molec. Biol.* **134**, 265–303.
Cochet, M., Gannon, F., Hen, R., Maroteaux, L., Perrin, F. and Chambon, P. (1979). *Nature, Lond.* **282**, 567–574.
Doherty, P. C. and Zinkernagel, R. M. (1976). *Immunology* **31**, 27–32.
Doherty, P. C., Blanden, R. V. and Zinkernagel, R. M. (1976). *Transplantn. Rev.* **29**, 98–123.
Gardner, I. D., Bowern, N. A. and Blanden, R. V. (1975). *Eur. J. Immunol.* **5**, 122–127.
Garoff, H., Simons, K. and Dobberstein, B. (1978). *J. molec. Biol.* **124**, 587–600.
Hale, A. H., Witté, O. N., Baltimore, D. and Eisen, H. N. (1978). *Proc. natn. Acad. Sci. U.S.A.* **75**, 974–980.
Harter, M. L. and Lewis, J. B. (1978). *J. Virol.* **26**, 736–749.
Henning, R., Schrader, J. W. and Edelman, G. M. (1976). *Nature, Lond.* **263**, 689–691.
Hérissé, J., Courtois, G. and Galibert, F. (1980). *Nucl. Acids Res.* **8**, 2173–2191.
Jokinen, M., Gahmberg, C.-G. and Andersson, L. C. (1979). *Nature, Lond.* **279**, 604–607.
Kitchingman, G. R. and Westphal, H. (1980). *J. molec. Biol.* **137**, 23–48.
Koszinowski, U. and Thomsen, R. (1975). *Eur. J. Immunol.* **5**, 245–251.
Kvist, S., Östberg, L., Persson, H., Philipson, L. and Peterson, P. A. (1978). *Proc. natn. Acad. Sci. U.S.A.* **75**, 5674–5678.
Lewis, J. B., Atkins, J. F., Baum, P. R., Solem, R., Gesteland, R. F. and Anderson, C. W. (1976). *Cell* **7**, 141–151.
Paterson, B. M., Roberts, B. E. and Kuff, E. L. (1977). *Proc. natn. Acad. Sci. U.S.A.* **74**, 4370–4374.
Persson, H., Öberg, B. and Philipson, L. (1978). *J. Virol.* **28**, 119–139.
Persson, H., Perricaudet, M., Tolun, A., Philipson, L. and Pettersson, U. (1979a), *J. biol. Chem.* **254**, 7999–8003.
Persson, H., Signäs, C. and Philipson, L. (1979b). *J. Virol.* **29**, 938–948.
Persson, H., Jansson, M. and Philipson, L. (1980). *J. molec. Biol.* **136**, 375–394.
Philipson, L. (1979). *Adv. Virus Res.* **25**, 357–405.
Schrader, J. W. and Edelman, G. M. (1976). *J. exp. Med.* **143**, 601–614.
Springer, T. A. and Strominger, J. L. (1976). *Proc. natn. Acad. Sci. U.S.A.* **73**, 2481–2485.
Uetake, H. and Inada, T. (1977). *Ann. Microbiol. (Inst. Pasteur)* **128B**, 517–530.

Wheelock, E. F. and Toy, S. T. (1973). *Adv. Immunol.* **16**, 123–184.
Zinkernagel, R. M. and Doherty, P. C. (1974). *Nature, Lond.* **248**, 701–702.
Zinkernagel, R. M. and Doherty, P. C. (1977). *Cold Spring Harb. Symp. quant. Biol.* **41**, 505–510.
Zinkernagel, R. M., Callahan, G. N., Althage, A., Cooper, S., Klein, P. A. and Klein, J. (1978). *J. exp. Med.* **147**, 882–896.

The Control of Adenovirus Early Gene Expression

HÅKAN PERSSON, HANS-JÜRG MONSTEIN and
LENNART PHILIPSON

Department of Microbiology, The Biomedical Center, University of Uppsala, S-751 23
Uppsala, Sweden

INTRODUCTION

Five regions of the adenovirus DNA are expressed early after infection (Pettersson *et al.*, 1976; Berk and Sharp, 1978; Chow *et al.*, 1979). These regions encode at least ten different early proteins (Lewis *et al.*, 1976; Harter and Lewis, 1978; Persson *et al.*, 1979b). A few of these have been purified and characterized (Linné *et al.*, 1977; Sugawara *et al.*, 1977; Persson *et al.*, 1978b, 1979a, 1980). There appears to be a regulated expression of the viral genome following virus infection because the different early proteins start to be synthesized and reach maximal rate of synthesis at different times after infection (Persson *et al.*, 1978a; Ross *et al.*, 1980). The control of the early protein cascade is obviously at the transcriptional level since the transcripts and the mRNAs from the early regions also appear sequentially (Nevins *et al.*, 1979). Late in the early phase of adenovirus gene expression there appears to exist a negative control on early gene transcription mediated by the DNA-binding protein encoded in the E2 region of the genome (Carter and Blanton, 1978).

Recently Jones and Shenk isolated mutants of adenovirus type 5 that contained defined deletions in the left-hand end of the genome (Jones and Shenk, 1979a). These mutants are defective in the transformation of rat embryo cells and one of them, ad5 dl 312, containing a deletion of viral DNA spanning map positions 1.5 to 4.5, does not induce cytoplasmic viral RNA from any of the early regions (Jones and Shenk, 1979a,b). Host-range mutants (hr) of adenovirus type 5, selected for their ability to grow in adenovirus-transformed cells (Harrison *et al.*, 1977) have also been used to study the regulation of viral gene expression. Berk *et al.* (1979a) showed that group I hr-mutants with mutations between 0 and 4.4 map units on the viral DNA are defective for the synthesis of viral RNA from all early regions except region E1A (map position 1.5–4.5). These results are consistent with the hypothesis that an adenovirus early gene function specified by the E1A region regulates the expression of other early genes. In conflict with this hypothesis Lassam *et al.*, (1979) found that group I hr-mutants induce the synthesis of early proteins encoded in early regions other than E1A.

The controversy may be due to the fact that the host-range mutants are extremely leaky at high multiplicity of infection, suggesting either that the mutant virus stocks contain defective recombinants which can complement each other at high multiplicity or that cellular products may complement (Shenk *et al.*, 1979). Irrespective of the cause of the multiplicity-dependence it may be anticipated that the gene products initiating early adenovirus gene expression are only required in catalytic amounts and are effective in *trans* since the transformed human cell line 293 containing limited copies of these genes complement the mutants. We used protein synthesis inhibitors to establish whether virus-coded or virus-induced protein products are required for the control events regulating early adenovirus gene expression.

A TRANSLATION BLOCK

The selection of protein-synthesis inhibitors must be made with caution since some of them may either not prevent protein synthesis completely or drastically affect RNA synthesis. The inhibitors should preferably be reversible to allow release of the block in macromolecular synthesis under *in vivo* conditions. Berk *et al.* (1979b) used emetine, an irreversible inhibitor of protein synthesis which affects RNA synthesis, to study the same events and

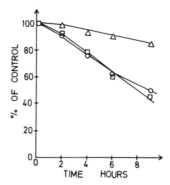

FIG. 1. Effect of puromycin on RNA synthesis in uninfected HeLa cells. Puromycin was added to uninfected HeLa cells in a final concentration of 100 μM. At the indicated times 10^7 cells were withdrawn and pulse-labeled with ^{3}H uridine (20 μCi/ml) for 30 min at 37°C. Cytoplasmic RNA was extracted and acid-insoluble radioactivity was determined before and after oligo(dT)-chromatography. Pulse-labeled HeLa cells incubated without puromycin served as control. All values have been correlated for the 50% inhibition of ^{3}H-uridine uptake caused by the inhibitor. (□—□) Total cytoplasmic RNA; (○—○) poly(A)$^-$ RNA; (△—△) poly(A)$^+$ RNA.

came to the conclusion that a protein gene product from region E1A of the adenovirus genome is required for transcription of other early regions. Puromycin and anisomycin may, however, be better suited for these studies since they inhibit protein synthesis effectively ($>99\%$) and do not suppress mRNA synthesis more than 15% over an 8-hour period in uninfected cells (Fig. 1). Cycloheximide on the other hand is less effective as a protein-synthesis inhibitor (95% inhibition) but it also inhibits RNA synthesis feebly. We therefore selected puromycin for further studies of the control events in adenovirus early gene expression, but the same results were obtained with anisomycin and cycloheximide. In the latter case it was necessary to inhibit protein synthesis 2 hours before virus infection to obtain comparable results. In contrast to the previously observed inhibition of a transcriptional or post-transcriptional step (Jones and Shenk, 1979b; Berk *et al.*, 1979a) we could establish a translational control *in vivo* early in adenovirus infection. When puromycin was added at the time of infection and the drug removed at 5 hours post infection essentially no adenovirus-specific

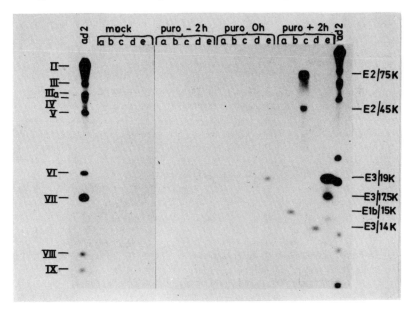

FIG. 2. Synthesis of ad2 early proteins after puromycin treatment. HeLa cells were treated with puromycin (100 μM) 2 hours before, at the onset, or 2 hours after ad2 infection. The drug was removed at 5 hours post infection, the cells were washed and pulse-labeled with ^{35}S methionine for 30 min at 37°C. Cell extracts were prepared and immunoprecipitated with a set of monospecific antisera directed against ad2 early proteins. The immunoprecipitates were analyzed by SDS-PAGE followed by fluorography. The following antisera were used: (a) pre-immune serum from rabbits; (b) an antiserum prepared in syngeneic rats against the ad2-transformed rat cell line F17 (Gilead *et al.*, 1976); (c) anti-DNA binding protein (E2/75K); (d) anti-E3/14K serum; (e) anti-E3/19K serum. Here and in the subsequent figures "ad2" denotes a ^{35}S-methionine-labeled ad2 virus marker.

early proteins could be detected after a 30 min pulse with ^{35}S methionine (Fig. 2). If, however, the puromycin block was introduced from 1–5 hours after infection, all identified early polypeptides were synthesized in large amounts at reversion, indicating that pretreatment with puromycin did not by itself inhibit translation of viral mRNAs. We have established with labeled puromycin that this difference is not due to differential accumulation of puromycin in infected and uninfected cells (not shown). The puromycin block was overcome within 1–2 hours after reversion and all early viral polypeptides (E2/76K, E3/19K, E1B/15K and E3/14K) were synthesized concurrently without a cascade appearance (Fig. 3) in sharp contrast to the situation early in adenovirus infection in untreated cells where expression is

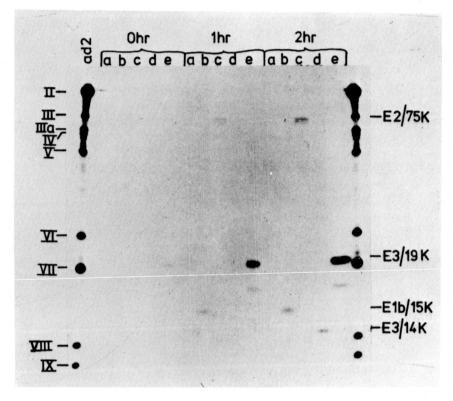

FIG. 3. Reversion of the puromycin block. HeLa cells infected with ad2 in the presence of puromycin (100 µM) were washed free of the drug at 5 hours post infection. At the times indicated at the top, 10^7 cells were withdrawn and pulse-labeled with ^{35}S methionine. Cell extracts were prepared and immunoprecipitated with the following antisera: (a) pre-immune serum from rabbits; (b) anti-F-17 serum; (c) anti-E2/75K serum; (d) anti-E3/14K serum; and (e) anti-E3/19K serum. The immunoprecipitates were analyzed by SDS-PAGE followed by fluorography of the dried gel.

regulated (Persson *et al.*, 1978a; Ross *et al.*, 1980). These results suggest that the products controlling expression are either synthesized during the puromycin treatment or shortly after withdrawal of the drug.

It was also established that viral mRNA was synthesized in the presence of puromycin from all five early regions of the genome by two different techniques. Quantitative hybridization to separated strands of restriction enzyme fragments revealed that between 15–80% of viral mRNA from untreated control cells was present in the cytoplasm of cells treated with puromycin from the onset of infection (Fig. 4). Messenger RNA from the E2 region encoding the DNA-binding protein was suppressed the most and mRNA from the E1A region was least sensitive to drug treatment. The same

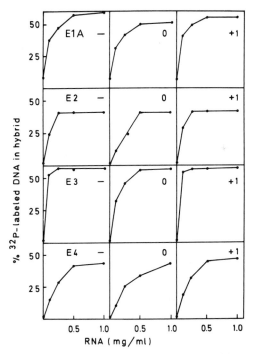

FIG. 4. Detection of early viral-specific mRNA from cells treated with puromycin. HeLa cells were infected with ad2 and cytoplasmic RNA was prepared at 5 hours after infection from cells not treated with puromycin (−), puromycin (100 μM) added at the onset of infection (0) or puromycin added 1 hour post infection (+1). RNA was hybridized in liquid to single-stranded ad2 ^{32}P-labeled restriction enzyme fragments and the RNA-DNA hybrids were isolated by hydroxylapatite chromatography. The abscissa shows the RNA concentrations used for hybridization and the ordinate percentage of the probe in hybrid. The concentration of RNA required to yield 50% of the observed saturation of the probe was used for quantitative comparison of the RNA. The ad2 restriction enzyme fragments used for hybridization were: E1A probe Sma I-J (0–2.9 map units); E2 probe Eco RI-B (58.5–70.7 map units); E3 probe Eco RI-D (75.9–84.0 map units); E4 probe Eco RI-C (89.7–100 map units).

52   HÅKAN PERSSON *et al.*

results were obtained when mRNA was selected by hybridization to
adenovirus DNA and translated in a cell-free system followed by assaying
radioactivity accumulating in polypeptides from each region (Fig. 5). The
two methods gave essentially the same quantitative results. When puromycin was introduced at 1 hour post infection the amount of viral mRNA
accumulating in the cytoplasm was increased by 20–150 % compared with

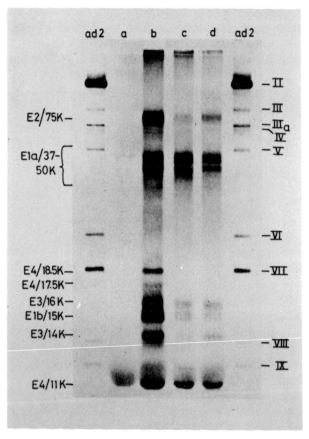

FIG. 5. Cell-free translation with viral mRNA prepared from puromycin-treated cells.
Cytoplasmic RNA was prepared at 5 hours after ad2 infection from cells treated with
puromycin (100 μM) from 2 hours before to 1 hour after infection. The cytoplasmic RNA was
purified by hybridization to ad2 DNA and the purified mRNA was translated *in vitro* and the
translation products were analyzed by SDS-PAGE. The location of the early proteins on the
genome was established by cell-free translation of ad2 mRNA purified by hybridization to
selected restriction fragments of ad2 DNA (not shown). (a) No RNA added. The RNA used for
hybridization and translation was prepared from infected cells 5 hours post infection, treated
with (b) puromycin from 1 hour post infection, (c) puromycin from the onset of infection, (d)
no puromycin.

untreated controls when assayed by *in vitro* translation. The mRNA made in the presence of puromycin was functional for translation irrespective of whether the drug was introduced at the onset or 1 hour post infection. Heteroduplex mapping, analytic cap analysis and size measurements of the poly(A) tail of the viral mRNA made in the presence of puromycin revealed no difference from mRNA of untreated control cells. We therefore conclude that puromycin unravels a translation control *in vivo* and a virus-coded or virus-induced cell product is required to allow selective translation of viral mRNA among the abundant cellular mRNA. Since puromycin also affects RNA synthesis we cannot at this stage claim that the controlling element is a protein, although this appears likely.

GENE PRODUCTS CONTROLLING THE BLOCK

A complementation assay was introduced using a low multiplicity infection with ad5 followed within 2 hours by a challenge with ad2 virus at high multiplicity. Some early polypeptides from these two viruses have the same antigenic characteristics but they vary in size (Fig. 6). If complementation is successful both polypeptides can be observed after immunoprecipitation and SDS-PAGE. When puromycin was introduced together with the complementing virus no products were observed after reversion of the *in vivo* block. On the other hand, when puromycin was introduced together with the challenge ad2 virus, both products were made after reversion, suggesting that ad5 at low multiplicity can complement ad2 for the control element regulating *in vivo* translation. When deletion mutants in the E1A or the E1B region were used as complementing viruses no complementation was observed (Fig. 7). These results suggest that a virus product from the E1B region is either directly or by induction involved in the expression of the translational control element since the mutants in both the E1A and the E1B regions were defective in expressing E1B products (Jones and Shenk, 1979b; Berk *et al.*, 1979b). Our results showing that a mixture of region E1A and E1B mutants (hr1 + hr7) partially overcomes the translation block suggest that the puromycin block requires a functional early region E1B. A similarity may therefore exist between the phenotype of the region E1B mutants (dl 313 and hr7) and the translation block described here. Both puromycin pretreatment and infection with region E1B mutants yield early viral mRNAs from all regions of the genome. The region E1B mutants which fail to grow in HeLa cells may therefore have a block at the translation level. If so, these mutants should have an impaired ability to synthesize early proteins *in vivo*. Lassam *et al.* (1979) recently showed that the group II mutants do not synthesize the E1B/58K protein but the other early

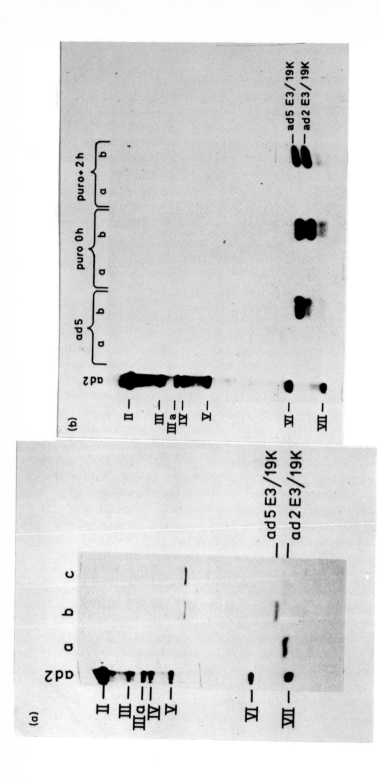

FIG. 6. Complementation with adenovirus type 5. (A) HeLa cells infected with ad2 or ad5 were treated with cycloheximide (25 µg/ml) from 2 to 5 hours post infection and then pulse-labeled for 30 min with ^{35}S methionine. Cell extracts were prepared and immunoprecipitated with the ad2 anti-E3/19K serum. (a) Ad2-infected cells; (b) ad5-infected cells; (c) a pre-immune serum with ad5 extracts. The immunoprecipitates were analyzed by SDS-PAGE and fluorography.

(B) HeLa cells were infected with ad5 (1 PFU/cell) and at 2 hours post infection the cells were divided into three aliquots. One was mock-infected (ad5), the second infected with ad2 in the presence of puromycin (puro 0h) and the third infected with ad2 and puromycin was added 2 hours after the ad2 infection. The cells were washed and pulse-labeled with ^{35}S methionine for 30 min at 5 hours after the ad2 infection. Cell extracts were prepared and immunoprecipitated with a normal rabbit serum (a) or the ad2 anti-E3/19K serum (b). The samples were analyzed by SDS-PAGE and fluorography.

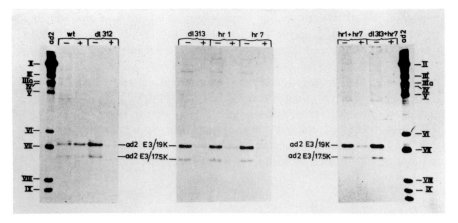

FIG. 7. Complementation with ad5 mutants. HeLa cells were infected with ad5 wild-type virus (wt) or the indicated mutants of ad5. Each virus was added at a multiplicity of infection corresponding to 0.1 PFU/cell. The samples were divided into two aliquots at 2 hours post ad5 infection. Both samples were infected with ad2 either in the presence (+) or in the absence (−) of puromycin (100 μM). The cells were washed and pulse-labeled with ^{35}S methionine for 30 min at 5 hours post ad2 infection. Cell extracts were prepared and immunoprecipitated with the ad2 anti-E3/19K serum. The immunoprecipitates were analyzed by SDS-PAGE and fluorography. When mixtures of different mutants were used in the pre-infection step each mutant was used at 0.05 PFU/cell.

polypeptides are accumulated normally. They used, however, a high multiplicity of infection (35 PFU/cell) and the translation block may have been masked since the mutants are leaky under these conditions (Shenk *et al.*, 1979). A multiplicity-dependence was also observed in our complementation assay since the translation block was partially complemented by all mutants at a multiplicity of 10 PFU/cell (not shown). Additional experiments established that the human 293 cells transformed with adenovirus type 5 which contains the E1A and E1B regions of the genome (Aiello *et al.*, 1979) could complement the *in vivo* translation block (not shown).

MULTIPLE CONTROLS IN EARLY ADENOVIRUS GENE EXPRESSION

There appears to be a contradiction between the results obtained with the host-range mutants and those obtained with puromycin. In the former case a transcriptional or a post-transcriptional block was identified (Jones and Shenk, 1979b; Berk *et al.*, 1979b), while puromycin only revealed a translational control under *in vivo* conditions. One explanation of this discrepancy might be that *two* control elements reside in the immediate early regions of the adenovirus genome, one governing a transcriptional step and the other translation. Since the adenovirus genome, like other virus DNAs, is compressed with at least three mRNAs from the E1A region (Spector et al., 1978; Chow et al., 1979), even single mutations may affect more than one product and the transcriptional block may obscure a translational control element from a neighboring region. If two control events are involved, the present results with puromycin where only a minor inhibition of viral mRNA was observed, suggest that the controlling element for the transcriptional or post-transcriptional control is an RNA moiety and experiments are in progress to delineate the details of these important control elements.

REFERENCES

Aiello, L., Giulfole, R., Huebner, K. and Weinman, R. (1979). *Virology* **94**, 460–469.
Berk, A. J. and Sharp, P. A. (1978). *Cell* **14**, 695–711.
Berk, A. J., Lee, F., Harrison, T., Williams, J. and Sharp, P. A. (1979a). *Cell* **17**, 935–944.
Berk, A. J., Lee, F., Harrison, T., Williams, J. and Sharp, P. A. (1979b). *In* "Eukaryotic Gene Regulation" (R. Axel, T. Maniatis and C. F. Fox, eds) Vol. XIV. ICN-UCLA Symposium on Molecular and Cellular Biology.
Carter, T. H. and Blanton, R. A. (1978). *J. Virol.* **28**, 450–456.
Chow, L. T., Broker, T. R. and Lewis, J. B. (1979). *J. molec. Biol.* **134**, 265–303.
Gilead, Z., Jeng, Y., Wold, W. S. M., Sugawara, H., Rho, M., Harter, M. L. and Green, M. (1976). *Nature, Lond.* **264**, 263–266.
Harrison, T., Graham, F. and Williams, J. (1977). *Virology* **77**, 319–329.
Harter, M. L. and Lewis, J. B. (1978). *J. Virol.* **26**, 736–749.
Jones, N. and Shenk, T. (1979a). *Cell* **17**, 683–689.
Jones, N. and Shenk, T. (1979b). *Proc. natn. Acad. Sci. U.S.A.* **76**, 3665–3669.
Lassam, N. J., Bayley, S. T. and Graham, F. L. (1979). *Cell* **18**, 781–791.
Lewis, J. B., Atkins, J. F., Baum, P. R., Solem, R., Gesteland, R. F. and Anderson, C. W. (1976). *Cell* **7**, 141–151.
Linné, T., Jörnvall, H. and Philipson, L. (1977). *Eur. J. Biochem.* **76**, 481–491.
Nevins, J. R., Ginsberg, H. S., Blanchard, J. M., Wilson, M. C. and Darnell, J. E. (1979). *J. Virol.* **32**, 727–733.
Persson, H., Pettersson, U. and Mathews, M. B. (1978a). *Virology* **90**, 67–79.
Persson, H., Öberg, B. and Philipson, L. (1978b). *J. Virol.* **28**, 119–139.
Persson, H., Signäs, C. and Philipson, L. (1979a). *J. Virol.* **29**, 938–948.
Persson, H., Perricaudet, M., Tolun, A., Philipson, L. and Pettersson, U. (1979b). *J. biol. Chem.* **254**, 7999–8003.
Persson, H., Jansson, M. and Philipson, L. (1980). *J. molec. Biol.* **136**, 375–394.
Pettersson, U., Tibbetts, C. and Philipson, L. (1976). *J. molec. Biol.* **101**, 479–501.
Ross, R. S., Flint, S. J. and Levine, A. J. (1980). *Virology* **100**, 419–432.
Shenk, T., Jones, N., Colby, W. and Fowlkes, D. (1979). *Cold Spring Harb. Symp. quant. Biol.* **44**, 367–375.
Spector, D. J., McGrogan, M. and Raskas, H. J. (1978). *J. molec. Biol.* **126**, 395–414.
Sugawara, K., Gilead, Z. and Green, M. (1977). *J. Virol.* **21**, 338–346.

Mutational Analysis of Regulatory Elements of Simian Virus 40

DANIEL NATHANS

Department of Microbiology, Johns Hopkins University School of Medicine, Baltimore, Maryland 21205, U.S.A.

INTRODUCTION

Simian Virus 40 (SV40) and similar small viruses are useful models for studying genetic regulation in mammalian cells. The genome of SV40, consisting of a duplex ring of DNA of 5243 nucleotide pairs (Fiers *et al.*, 1978; van Heuverswyn and Fiers, 1979; Reddy *et al.*, 1978) replicates in nuclei of infected monkey cells and undergoes regulated expression of its genes in the course of virus development. Since the nucleotide sequence of SV40 DNA has been determined and extensive genetic and biochemical studies have been carried out the genetic organization of the DNA is known in considerable detail (Tooze, 1980) (see Fig. 1). Of particular note for the purpose of this chapter is the segment of DNA lying between the start of early and late genes (the "regulatory segment") that contains the origin of replication of viral DNA (*ori*), sequences corresponding to the 5' ends of mRNAs, and T-antigen binding sites, but lacks protein coding sequences (Fig. 2). My colleagues and I set out to determine what sequences of nucleotides in the regulatory segment make up the replication origin and which are involved in regulating gene transcription. We were also interested in delineating possible domains of SV40 T-antigen involved in these regulatory phenomena. Our approach has been a genetic one, namely, the isolation of mutants with nucleotide changes in this region that are defective in DNA replication and/or transcription, thereby allowing a correlation between sequence changes and regulation of these events. Since no direct selection for the desired phenotypes was available, we applied a local mutagenesis procedure to construct viral DNA molecules with base substitutions or small deletions in the segment of interest, concentrating initially on partially or conditionally defective mutants that can be tested for regulatory activities directly in infected cells in culture.

57

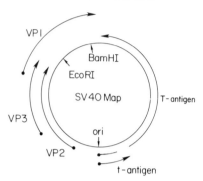

FIG. 1. A functional map of the SV40 genome. The duplex circular genome is represented by the inner circle showing the unique origin of DNA replication (*ori*) and the positions of the single cleavage sites for Eco RI and Bam HI. Around the outside is depicted the approximate coding regions for the viral late proteins VP1, VP2, VP3 and the early gene products large T-antigen and small t-antigen; the direction of their transcription 5' to 3' is given by the arrows. The T-antigens are synthesized early after infection of monkey cells and are found in SV40-transformed cells. T-antigen, a multifunctional protein of about 84 000 daltons, is required for the initiation of viral DNA replication, stimulation of cellular DNA and RNA synthesis, neoplastic transformation, negative control of early region transcription and possibly turn-on of late transcription, as well as supplying an adenovirus-helping function to enable adenoviruses to grow on normally non-permissive monkey cells. The other early protein, t-antigen, is about 17 000 in molecular weight and its function is at present unclear since it is not required for lytic growth. The gene products VP1, VP2 and VP3 are the structural proteins of the virus capsid and are synthesized late in infection. For reviews see Tooze, 1980.

ISOLATION OF REGULATORY MUTANTS

The local mutagenesis procedure developed by Shortle and Nathans (1979) allows the construction of DNA molecules containing bisulfite-induced base substitutions at preselected restriction sites in the DNA. Shortle's first mutants were constructed at the single Bgl I site of SV40 DNA located in a long G·C-rich palindrome within the regulatory segment (Fig. 3). He created a target for bisulfite by nicking the DNA with Bgl I and extending the nick to a small gap by means of the 5' and 3' exonucleolytic activity of DNA polymerase I (Shortle and Nathans, 1978). To isolate mutants with changes adjacent to the palindrome, Daniel DiMaio positioned a small gap near the Bgl I site by controlled nick translation from the Bgl I site (DiMaio and Nathans, 1980). From such gapped molecules he constructed mutants with small deletions or base substitutions as shown in Fig. 4. Transfection of BSC-40 monkey cells with the variously modified DNAs allowed efficient isolation of non-defective mutants or mutants with partial or conditional defects in growth, evidenced by changes in plaque size or character (Fig. 5). More recently, non-viable mutants have been isolated as plasmid recombinants in *E. coli* (Peden *et al.*, 1980).

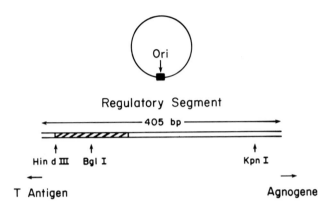

FIG. 2. The regulatory segment of the SV40 genome between the start of early and late genes, containing the origin of replication of viral DNA (Danna and Nathans, 1972; Sebring *et al.*, 1971); the 5' ends of early and late mRNAs (Ghosh *et al.*, 1978); and T-antigen binding sites (Tjian, 1978a; Shalloway, 1980). The nucleotide sequence is derived from the hatched region of the regulatory segment (Reddy *et al.*, 1978). Dark letters represent a palindromic sequence containing the Bgl I cleavage sites (arrows). "Agnogene" refers to a putative late gene identified by nucleotide sequence analysis only (Thimmappaya *et al.*, 1977).

Nucleotide Sequence Changes in Mutant DNAs

After isolation of a set of mutants by the procedures just outlined, their DNAs were screened by restriction followed by nucleotide sequence analysis of the relevant DNA fragment. The results are summarized in Fig. 6. Within the G·C-rich palindrome seven different mutants, each with a single base-pair change, were isolated. A set of overlapping deletion mutants was identified with the sequence changes given in Fig. 6; as expected, some of the deleted segments were to the left ("early" side) and some to the right ("late" side) of the palindrome. Bisulfite-induced base substitutions were also seen in these regions; in all cases these were multiple, probably reflecting the detection methods employed, namely altered plaque size or altered restriction pattern.

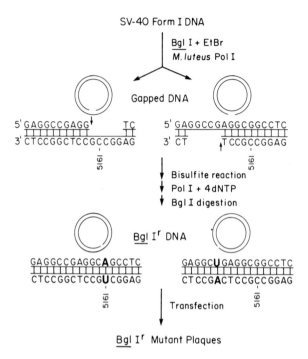

FIG. 3. Outline of the procedure used to construct mutants with base-pair substitutions within the G·C-rich palindrome of the regulatory segment. For details see Shortle and Nathans, 1979.

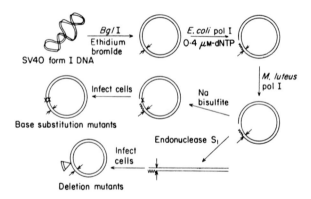

FIG. 4. Outline of the procedure used to construct mutants with deletions or base-pair substitutions adjacent to the Bgl I site of the G·C-rich palindrome. For details see DiMaio and Nathans, 1980.

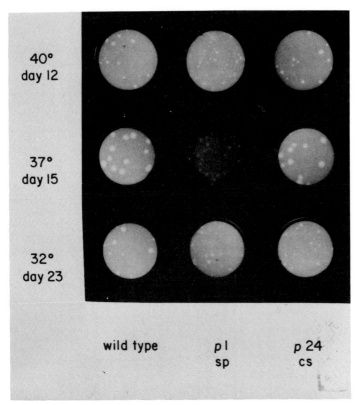

FIG. 5. Examples of plaque morphology changes resulting from mutations in the regulatory segment. Wild-type; (sp) small-plaque; (cs) cold-sensitive.

Phenotypes of Mutant Classes

As indicated in Fig. 5, many of the mutants with sequence changes in the regulatory region were partially or conditionally defective, as assessed by plaque morphology, whereas others were indistinguishable from wild-type virus by this criterion. Most common was cold sensitivity of plaque size relative to wild-type SV40. This characteristic could be due to alteration of a protein-binding site or to changes in sequences required for proper folding of an RNA transcript or mRNA. The cold-sensitive phenotype is especially common for mutants with changes in the palindrome and to the early side of the palindrome. Both sites are known to bind SV40 T-antigen preferentially (Tjian, 1978b).

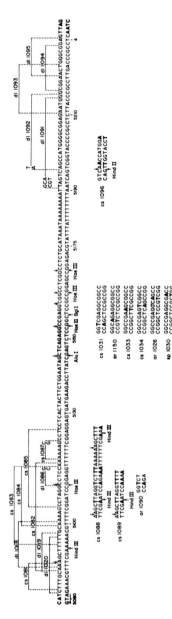

FIG. 6. Nucleotide sequence changes in mutants isolated by local mutagenesis as described in the text. (dl) Deletion mutant producing a wild-type plaque; (ar) altered restriction, wild-type plaque; (cs) cold-sensitive for plaque formation; (sp) small plaque. In some cases, new restriction sites resulted from mutational changes, as noted. Also included are 3 viable mutants (dl 1018, 1019, 1021) isolated earlier by deletion of the Hind III site preceeding the start of the gene for T-antigen and sequenced by S. Lazarowitz.

Shortle and DiMaio also measured rates of DNA replication in cells infected with several of the regulatory mutants. In general, most mutations in the palindrome led to decreased DNA replication or cold-sensitive DNA replication. (A notable exception was the mutant ar1026, in which the G·C base pair forming the axis of symmetry of the palindrome was changed to A·T). Especially defective was mutant sp1030 in which C·G at residue 5162 was changed to T·A; its rate of DNA replication was about 5% that of wild-type SV40 DNA. These defects were not *trans*-complementable and therefore were in a *cis*-acting element, presumably the origin signal of the SV40 replicon.

The mutants with deletions to the late side of the palindrome appeared to be non-defective. (Similar deletion mutants analyzed earlier by Subramanian and Shenk (1978) were also non-defective.) We interpret this to mean that the deleted sequences, which fall in the "third" T-antigen binding-site described by Tjian (1978b), are not essential parts of the origin signal, and indeed are not required for virus growth.

The defective mutants with deletions or base-pair changes on the early side of the palindrome that were analyzed by DiMaio showed normal or only slightly reduced rates of DNA replication. However, they over-produced early RNA and T-antigen (DiMaio, unpublished observations). DNA from these mutants showed greatly reduced binding of SV40 T-antigen *in vitro* (McKay and DiMaio, 1980). We infer that the mutations are probably within the postulated autoregulatory site where T-antigen binding represses early transcription (Tegtmeyer *et al.*, 1975; Read *et al.*, 1976; Khoury and May, 1978).

Pseudorevertants of Origin-defective Mutants

There is genetic evidence that T-antigen is the "initiator protein" (Jacob and Brenner, 1963) of the SV40 replicon (Tegtmeyer, 1972). To test this hypothesis further and to try to identify the region of T-antigen that interacts with the replication origin, Shortle and Margolskee (1979) isolated second-site revertants (pseudo-revertants) of origin-defective mutants. In all cases tested, the second-site mutations were located within the gene for T-antigen. We infer from this result that T-antigen interacts with the origin sequence in infected cells, as it does *in vitro*, and that this is essential for initiating DNA replication. The mutational sites of two pseudorevertant mutants have been sequenced by Margolskee; the base substitutions are within 30 base-pairs of each other in the large T-antigen coding sequence. Possibly this region of T-antigen is involved in recognizing the origin signal. DiMaio is now attempting to isolate similar second-site revertants for the putative autoregulation-defective mutants.

CONCLUSION

Local mutagenesis has been used to isolate a series of mutants with nucleotide deletions or substitutions in the regulatory segment of SV40 DNA. By correlating base-sequence changes with changes in viral DNA replication, transcription and binding of T-antigen to the regulatory segment, we have defined part of the replication origin sequence, and possibly also the site where T-antigen represses early gene transcription. We are now extending these studies by isolating additional mutants of SV40 as plasmid recombinants in *E. coli* (Peden *et al.*, 1980). Eventually we hope to define the full extent of each viral regulatory element at the sequence level. Such structurally and functionally characterized mutants should be useful in *in vitro* biochemical studies of DNA replication and transcription.

REFERENCES

Danna, K. J. and Nathans, D. (1972). *Proc. natn. Acad. Sci. U.S.A.* **69**, 3097–3100.
DiMaio, D. and Nathans, D. (1980). *J. molec. Biol.* **140**, 129–142.
Fiers, W., Contreras, R., Haegeman, G., Rogiers, R., Van de Voorde, A., Van Heuverswyn, H., Van Herreweghe, J., Volckaert, G. and Ysebaert, M. (1978). *Nature, Lond.* **273**, 113–120.
Ghosh, P. K., Reddy, V. B., Swinscoe, J., Lebowitz, P. and Weissman, S. M. (1978). *J. molec. Biol.* **126**, 813–846.
Jacob, F. and Brenner, S. (1963). *Compt. Rend. Acad. Sci.* **256**, 298–399.
Khoury, G. and May, E. (1978). *J. Virol.* **23**, 167–176.
McKay, R. and DiMaio, D. (1980). *Nature, Lond.* In press.
Peden, K. W. C., Pipas, J. M., Pearson-White, S. and Nathans, D. (1980). *Science, N.Y.* **209**, 1392–1396.
Reddy, V. B., Thimmappaya, B., Dhar, R., Subramanian, K. N., Zain, B. S., Pan, J., Ghosh, P. K. and Weissman, S. B. (1978). *Science, N.Y.*, **200**, 494–502.
Reed, S. L., Stark, G. S. and Alwine, S. (1976). *Proc. natn. Acad. Sci. U.S.A.* **73**, 3083–3087.
Sebring, E. D., Kelly, T. J., Jr., Thoren, M. M. and Salzman, N. P. (1971). *J. Virol.* **8**, 479–485.
Shalloway, D., Kleinberger, T. and Livingston, D. M. (1980). *Cell* **20**, 411–422.
Shortle, D. and Nathans, D. (1978). *Proc. natn. Acad. Sci. U.S.A.* **75**, 2170–2174.
Shortle, D. and Nathans, D. (1979). *Cold Spring Harb. Symp. quant. Biol.* **43**, 663–668.
Shortle, D., Margolskee, R. F. and Nathans, D. (1979). *Proc. natn. Acad. Sci. U.S.A.* **76**, 6128–6131.
Subramanian, K. N. and Shenk, T. (1978). *Nucl. Acids Res.* **5**, 3635–3642.
Tegtmeyer, P. (1972). *J. Virol.* **10**, 591–598.
Tegtmeyer, P., Schwarz, M., Collins, J. K. and Rundell, K. (1975). *J. Virol.* **16**, 168–178.
Thimmappaya, B., Reddy, V. B., Dhar, R., Celma, M., Subramanian, K. N., Zain, B. S., Pan. J. and Weissman, S. M. (1977). *Cold Spring Harb. Symp. quant. Biol.* **42**, 449–456.

Tjian, R. (1978a). *Cell* **13**, 165–179.
Tjian, R. (1978b). *Cold Spring Harb. Symp. quant. Biol.* **63**, 655–662.
Tooze, J. (ed.) (1980). "DNA Tumor Viruses", 2nd edition, part 2. Cold Spring Harbor Laboratory, New York.
Van Heuverswyn, H. and Fiers, W. (1979). *Eur. J. Biochem.* **100**, 51–60.

II

Structure and Expression of Retrovirus Genes

OK 10 Virus-transformed Bone-marrow-derived Cell Line Produces Infectious Transforming Virus without Detectable Associated Virus

SUSAN PFEIFER, RALF F. PETTERSSON, ISMO ULMANEN,
ANTTI VAHERI and NILS OKER-BLOM

*Department of Virology, University of Helsinki, Haartmaninkatu 3,
00290 Helsinki 29, Finland*

INTRODUCTION

Acute leukemia viruses, which cause rapidly progressive leukemias and various other types of neoplasms *in vivo*, and transformation of specific hematopoietic target cells *in vitro* (Graf and Beug, 1978) have recently received much attention. The discovery of three new types of viral oncogenes of cellular origin, *erb* (erythroblast-specific), *mac* (macrophage) and *myb* (myeloblast), specifically involved in transformation of hematopoietic cells by avian acute leukemia viruses (Roussell *et al.*, 1979; Beug *et al.*, 1979), has suggested a possible mechanism by which these viruses transform their target cells. It has been proposed that avian acute leukemia viruses arose by recombination of a common vector virus related to endogenous retroviruses and one of the three types of cellular genes possibly involved in hematopoietic differentiation (Roussell *et al.*, 1979). Cellular sequences related to these three newly discovered oncogenes have been found in the non-repetitive DNA of normal cells of all higher vertebrates studied. These genes have been strongly conserved during evolution and presumably have an essential function in the life-cycle of the cell in which they are expressed. According to one model, avian acute leukemia viruses transform their hematopoietic target cells through the action of the oncogene-coded transforming protein, which blocks the differentiation of the target cell by competitively inhibiting the action of a hypothetical homologous cellular differentiation protein (Graf *et al.*, 1978; Gazzolo *et al.*, 1979; Graf *et al.*, 1980).

All known acute leukemia viruses described so far are replication-defective,

69

requiring a helper virus, usually a lymphoid leukosis virus, to provide functions necessary for virus replication and assembly (Graf and Beug, 1978). In this report we have characterized the avian acute leukemia virus OK 10, produced by the continuous macrophage-like cell line, OK-BM (Pfeifer *et al.*, 1980). These cells produce transforming infectious OK virus, but we have been unable to find biological or biochemical evidence for the presence of an associated helper virus.

BIOLOGICAL PROPERTIES OF OK 10 VIRUS

The OK 10 virus complex was originally isolated from the embryo of a leukemic hen with lymphoid leukosis. The virus was clone-purified and shown to contain a transforming virus OK 10 and an associated virus OK 10 AV belonging to subgroup A (Oker-Blom *et al.*, 1978).

OK 10 virus has several properties of an acute leukemia virus both *in vivo* and *in vitro*. It induces multiple tumors within 2–6 weeks, characteristically located in the mesenterium of the gut and in the liver and testes (Hortling, 1978a). The associated, non-transforming OK 10 AV, like lymphatic leukemia viruses in general, induces lymphatic leukemia in chickens after a long 6- to 12-month lag period (Hortling, 1978a).

In vitro OK 10 virus transforms bone-marrow cells and chicken-embryo fibroblasts and under appropriate conditions, induces foci and soft agar colonies of transformed cells in cultures of chick-embryo cells (Oker-Blom *et al.*, 1978; Graf *et al.*, 1979). Like the acute leukemia viruses MC 29 and MH 2 (Hu *et al.*, 1978, 1979a; Gazzolo *et al.*, 1979), OK 10 virus transforms macrophages *in vitro* (Hortling, 1978b; Graf *et al.*, 1979). Non-producer clones of transformed cells containing replication-defective OK 10 virus have been isolated (Graf *et al.*, 1979; our unpublished results).

ESTABLISHMENT OF A STABLE BONE-MARROW-DERIVED CELL LINE, OK- BM, CONTINUOUSLY PRODUCING TRANSFORMING OK 10 VIRUS

The OK-BM cell line was established from the bone marrow of chicks inoculated with cells transformed by the original OK 10 complex containing both the transforming virus OK 10 and the associated virus OK 10 AV. The OK 10 virus-transformed cell line, OK-BM (Fig. 1), has now been maintained in culture for 444 passages and more than 4 years. This "immortality" as such is an unusual property for avian cell lines, which mostly have a limited life-span. Characterization of the cells has shown that they are of

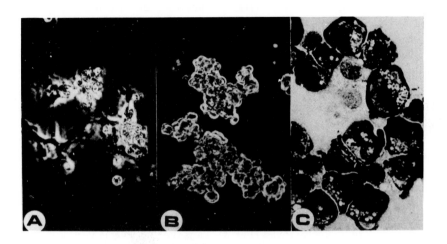

FIG. 1. OK-BM cells. (A) Early passage showing adherent macrophage-like cells (× 500). (B) During later passages the cells grew exclusively as a suspension culture, characteristically in clusters of a few hundred cells (× 500). (C) May-Grünwald-Giemsa-stained cytocentrifuged cells. Note abundant vacuolization, large eccentric nucleus and enlarged nucleolus (× 1000). (From Pfeifer *et al.* (1980) with permission.)

myeloid origin and consist of proliferating transformed macrophage-like cells (Pfeifer *et al.*, 1980) (Table I). The OK-BM cell line continuously produces infectious transforming OK 10 virus (OK 10-BM virus) with biological properties indistinguishable from those of the original OK 10 virus complex.

TABLE I

Cellular characteristics of the OK-BM cell line

Property	Finding
Histone H 5	Negative
Surface IgG	Negative
Retrovirus core protein p27	Present
Phagocytic capacity	Strong
Rosette assay for Fc receptors	Strong
Esterase activity	Strong
Intracellular fibronectin	Present

CHARACTERIZATION OF THE VIRUS PRODUCED BY THE OK-BM CELL LINE

The morphology of the OK 10 virus produced by the OK-BM cell line is typical of C-type viruses as judged by electron microscopy (Fig. 2).

Virion RNA. RNA extracted from the OK 10-BM virus had a sedimentation coefficient of 60–70S in sucrose density gradients. The RNA of the virus has been analyzed in either non-denaturing or denaturing agarose gels (Bailey and Davidson, 1976) containing 5 mM methylmercuryhydroxide, and compared to the RNAs of RSV strain PR-A, of the avian acute leukemia viruses strains MC 29 and MH 2, and of the lymphoid leukosis viruses RAV_1, RAV_2, RAV_3, and the associated virus OK 10 AV of the original OK 10 complex. Under non-denaturing conditions the ^{32}P-labeled RNA of the OK 10-BM virus migrated as a single peak comigrating both with the ^{3}H RNAs of the transformation-defective (td)PR-A(RSV) and with the ^{3}H RNA of leukosis viruses (Fig. 3). Thus the OK 10-BM virus contains only one size-class of RNA (8000–8500 nucleotides); that is, an RNA

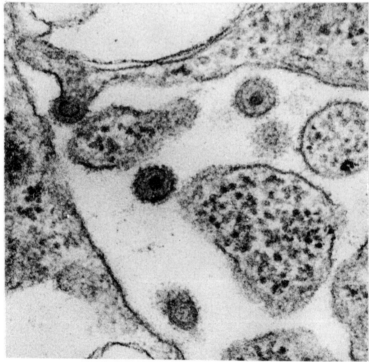

FIG. 2. Thin section of OK-BM cell. Budding and mature C-type viruses are seen (× 120 000).

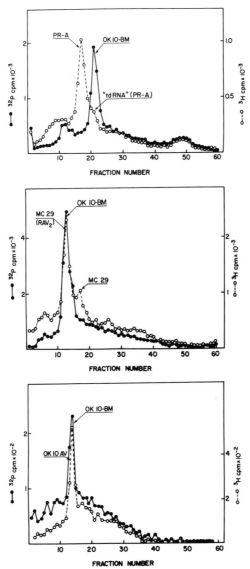

FIG. 3. Analysis of OK 10-BM RNA by agarose gel electrophoresis. OK-BM cells were labeled with ^{32}P, 400 μCi/ml medium, and cultures of chicken fibroblasts infected with PA-A, MC 29 (RAV$_2$), or OK 10 AV with 50 μCi ^{3}H uridine/ml medium for 6 hours. Culture medium was collected twice daily during 2–3 days and used for virus purification according to a published procedure (Pettersson *et al.*, 1971). Viral 60–70S RNA was isolated by sucrose sedimentation in 15–30 % density gradients, denatured for 1 min at 78°C and oligo-dT-purified. The oligo-dT-purified 35S ^{32}P and ^{3}H RNAs were mixed in the combination shown in the three lanes of the figure and subjected to electrophoresis in 1 % agarose gels. The gels were sliced in 2 mm segments and the radioactivity determined.

considerably larger than that of the other avian acute leukemia viruses, which are 5500–6000 nucleotides in size (Duesberg *et al.*, 1977, 1979; Hu *et al.*, 1979b).

The RNA of OK 10-BM virus was also analyzed by oligonucleotide fingerprinting (De Wachter and Fiers, 1972; Pettersson *et al.*, 1977).

The RNA from the OK 10-BM virus showed four unique oligonucleotides not found in the fingerprint of the OK 10 AV RNA, whereas the fingerprint of the OK 10 AV RNA displayed three unique oligonucleotides absent from the OK 10-BM virus RNA (Fig. 4, lower panels, arrow). Unanswered is the question of what the faint, submolar spots (Fig. 4, lower right panel, dotted circles) in the fingerprints of OK 10-BM represent. They do not seem to be

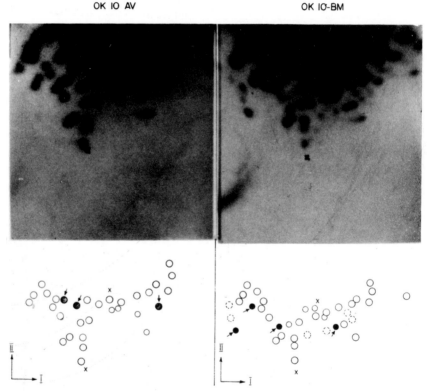

FIG. 4. Two-dimensional polyacrylamide gel analysis of T_1-oligonucleotides from OK 10 AV and OK 10-BM RNA. OK 10-BM and OK 10 AV RNA, labeled with ^{32}P and purified as described in Fig. 3, was digested with RNase T_1 and the oligonucleotides fractionated on a two-dimensional gel (De Wachter and Fiers, 1972; Pettersson *et al.*, 1977). The arrows indicate OK 10 AV and OK 10-BM specific spots.

derived from contaminating cellular RNAs, since they appear to be present regardless of whether the OK 10 virus is produced by OK-BM cells, chicken fibroblasts or quail fibroblasts.

Structural proteins. In the preliminary protein analysis carried out so far, we have detected distinct changes in the glycoproteins of the OK 10-BM virus. ^{35}S-methionine-labeled proteins of the OK 10-BM virus were analyzed in SDS-polyacrylamide slab gels and compared to the proteins of OK 10 AV, the associated virus of the original OK 10 virus complex (Fig. 5). It is evident that the polypeptide in the position of gp85 is not detected in the OK 10-BM virus using ^{35}S-methionine labeling. When the proteins were labeled with ^{3}H mannose, a polypeptide migrating with an apparent molecular weight of 78 000 daltons was seen in the OK 10-BM virus but not in the OK 10 AV virus (Fig. 6). This gp78 was immunoprecipitable with anti-gp85 serum (kindly provided by Dr H. Diggelmann, Swiss Institute for Experimental Cancer Research, Lausanne, Switzerland) (data not shown).

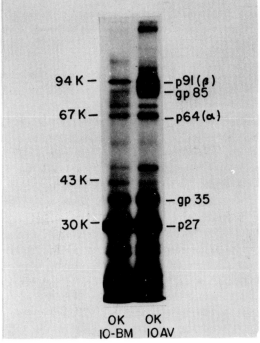

FIG. 5. Structural proteins of OK 10-BM and OK 10 AV viruses. OK-BM cells and chicken fibroblasts infected with OK 10 AV were labeled overnight with 50 µCi ^{35}S methionine/ml medium and virus purified from supernatant fluid. Electrophoresis was in 10% SDS-polyacrylamide gels (Laemmli, 1970). The figures on the left show positions of molecular weight marker proteins in kilodaltons and those on the right tentative identification of the major structural proteins.

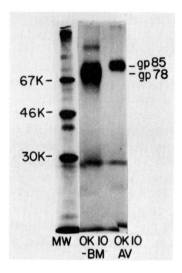

FIG. 6. Glycoproteins of OK 10-BM and OK 10 AV viruses. D-2-H³-mannose-labeled OK 10-BM and OK 10 AV viruses were analyzed in 10% SDS-polyacrylamide slab gels according to Laemmli (1970). For further explanation see legend of Fig. 5.

With respect to the other structural proteins, tentatively identified according to their apparent molecular weight, we found no changes in the OK 10-BM virus as compared to the OK 10 AV virus (Fig. 5).

In vitro *translation of the virion RNA*. In *in vitro* translation of the OK 10-BM virus RNA in reticulocyte extracts (Pelham and Jackson, 1976), a polypeptide of about 200 000 daltons was produced. This polypeptide was not detected in extracts programmed with OK 10 AV RNA. Products which based on their molecular weights could be pr76 *gag* and pr180 *gag-pol* (Eisenman and Vogt, 1978) were found in the *in vitro* translates of both viruses (Fig. 7).

All these products could be immunoprecipitated with anti-p27 rabbit serum (Fig. 7) (anti-p27 serum kindly provided by Dr H. Diggelmann), suggesting that the p200, pr180 and pr76 are polyproteins containing p27-specific determinants.

Other experiments carried out to detect the presence of a possible helper virus. The presence of only one size-class of RNA, the results obtained by fingerprint analysis, and the absence of gp85 in the OK 10-BM virus suggests that the OK 10-BM virus may contain little or no helper virus.

In agreement with these biochemical findings we have not been able to detect an associated virus by employing a number of other methods. These included interference tests using challenge virus representing subgroups A, B, C and D, all of which gave negative results even when 2-fold dilutions of

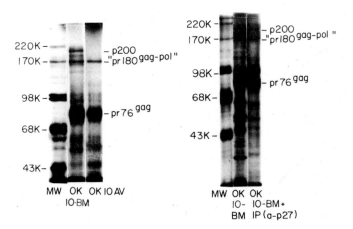

FIG. 7. Analysis by polyacrylamide gel electrophoresis of *in vitro* translation products of OK 10-BM virion RNA. Virus was purified from the media of OK-BM cells and OK 10 AV-infected chicken fibroblasts and the 60–70S RNAs isolated as described in Fig. 3. Oligo-dT-purified 35S RNA was used for *in vitro* translation in reticulocyte extracts according to Pelham and Jackson (1976) and in the case of OK 10-BM RNA also for subsequent immunoprecipitation with anti-p27 (track furthest on the right). MW = molecular weight marker proteins in kilodaltons.

OK 10-BM virus were used (Pfeifer *et al.*, 1980). Neither could we detect any non-transforming virus by assaying for reverse transcriptase activity in dilutions beyond the end-point titer of the focus-forming virus (Pfeifer *et al.*, 1980). Immunofluorescence tests using anti-p27 serum also speak against the presence of an associated non-transforming virus, since only the transformed foci, but not areas of non-transformed cells in cultures of chicken-embryo cells infected with OK-BM virus, were stained. Thus, on the basis of the evidence so far obtained (Table II), we conclude that OK 10-BM virus may consist of an infectious transforming virus without a helper virus or with amounts of helper virus which we are unable to detect by the methods employed.

GENERATION OF QUAIL CELL LINES PRODUCING DEFECTIVE NON-INFECTIOUS OK 10 VIRUS

The generation of non-producer cells at high frequency—or in fact defective producers—is the only strong argument in favor of a yet unidentified associated virus possibly present in the OK 10-BM virus.

We have isolated non-producer cell lines according to Graf *et al.* (1979) and tested for virus production using the focus-formation and the reverse-

TABLE II

Evidence in favor or against the presence of helper virus in the culture medium of OK 10 virus-transformed bone-marrow cell line

Against	In favor
Interference with subgroups A–D is negative	Easy to produce quail "non-producers" (defective producers)
Reverse transcriptase activity beyond end-point dilution of transforming virus is negative	Unidentified faint, submolar spots in OK 10-BM RNA fingerprints
Immunofluorescence using anti-p27 detected only in OK 10-BM foci	
Fingerprints of OK 10-BM RNA lack OK 10 AV-specific spots	
Nucleic acid hybridization of OK 10-BM RNA has failed to detect helper sequences	

transcriptase assays from culture media of quail-fibroblast clones infected and transformed by the OK 10-BM virus. About 60–70% became "non-producers". The non-producer cells had a transformed morphology and an expanded life-span *in vitro*. We have investigated the properties of two non-producer cell lines, OK 10 QDP 9C (isolated in our laboratory) and OK 10 QB5 (kindly supplied by T. Graf) in more detail. Both cell lines are actually defective producers as they produce large amounts of non-infectious virus particles into the culture medium. Both were negative in the focus-formation assay, but were rescuable with subgroup A helper viruses; viruses of the other subgroups grew poorly in our quail cells. Both were negative in the interference assay using challenge virus representing subgroups A–D. The 9C clone was strongly positive for reverse transcriptase but the B5 clone was reverse transcriptase negative (Table III). Both cell lines, when inoculated into animals *in vivo* produced tumors from which cell lines could be re-established. The cells from these tumors produced defective virus particles, with properties indistinguishable from those of virus produced by the inoculated cells. Specifically, no infectious virus was produced.

The defective virus particles produced by the cell lines OK 10 QDP 9C and OK 10 QB5, have also been analyzed by the biochemical methods described above for the OK 10-BM virus. The RNAs of both the 9C and B5 virus particles had a sedimentation coefficient of 60–70S in sucrose density gradients. Analysis of the RNAs in denaturing methylmercuryhydroxide agarose gels indicated that they consisted of only one size class of RNA with about 8000–8500 bases, thus being indistinguishable in size from RNA of the OK 10-BM virus as well as that of the OK 10 AV (Fig. 8).

TABLE III

Characteristics of virus particles produced by OK 10-BM-transformed quail "defective producers"

Property	QDP 9C	QB5
Infectivity (FFU/ml)	—	—
EM: particles	+ +	+
Genome size	8000–8500 nucleotides	8000–8500 nucleotides
Fingerprints	"Identical" to OK 10-BM	ND
Reverse transcriptase activity	100 000 cpm/5 µg viral protein	9000 cpm
Transforming virus rescuable with subgroup A helper	+	+

Oligonucleotide fingerprinting showed that the RNAs of the OK 10-BM virus and that of the OK 10 QDP 9C are very similar (Fig. 9). The four unique oligonucleotides of the OK 10-BM virus are also found in the OK 10 QDP 9C fingerprints. No OK 10 AV-specific oligonucleotides could be detected in either OK 10-BM or OK 10 QDP 9C. The RNA of the OK 10 QB5 virus particles has not yet been analyzed by oligonucleotide fingerprinting.

Protein analyses showed changes in the structural proteins of OK 10 defective particles. SDS polyacrylamide gel electrophoresis (Fig. 10) of ^{35}S-methionine-labeled OK 10 QDP 9C and B5 viral proteins has given the following results: (1) the polypeptide in the position of gp85 is not detected in either 9C or B5 defective particles and in addition also gp35 seems to be missing from B5; (2) polypeptides in the position of both the α and β components of the reverse transcriptase are found in 9C but not in B5. No differences were seen in the other structural proteins.

DISCUSSION

We conclude that in its biological properties the OK 10 virus closely resembles the replication-defective avian acute leukemia viruses. It is able to cause *in vivo* a severe leukemia and other malignancies after a short period of latency and to transform both fibroblasts and hematopoietic cells *in vitro*. In its target cell specificity (Gazzolo *et al.*, 1979; Beug *et al.*, 1979; Graf *et al.*, 1980) OK 10 virus is similar to the MC 29 group of viruses and nucleic acid hybridization has shown it to contain sequences of the *mac* gene, the oncogene of this group (Stehelin *et al.*, 1980).

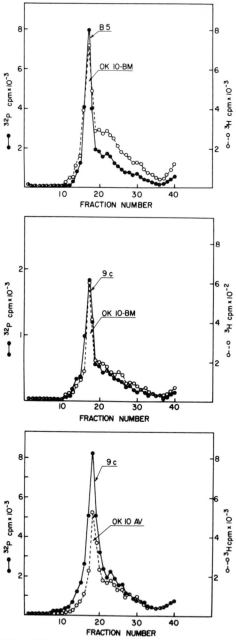

FIG. 8. Analysis of OK 10 QDP 9C and OK 10 QB5 RNAs in denaturing agarose gels. ^{32}P-labeled OK 10 QDP 9C and OK 10 QB5 35S RNAs were purified as in Fig. 3 from the supernatant fluid of quail cell lines producing these defective virus particles, and subjected to coelectrophoresis with H^3 OK 10-BM or H^3 OK 10 AV 35S RNAs. Electrophoresis was in a 1% agarose gel containing 5 mM CH$_3$HgOH according to Bailey and Davidson (1976).

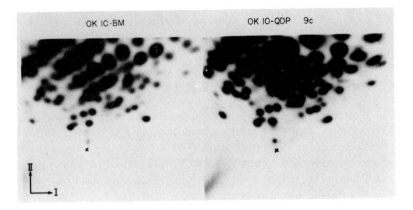

FIG. 9. Fingerprints of OK 10-BM and OK 10 QDP 9C RNAs. ^{32}P-labeled oligo-dT-purified OK 10-BM and 9C viral 35S RNAs were digested with RNase T_1 and analyzed by two-dimensional gel electrophoresis as in Fig. 4.

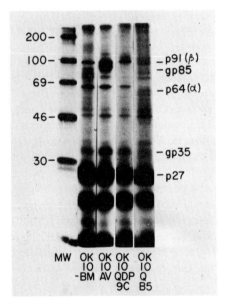

FIG. 10. Structural proteins of OK 10 QDP 9C and OK 10 QB5 viruses. ^{35}S-methionine-labeled OK 10 QDP 9C, B5, OK 10-BM and OK 10 AV viruses were prepared as in Fig. 5 and analyzed in 10% SDS-polyacrylamide gels. MW = molecular weight marker proteins in kilodaltons.

The virus produced by the continuous cell line, the OK 10-BM virus which also has *mac* gene sequences (Saule *et al.*, personal communication) differs, however, from other avian acute leukemia viruses in its much larger genome size and the lack of detectable associated helper virus. It has been reported that when the avian acute leukemia virus MH 2 replicates in macrophages, the transforming-to-helper virus ratio increases (Hu *et al.*, 1979a). It should be noted that the history of the virus produced by the OK 10-BM cell line is rather complex and thus this virus may have been generated during the prolonged growth (444 passages) of the virus in its transformed target cells. Thus the OK 10-BM virus may contain only a minimum amount of helper virus, below the level of detection by the methods used.

Analysis of the RNA, nucleic acid hybridization (S. Saule, personal communication) and oligonucleotide fingerprinting have failed to detect a helper virus. Circumstantial evidence is also given by other experiments such as negative interference assays and the immunofluorescence tests for p27. The fact that it is easy to obtain non-producer cells on the other hand argues in favor of the presence of a small amount of helper virus.

The analysis of the structural proteins as well as *in vitro* translation of virion RNA showed that the RNA encodes a p200, pr180$^{gag-pol}$, and pr76. This suggests the presence of a complete (or a nearly complete) *gag* and *pol* gene. Ramsay and Hayman (1980) have detected in OK 10-induced non-producer quail cells a polyprotein similar in size to the p200 found in our *in vitro* translates. They have shown that the polyprotein is related to the *gag* and *pol* gene products, since it contains all of the ^{35}S-methionine-labeled tryptic peptides found in pr76 and all but one of the *pol*-specific peptides. This suggests the presence of a complete *gag* gene and a nearly complete *pol* gene. In addition p200 contains unique peptides specific for OK 10 and not related to the *env* products. The fact that the OK 10-BM RNA directed the *in vitro* synthesis of both a 180 000 and a 200 000 dalton polypeptide in roughly equimolar amounts could mean (1) that a helper RNA, directing the synthesis of pr180, is present in the virus preparation, (2) that pr180 represents a premature translational termination of p200, or (3) that p200 can be occasionally cleaved *in vitro* to pr180.

We have been unable to detect a gp85 in the OK 10-BM virus but instead we found a smaller glycoprotein, gp78. One possible explanation for this could be that the envelope gene is deleted and codes for a truncated glycoprotein.

Since the genome of the OK 10-BM virus is of the same size as that of non-transforming leukosis viruses, about 8200 bases, it means that in order to accommodate a *mac* oncogene or part thereof there must be equivalent amounts of sequences deleted from the other genes. From our results and those of Ramsay and Hayman (1980) it appears that the major deletions are not in the *gag* or *pol* genes.

ACKNOWLEDGEMENTS

We would like to thank Annikki Kallio and Virpi Tiilikainen for expert technical assistance, and Drs P. Duesberg, T. Graf and D. Stehelin for valuable discussions. This work was supported by grants from the Finnish Cancer Foundation, the Jusélius Foundation, Helsinki, and the National Institutes of Health, NCI (grant no. CA 24605).

REFERENCES

Bailey, J. M. and Davidson, N. (1976). *Analyt. Biochem.* **70**, 75–85.
Beug, H., von Kirchbach, A., Döderlein, G., Conscience, J.-F. and Graf, T. (1979). *Cell* **18**, 375–390.
Bister, K., Vogt, P. K. and Hayman, M. J. (1978). *In* "Avian RNA Tumour Viruses" (S. Barlati and C. De Giuli-Morghen, eds) pp. 57–66. Piccin Medical Books, Padua.
De Wachter, R. and Fiers, W. (1972). *Analyt. Biochem.* **49**, 184–197.
Duesberg, P. H., Bister, K. and Vogt, P. K. (1977). *Proc. natn. Acad. Sci. U.S.A.* **74**, 4320–4324.
Duesberg, P. H. and Vogt, P. K. (1979). *Proc. natn. Acad. Sci. U.S.A.* **76**, 1633–1637.
Eisenman, R. N. and Vogt, V. M. (1978). *Biochim. biophys. Acta* **473**, 187–239.
Gazzolo, L., Moscovici, C., Moscovici, M. G. and Samarut, J. (1979). *Cell* **16**, 627–638.
Graf, T. and Beug, H. (1978). *Biochim. biophys. Acta* **516**, 269–299.
Graf, T., Royer-Pokora, B., Meyer-Glauner, W., Claviez, M., Götz, E. and Beug, H. (1977). *Virology* **83**, 96–109.
Graf, T., Ade, N. and Beug, H. (1978). *Nature, Lond.* **275**, 496–501.
Graf, T., Oker-Blom, N., Todorov, T. G. and Beug, H. (1979). *Virology* **99**, 431–436.
Graf, T. Beug, H. and Hayman, M. J. (1980). *Proc. natn. Acad. Sci. U.S.A.* **77**, 389–393.
Hortling, L. (1978a). *Acta path. microbiol. scand. Sect.* B **86**, 185–192.
Hortling, L. (1978b). Thesis, University of Helsinki.
Hu, S. S. F., Moscovici, C. and Vogt, P. K. (1978). *Virology* **89**, 162–178.
Hu, S. S. F., Duesberg, P. H., Lai, M. M. C. and Vogt, P. K. (1979a). *Virology* **96**, 302–306.
Hu, S. S. F., Lai, M. M. C. and Vogt, P. K. (1979b). *Proc. natn. Acad. Sci. U.S.A.* **76**, 1265–1268.
Laemmli, U. K. (1970). *Nature, Lond.* **227**, 680–685.
Oker-Blom, N., Hortling, L., Kallio, A., Nurmiaho, E.-L. and Westermarck, H. (1978). *J. gen. Virol.* **40**, 623–633.
Pelham, H. R. B. and Jackson, R. T. (1976). *Eur. J. Biochem.* **67**, 247–256.

84   SUSAN PFEIFER *et al.*

Pettersson, R. F., Kääriäinen, L., von Bonsdörff, C.-H. and Oker-Blom, N. (1971). *Virology* **46**, 721–729.

Pettersson, R. F., Hewlett, M. J., Baltimore, D. and Coffin, J. M. (1977). *Cell* **11**, 51–63.

Pfeifer, S., Kallio, A., Vaheri, A., Pettersson, R. F. and Oker-Blom, N. (1980). *Int J. Cancer* **25**, 235–242.

Ramsay, G. and Hayman, M. J. *Virology* **106**, 71–81.

Roussel, M., Saule, S., Lagrou, C., Rommens, C., Beug, H., Graf, T. and Stehelin, D. (1979). *Nature, Lond.* **281**, 452–455.

Stehelin, D., Saule, S., Roussel, M., Sergeant, A., Lagrou, C., Rommens, C. and Raes, M. B. (1980). *Cold Spring Harb. Symp. quant. Biol.* **44**, 1215–1223.

Origin of the Transforming Insert in the Abelson Murine Leukemia Virus Genome

DAVID BALTIMORE,[1] STEPHEN P. GOFF,[1] ELI GILBOA[2] and OWEN N. WITTE[3]

[1] Center for Cancer Research and
Department of Biology,
Massachusetts Institute of Technology,
Cambridge, Massachusetts 02139, U.S.A.

[2] Department of Biochemical Sciences,
Princeton University, Princeton,
New Jersey 08540, U.S.A.

[3] Molecular Biology Institute and
Department of Microbiology,
University of California,
Los Angeles, California 90024, U.S.A.

INTRODUCTION

Abelson murine leukemia virus (A-MuLV) is a replication-defective retro-virus capable of rapid *in vivo* and *in vitro* transformation of bone-marrow lymphoid target cells (Abelson and Rabstein, 1970b; Rosenberg and Baltimore, 1980). A-MuLV was originally isolated from a corticosteroid-treated BALB/c mouse that had been inoculated with the replication-competent virus, Moloney MuLV (M-MuLV) (Abelson and Rabstein, 1970a). The genome of A-MuLV is a hybrid containing portions of the parental M-MuLV genome joined to a large central substitution of novel genetic material (Shields *et al.*, 1979). This inserted DNA, referred to below as *abl*, presumably confers to A-MuLV its characteristic biological pro-perties, the ability to transform both cells of the B-lymphocyte lineage (Rosenberg and Baltimore, 1976) and fibroblastic NIH/3T3 cells (Scher and Siegler, 1975).

Only one protein product encoded by the A-MuLV has been demon-strated. It is of hybrid genetic origin, one segment reflecting a sequence of M-MuLV origin and the other being encoded as part of the *abl* sequence (Witte *et al.*, 1978; Reynolds *et al.*, 1978; Witte *et al.*, 1979b). A variety of A-MuLV strains have been characterized which make different-sized proteins; the prototype makes a protein of 120 000 molecular weight (P120) (Witte *et al.*, 1978; Rosenberg and Witte, 1980). The P120 protein is associated with a protein kinase activity capable of transferring the γ-phosphate of ATP to itself, forming a phosphotyrosine linkage (Witte *et al.*, 1980a); genetic evidence suggests that the protein is responsible for the ability of A-MuLV to transform target cells (Witte *et al.*, 1980b). A cellular protein has been detected in uninfected cells which shares antigenic determinants with the A-MuLV-specific portion of P120 (Witte *et al.*, 1979a). This protein, termed NCP150, is presumed to be the normal product of a cellular gene closely related to *abl*.

To examine the structure of the A-MuLV genome, we have isolated recombinant λ phages containing full-length DNA reverse transcripts from the viral RNA. With these cloned DNAs we have analyzed the structure of the viral genome and the origin of the *abl* DNA acquired during the formation of the genome. These sequences have shown that the *abl* sequences are homologous to DNA sequences in the normal mouse cell. In the mouse genome these sequences are broken up by introns but in the A-MuLV genome they consist of a contiguous coding sequence.

CLONING OF CIRCULAR A-MuLV DNA

To derive molecular clones of the A-MuLV genome, circular DNA was prepared from NIH/3T3 cells recently infected with a high titer stock of A-MuLV containing M-MuLV as the helper. We first ascertained that A-MuLV DNA is cleaved at a single site by Hind III and then cloned the DNA into λ Charon 21A at its Hind III site. The clones were identified by the Benton-Davis (1977) procedure using as probes either the cloned M-MuLV genome or a fragment of the M-MuLV genome that covers a region known to be missing from the A-MuLV genome. Plaques that hybridized with the full-length probe but not the restricted probe were picked as putative A-MuLV recombinant phages and were submitted to restriction enzyme analysis.

Four classes of recombinant phages were detected and one (λAb3) was selected for detailed analysis (Fig. 1). The region of homology with M-MuLV was easily recognized by the characteristic cleavage pattern and the long terminal redundancy (LTR) of the M-MuLV genome could be positioned by the unique combination of Xba I and Kpn I sites. The approximate region of *abl* was evident from cleavage sites that did not correspond to sites on M-MuLV DNA.

To map the *abl* insert more precisely it was heteroduplexed to a cloned M-MuLV DNA containing a single LTR (λMo8, Shoemaker *et al.*, 1980). The heteroduplexes showed a central homology region measuring 2.1 kb

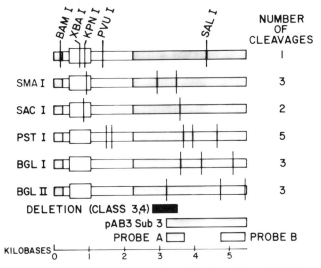

FIG. 1. Map of A-MuLV-cloned DNA. A member of the class 1 clones (λAb3) was mapped in detail. The A-MuLV-specific region is shaded; the single LTR is boxed. The cleavage sites for each enzyme are marked by vertical lines. The 700 bp deletion in classes 3 and 4 is localized at the bottom of the map. The region subcloned into plasmid pAB3Sub3, as well as the two smaller fragments used as 3′ and 5′ A-MuLV-specific probes, are shown aligned with the map.

and representing the sequence held in common between M-MuLV and A-MuLV. The *abl* sequences were evidenced as a bubble and the position of this sequence is shown as the gray area in Fig. 2.

With the knowledge of the structure of the λAb3 clone, the four classes of recovered inserts could be analyzed (Fig. 2). Class 1 inserts, typified by λAb3, were 5.5 kb long and had a single LTR and a 3.5 kb *abl* region. Class 2 inserts were 6.1 kb in length and were like the class 1 inserts except that they had two adjacent LTRs. These clones corresponded to the large circle detected in other retroviral systems (Shank *et al.*, 1978; Yoshimura and Weinberg, 1979). All such clones containing tandem duplications of the LTR were unstable and generated class 1 inserts presumably by homologous recombination (Bellet *et al.*, 1971).

Class 3 inserts, 4.5 kb in length, were similar to the class 1 inserts but lacked a 700 bp region within the *abl* region. Class 4 inserts contained the same deletion but included two copies of the LTR. Further work has shown that these latter two classes derived from a deleted form of A-MuLV found in the cells used to produce the viral stocks for this study. The deleted virus has been cloned away from its parental virus and shown to represent a transformation-defective A-MuLV (Witte *et al.*, 1980b).

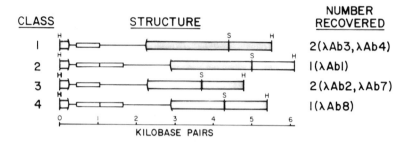

FIG. 2. The structure of the four classes of A-MuLV-cloned DNA. Classes 1 and 2 are full-length A-MuLV clones containing 1 and 2 LTRs, respectively; classes 3 and 4 contain an identical deletion of 700 bp in the A-MuLV-specific region, and also have 1 and 2 LTRs, respectively. The region homologous to M-MuLV is shown as a single line; the LTRs contained within this region are boxed. The unique A-MuLV regions are shaded. The names of individual clones are indicated.

PREPARATION OF AN A-MuLV-SPECIFIC PROBE

The detailed restriction map allowed the preparation of a cloned probe for the *abl* sequences. The position of the cloned probe on the A-MuLV genome is shown in Fig. 1 (pAB3Sub3). When this cloned probe was nick-translated with high specific activity α-^{32}P deoxyribonucleoside triphosphates, it could be used as a probe for the origin of the *abl* sequence.

To examine whether the *abl* probe was homologous to normal cellular DNA, DNA from BALB/c liver, C57Bl liver and the NIH/3T3 cell line was isolated, cleaved with one of several restriction enzymes and fractionated by agarose gel electrophoresis. The DNA was transferred to nitrocellulose paper (Southern, 1975) and hybridized with the *abl* probe. Autoradiography revealed that several fragments of different sizes were labeled (Fig. 3). Restriction enzymes used in this experiment did not cleave the viral probe and thus each fragment must have contained an isolated region of homology to this probe with non-homologous segments on both sides. Three strains of mice showed identical patterns with all tested enzymes. Most of the homology was found in the single 28 kb Eco RI band indicating that the *abl*-specific sequence must be located largely in one region of the mouse genome.

Further experiments using subfragments of the cloned probe (Fig. 2) have shown that the cellular sequence related to the *abl* probe is distributed over 11–20 kb of DNA. The probe itself, however, is only 2 kb in length and thus there must be intervening sequences within the normal counterpart of the *abl* sequence.

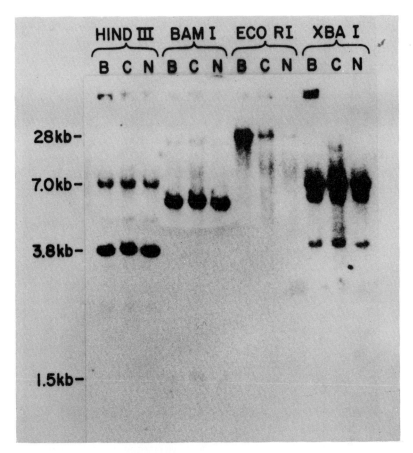

FIG. 3. Analysis of normal mouse-cell DNAs homologous to *abl*. Total cell DNA from BALB/c liver (lanes B), C57/BL6 liver (lanes C) or NIH/3T3 cells (lanes N) were digested with the designated restriction enzyme and applied to a 0.6% agarose gel. After electrophoresis the DNA was blotted and hybridized with the *abl*-specific probe, ^{32}P-labeled pAB3Sub3 DNA (see Fig. 1). The positions of the marker DNA fragments were determined directly from a photograph of the stained gel.

CELLULAR DNAs OF OTHER SPECIES ARE HOMOLOGOUS TO *abl*

To examine the homology of *abl* to DNA from other species, total DNAs of rat liver, Chinese hamster ovary cells and human HeLa cells were cleaved with various restriction enzymes, electrophoretically fractionated, and analyzed for hybridization to the *abl* probe. Rat DNA generated several fragments with homology to the probe and these were different in size from those of mouse DNA (Fig. 4). The intensity of the labeling of the most intense rat DNA bands was essentially identical to that of mouse DNA. Thus the A-MuLV gene is apparently very closely conserved between rat and mouse but the surrounding cleavage sites are different. Chinese hamster DNA showed weaker homology to the probe, and human DNA showed extremely weak but detectable homology. In similar analyses, NZW rabbit

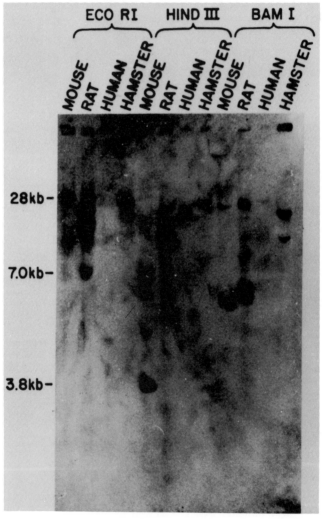

FIG. 4. Analysis of heterologous cellular DNAs for sequences homologous to *abl*. DNA from BALB/c mouse liver, rat liver, Chinese hamster ovary cells in culture or HeLa cells in culture was cleaved, displayed on a 0.6% agarose gel, and hybridized with ^{32}P-labeled pAB3Sub3 DNA.

liver DNA and chicken DNA also showed weak homology and individual patterns of bands.

Because of the difference between the restriction digest patterns of mouse DNA and Chinese hamster ovary DNA it has been possible to use hybrid cell lines containing a small number of mouse chromosomes on a Chinese hamster cell background to map the location of the *abl*-related sequences in mouse DNA. Preliminary evidence from collaborative experiments with Drs D'Eustachio and Ruddle have indicated that the *abl*-related sequence is located on either chromosome 2 or 19 of the mouse.

SIGNIFICANCE

These data imply that the A-MuLV genome arose by a recombination event between a retrovirus (M-MuLV) and normal cellular genetic information. In this way the viral genome appears to have acquired its transforming ability because deletion of part of the acquired sequence makes the virus transformation-defective. In many retroviral systems, normal cell genetic information plays a crucial role in the acquisition of transforming activity (Roussel *et al.*, 1979; Erikson, 1980; Shih and Scolnick, 1980). Thus A-MuLV confronts us with the same problem that is inherent in the structure of other transforming retroviruses: how does normal cellular genetic information make a virus able to transform cells? This question is similar to the puzzle of how simple chemically or radiation-induced mutations can transform cells and implies that small differences in the structure of genes or in their expression are sufficient to change a normal cell into a tumor cell.

ACKNOWLEDGEMENTS

This work was supported by grants from the National Cancer Institute and a contract from the Virus Cancer Program of the National Cancer Institute. D.B. is an American Cancer Society Research Professor. S.P.G. was a post-doctoral fellow of the Jane Coffin Childs Memorial Fund for Medical Research. O.N.W. was a post-doctoral fellow of the Helen Hay Whitney Foundation.

REFERENCES

Abelson, H. T. and Rabstein, L. S. (1970a). *Cancer Res.* **30**, 2208–2212.
Abelson, H. T. and Rabstein, L. S. (1970b). *Cancer Res.* **30**, 2213–2222.

Bellet, A. J. D., Busse, H. G. and Baldwin, R. L. (1971). *In* "The Bacteriophage Lambda" (A. D. Hershey, ed.), p. 501. Cold Spring Harbor Laboratory, New York.

Benton, W. D. and Davis, R. W. (1977). *Science, N.Y.* **196**, 180–181.

Erikson, R. L. (1980). *In* "Viral Oncology" (G. Klein, ed.), pp. 39–53. Raven Press, New York.

Reynolds, F. H., Sacks, T. L., Deobaghar, D. N. and Stephenson, J. R. (1978). *Proc. natn. Acad. Sci. U.S.A.* **75**, 3974–3978.

Rosenberg, N. and Baltimore, D. (1976). *J. Exp. Med.* **143**, 1453–1463.

Rosenberg, N. and Baltimore, D. (1980). *In* "Viral Oncology" (G. Klein, ed.), pp. 187–203. Raven Press, New York.

Rosenberg, N. and Witte, O. N. (1980). *J. Virol.* **33**, 340–348.

Roussel, M., Saule, S., Lagrou, C., Rommens, C., Beug, H., Graf, T. and Stehelin, D. (1979). *Nature, Lond.* **281**, 452–455.

Scher, C. D. and Siegler, R. (1975). *Nature, Lond.* **253**, 729–738.

Shank, P. R., Hughes, S. H., Kung, H., Majors, J. C., Quintrell, N., Guntaka, R. V., Bishop, J. M. and Varmus, H. E. (1978). *Cell* **15**, 1383–1395.

Shields, A., Goff, S., Paskind, M., Otto, G. and Baltimore, D. (1979). *Cell* **18**, 955–962.

Shih, T. Y. and Scolnick, E. M. (1980). *In* "Viral Oncology" (G. Klein, ed.), pp. 135–160. Raven Press, New York.

Shoemaker, C., Goff, S., Gilboa, E., Paskind, M., Mitra, S. W. and Baltimore, D. (1980). *Proc. natn. Acad. Sci. U.S.A.* **77**, 3932–3936.

Southern, E. M. (1975). *J. molec. Biol.* **98**, 503–517.

Witte, O. N., Rosenberg, N., Paskind, M., Shields, A. and Baltimore, D. (1978). *Proc. natn. Acad. Sci. U.S.A.* **75**, 2488–2492.

Witte, O. N., Rosenberg, N. E. and Baltimore, D. (1979a). *Nature, Lond.* **281**, 396–398.

Witte, O. N., Rosenberg, N. and Baltimore, D. (1979b). *J. Virol.* **31**, 776–784.

Witte, O. N., Dasgupta, A. and Baltimore, D. (1980a). *Nature, Lond.* **283**, 826–831.

Witte, O. N., Goff, S., Rosenberg, N. and Baltimore, D. (1980b). *Proc. natn. Acad. Sci. U.S.A.* **77**, 4993–4997.

Yoshimura, F. K. and Weinberg, R. A. (1979). *Cell* **16**, 323–332.

Structural and Functional Properties of DNA Intermediates in the Replication of Retroviruses

HAROLD E. VARMUS, JOHN E. MAJORS, RONALD
SWANSTROM, WILLIAM K. DeLORBE, GREGORY S. PAYNE,
STEPHEN H. HUGHES, SUZANNE ORTIZ, NANCY
QUINTRELL and J. MICHAEL BISHOP

*Department of Microbiology and Immunology, University of California,
San Francisco, California 94143, U.S.A.*

INTRODUCTION

The Replication of Retroviruses

RNA tumor viruses exhibit diverse biological behavior, and representatives have been isolated from many species of animals, but the agents are united by their common mode of replication, an unusual process which requires production of a DNA copy of the viral RNA genome and insertion of that DNA into host chromosomes. Once established in the host genome, viral genes are regulated and expressed in a fashion similar to that of cellular genes. Hence retroviruses provide useful models for understanding fundamental cellular processes.

A schematic representation of our current view of the replication cycle of retroviruses is presented in Fig. 1. Some older aspects of this scheme have been recently reviewed (Bishop, 1978; Coffin, 1979; Varmus et al., 1979a; Taylor, 1979); new facets will be illustrated during a discussion of our studies of the DNA of avian sarcoma virus (ASV) and mouse mammary tumor virus (MMTV) and are supported by work on a variety of retroviruses in several other laboratories (Van Beveren et al., 1980; Sutcliffe et al., 1980; Dhar et al., 1980; Shimotohno et al., 1980; Shoemaker et al., 1980).

Repeated sequences, a major theme, exist even at the most fundamental level, in that the viral RNA genome itself is diploid, composed of two single-stranded, identical subunits of c. 5–10 kb. For simplicity, only a single subunit has been drawn, but it is possible that both subunits are normally

93

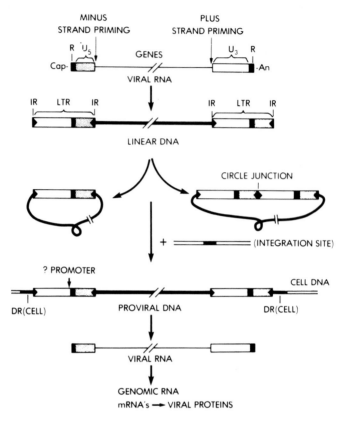

FIG. 1. Replication of the genomes of retroviruses. A detailed explanation of the scheme is provided in the text. R = direct repeat at ends of viral RNA; U_3 = sequences unique to the 3′ end of viral RNA which are present twice in linear DNA; U_5 = sequences unique to the 5′ end of viral RNA which are present twice in linear DNA; LTR = long-terminal direct repeat in linear DNA, composed of U_3, R and U_5; DR = direct repeat; IR = inverted repeat. The 5′ end of viral RNA is shown on the left, the 3′ end on the right.

used during the synthesis of viral DNA; diploidy may also contribute important structural features to the virion and facilitate genetic flexibility (e.g. heterozygosis and recombination).

The genomic RNA has features of eukaryotic mRNAs, being capped at the 5′ end and polyadenylated at the 3′ end; but these aspects of the RNA and the structural genes themselves are of less importance to the scheme under discussion than more unusual features indicated by the solid squares and rectangular boxes at the 3′ and 5′ termini. Each subunit ends with a short direct repeat (solid squares labeled "R") used in an early phase of

DNA synthesis to expedite a necessary transfer of reverse transcriptase between templates. The rectangular boxes (U_5 and U_3) are defined on one side by the ends of the viral genome (the "cap" and the poly(A) sequences are added following RNA synthesis and are not virus-coded), and on the other side by the sites at which the first (minus) and second (plus) strands of viral DNA are initiated. Priming of the minus strand is performed by a cellular tRNA, the 3' end of which is hydrogen-bonded to viral RNA 100–180 nucleotides from its 5' terminus (Taylor, 1977). Priming of the plus strands may occur at several sites during synthesis of the DNA of some retroviruses, but all retroviruses appear to have at least one highly preferred site for initiation of plus-strand DNA (Varmus et al., 1978a; Mitra et al., 1979; Kung et al., 1981). This site is located in a purine-rich region positioned 250 to 1200 bases from the poly(A) tract, but the nature of the primer is not known.

The completed product of DNA synthesis in the cytoplasm of infected cells is a linear duplex molecule; the linear DNA is slightly longer than a subunit of viral RNA and contains long terminal repeats (LTRs) composed of the boxed sequences (U_3, R, U_5) present at both ends of viral RNA (Shank et al., 1978a; Hsu et al., 1978). Although the nucleotide sequences at the absolute ends of linear DNA have not been determined directly, the structure has been closely approximated by restriction mapping of DNA from infected cells and by sequencing of minus strands synthesized in vitro and of cloned circular and proviral forms. The sequencing studies indicate that the LTRs conclude with short and often imperfect inverted repeats (shown as solid triangles labeled "IR").

Some of the linear DNA migrates to the nucleus where a fraction of it is converted to closed circular species (Shank and Varmus, 1978). Circularization appears to occur either by homologous recombination between the redundant ends (resulting in the loss of one copy of the LTR unit) or by direct joining of the ends of linear (Shank et al., 1978a,b; Hsu et al., 1978; Yoshimura and Weinberg, 1979). In some cases, the joining may be accompanied by the loss of sequence from one or both ends, but in at least some other cases, the sequence of the junctional region of cloned circular DNA suggests that no nucleotides have been lost from linear DNA (Swanstrom et al., 1981).

Integration of viral DNA into the host genome can occur in many different regions of cellular DNA (Hughes et al., 1978). Thus far, these sites appear to lack homology with each other or with the ends of viral DNA. However, it remains possible that preferred integration sites exist in host DNA. The integrated (proviral) DNA appears to be precisely colinear with the linear form of unintegrated DNA (Hughes et al., 1978; Sabran et al., 1979), thus ending with inverted repeats contained within the large direct repeats;

however, the proximal precursor to the provirus has not been ascertained (hence the ambiguous arrow in Fig. 1). The available sequence data from host-viral junctions reveal two additional, mechanistically significant attributes of proviruses: (i) the ends of proviruses appear to lack a few (generally two) nucleotide pairs present at the ends of linear DNA (and thus at the junction point in circular DNA bearing two complete copies of the LTR); (ii) proviruses are flanked by direct repeats of a short cellular sequence (4–6 nucleotide pairs) present once in unoccupied integration sites.

Proviruses normally harbor the complete sequence coding for viral RNA, stretching from the point within the left-hand LTR corresponding to the cap site to a point within the other LTR corresponding to the poly(A) addition site(s). Thus there are additional viral sequences (the U_3 region) upstream (in a transcriptional sense) from the complete copy of viral RNA, and additional sequences (the U_5 region) downstream (Fig. 1). Sequences likely to promote initiation of RNA synthesis at the cap site and to direct polyadenylation of viral RNA appear to be present at suitable locations within the LTRs. The primary transcript initiated within the leftward LTR is presumably equivalent to an unmodified subunit of viral RNA; after capping and polyadenylation it may be packaged into progeny virions, used as messenger RNA, or converted (by "splicing") into subgenomic messenger RNAs for internal cistrons.

STRUCTURAL FEATURES OF THE LTR OF AVIAN SARCOMA VIRUS

Molecular cloning of retroviral DNA now permits definitive sequencing of its interesting regions. To help understand the role of the long terminally repeated unit of viral DNA in DNA synthesis, integration and transcription, we determined the sequence of the LTRs present in molecularly cloned ASV DNA (Swanstrom *et al.*, 1981). Ideally, the complete sequence of the LTR should be determined from the ends of linear DNA, since the LTR could be altered by subsequent events in the life-cycle (e.g. circularization or integration). However, intact linear DNA has yet to be cloned from infected cells, and we have relied instead upon closed circular DNA isolated from quail cells acutely infected by the Schmidt-Ruppin A (SR-A) strain of ASV (DeLorbe *et al.*, 1980). We believe that one of our cloned molecules (SRA-2) contains two *complete* copies of the LTR for the following reasons: (i) the sequence very likely to be present at the right end of linear DNA (U_5; the sequence beginning with the first base after the minus strand priming site; Fig. 1) has been previously determined by analysis of cDNA synthesized *in vitro* (Shine *et al.*, 1977; Haseltine *et al.*, 1977), and this sequence is present in its entirety at the point in circular DNA at which the ends of linear DNA have been joined (the "circle junction"); (ii) the sequence derived from the 3′ end of the RNA at the circle junction most likely, for reasons discussed elsewhere (Swanstrom et al., 1981), represents all (or almost all) of the U_3 sequence expected in the LTR; and (iii) the sequence at the circle junction forms a 30-base, slightly imperfect palindrome (GCAGAAGGCTTCATT AATGTAGTCTTATGC), implying that each LTR in ASV DNA concludes with a 15-base-pair inverted repeat sequence (of which 3 bases are mismatched). Inverted repeat sequences of variable size (5 to 23 base pairs) have been found at or near the presumptive ends of LTRs of other retroviruses (Shimotohno et al., 1980; Sutcliffe et al., 1980; Van Beveren et al., 1980; Dhar et al., 1980; Shoemaker et al., 1980; Majors and Varmus, 1980, 1981), supporting the idea that the circle junction in this case corresponds to the junction between two complete LTRs.

On the other hand, we and others (Shoemaker et al., 1980; Ju et al., 1980; Highfield et al., 1980) have cloned circular DNAs which exhibit aberrations at the circle junction. In one case of ASV DNA which we have examined carefully (Swanstrom et al., 1981), the junction region is likely to reflect an event analogous to integration, since 2 base-pairs have been lost from the U_5 sequence and 61 from the U_3 sequence; it is characteristic of integrated termini to exhibit the loss of 2 bases from the LTR (see below).

The sequence of the ASV LTR reveals other significant features: (i) as predicted from current models for viral DNA synthesis (see Coffin, 1979, for review), only one copy of the R sequence is present in each copy of the LTR; (ii) an AT-rich sequence (TATTTAA), similar to other putative eukaryotic promoters (Ziff and Evans, 1978), is present from 24 to 30 bases preceding the probable start site for transcription (the point corresponding to the capping site at the 5′ end of genomic RNA); (iii) the sequence AATAAA, implicated as a recognition signal for polyadenylation (Proudfoot and Brownlee, 1974), precedes by 17–22 bases the region encoding the three adjacent sites which appear to be polyadenylated in ASV virion RNA (Schwartz et al., 1977).

PORTRAITS OF MMTV PROVIRUSES

Our recent studies of cloned proviruses of the mouse mammary tumor virus (MMTV) support the description of integrated DNA in the scheme shown in Fig. 1; in this section we will briefly review the rationale, difficulties and results of that work.

Although similar in many respects to other retroviruses which are more malleable experimentally, MMTV invites the study of proviral structure and function in view of its dramatic regulation at the transcriptional level by glucocorticoid hormones, its poorly understood capacity to induce mammary carcinomas and its presence in the germ line of many mice in the form of endogenous proviruses (for reviews of these topics, see Varmus *et al.*, 1978b; Hilgers and Bentvelzen, 1978; Varmus *et al.*, 1979b; Cohen and Varmus, 1979). However, our efforts to obtain molecular clones of MMTV DNA, starting either with circular or proviral forms, have been fraught with difficulty, regardless of whether we used plasmid or bacteriophage vectors. Judging from the variably deleted molecules isolated during attempts to clone circular DNA, it appears that a small region of the viral genome (less than 1 kb), near or in the *gag* gene encoded near the 5' end of MMTV RNA, is relatively or absolutely unclonable in *E. coli* K12 host-vector systems (Majors and Varmus, 1980, 1981). We refer to this sequence as the "poison sequence".

Despite this obstacle, we have been able to clone an intact, steroid-responsive proviral unit from MMTV-infected rat cells because we have fortuitously isolated a cell containing a single provirus lacking the "poison sequence". In addition, we have cloned interesting regions of other proviruses, particularly host-viral junctions, by using restriction enzymes which generate fragments outside the "poison sequence". The proviruses were isolated from rat XC cells infected with MMTV produced by a cultured C3H mammary tumor line (Fine *et al.*, 1974). Approximately two-thirds of cells cloned non-selectively after infection at a low multiplicity contained MMTV proviruses, and about half of these had a single provirus. As expected from previous studies of MMTV DNA (Cohen *et al.*, 1979; Ringold *et al.*, 1979), the proviruses are located in different regions of host DNA and are generally colinear with the unintegrated linear species. However, the provirus in one cell line, line 8, lacks most of the coding region for viral proteins (*gag* and *pol*), as well as the solitary Eco RI site present in the *pol* region of wild-type MMTV DNA. The line 8 provirus is transcribed into viral RNA, and the concentration of viral RNA is markedly stimulated by addition of dexamethasone to the culture medium. (The presence of a complete *env* gene in the line 8 provirus has been confirmed by the findings of normal *env* mRNA and *env* glycoprotein in line 8 cells (unpublished results of D. Robertson *et al.*).) The provirus in another cell line, line 34, is complete but not expressed at detectable levels, even in the presence of hormones.

All or parts of these proviruses were cloned initially in lambda phage vectors and subcloned in pBR322 (see Majors and Varmus, 1981, for details). In addition, DNA including the flanking regions from the line 8 provirus was in turn used as a hybridization probe to identify and clone an Eco RI fragment bearing the unoccupied integration site for the line 8

provirus from another clone of XC cells. Analyses during these manipulations indicated that the flanking region is composed mostly if not
completely of unique sequence DNA, and that no gross rearrangements or
deletions had occurred in this region during the integration events.

The sequencing strategy was designed to explore primarily the host-viral
junctions and possible regulatory signals encoded within the terminal
redundancies. Precise definition of the ends of the provirus was facilitated by
comparison of each junction with the corresponding internal region (i.e. the
priming sites for plus and minus strands with left and right junctions,
respectively; Fig. 1). Identification of the right end was also aided by
independent sequencing of the U_5 region from "strong stop" cDNA, primed
by the native tRNA primer in an *in vitro* reaction (J.E.M., unpublished).
Alterations of the host sequence were determined by comparison of the host
sequence at the right and left ends of the line 8 provirus with the
corresponding sequence from the unoccupied integration site.

The following observations pertinent to the scheme drawn in Fig. 1 were
made during the sequencing studies:

(i) The LTRs (*c.* 1350 bp) at the ends of the provirus are concluded with a
perfect inverted repeat of six nucleotide pairs $\left(\begin{array}{l} \text{TGCCGC ... GCGGCA} \\ \text{ACGGCG ... CGCCGT} \end{array}\right)$.

(ii) At least one and probably two of the nucleotide pairs from the U_5
sequence are missing from the right ends of proviral DNA. By positioning
the probable end of U_3 by indirect arguments (see (v) below), we believe two
nucleotide pairs from the U_3 sequence are missing from the left ends of
proviral DNA. The missing bases would not contribute to the length of the
inverted repeats in unintegrated DNA.

(iii) A cellular sequence of 6 nucleotide pairs present once in the
unoccupied integration site is present as a direct repeat on both sides of the
provirus. This hexanucleotide reads $\left(\begin{array}{l} \text{GTAAGG} \\ \text{CATTCC} \end{array}\right)$ in line 8 and $\left(\begin{array}{l} \text{GAGGTT} \\ \text{CTCCAA} \end{array}\right)$
in line 34.

(iv) There is no evidence of homology between the ends of the proviruses
and the immediately surrounding cellular DNA. Moreover, there appear to
be no similarities among the host sequences flanking the ends of MMTV
proviruses from our infected cells or from a mouse mammary tumor line
(Hager and Donehower, 1980).

(v) The site at which the critical plus strand is primed includes a purine-
rich plus-strand sequence (AAAAAGAAAAAGGGGGAAATG), the final
11 bases of which are identical to the functionally equivalent sequence in
avian sarcoma virus (ASV) DNA (AGGGAGGGGGAAATG) (Swanstrom
et al., 1981; Czernilofsky *et al.*, 1980). This finding is particularly striking

since the genomes of ASV and MMTV are not homologous in conventional annealing tests, and their plus-strand priming sites are different distances from the poly(A) sites.

(vi) The minus-strand priming site is preceded by an 18-base sequence which would provide a completely matched binding site for tRNA$_3^{lys}$, recently identified as the minus-strand primer for MMTV (Peters and Glover, 1980).

(vii) A sequence positioned 24–32 nucleotides upstream from the cap site (TATAAAAGA) is similar to many other putative eukaryotic promoters and identical to a sequence similarly positioned upstream from the initiation site for the late RNAs of adenovirus 2 (Ziff and Evans, 1978). A sequence of 10 nucleotides (CTTATGTAAA) is directly repeated in the region 40–60 nucleotides upstream from the cap site, but the significance of this repetition is obscure.

IMPLICATIONS OF PROVIRAL STRUCTURE FOR MECHANISMS OF INTEGRATION AND TRANSCRIPTIONAL REGULATION

Although the available data are insufficient to determine the mechanism of integration—or even to decide which of the three well-defined species of unintegrated DNA is the proximal precursor to the provirus—several constraints must now be placed upon any proposed mechanisms. (i) Viral DNA must enter host DNA either randomly or by recognition of signals more subtle than simple homology of the type used for integration of lambda phage DNA (Landy and Ross, 1977). (ii) The provirus must be colinear with unintegrated linear DNA. This implies either that linear DNA itself is the substrate for integrative recombination (an attractive possibility in view of its free ends) or that host or viral enzymes with suitable specificity are available to open circular DNA at precisely defined sites. (iii) The absence of a deletion in host DNA at the site of integration suggests that the ends of viral DNA must be juxtaposed prior to integration, either as part of a circular intermediate or by some means for bringing the ends of the linear species together. (iv) A duplication of a short region of host DNA must occur during integration; this would be accomplished most simply by the production of a staggered cleavage of cellular DNA to create ends for ligation as proposed for insertion of transposable elements (Grindley and Sherratt, 1978; Shapiro, 1979). Resulting single-stranded gaps on both sides of the provirus could then be filled by DNA polymerase, thereby completing the duplication. (v) It appears that a small number of base-pairs, generally two, is missing from the right and probably from the left end of integrated viral DNA. A mechanism for eliminating this limited amount of viral DNA between synthesis and integration is required of any model for integration.

Whatever the mechanisms of integration, it is apparent that the provirus is admirably suited to help regulate its own transcription and the processing of transcripts. In this light it seems probable to us that the response of MMTV gene expression to glucocorticoid hormone is controlled by a viral sequence (e.g. a binding site for activated hormone receptor) located within the c. 1200 nt of U_3 sequence upstream from the probable initiation site for RNA synthesis.

A virus-coded regulatory sequence would be the simplest answer to the problem of how MMTV proviruses at many different sites in several different cell types may be strongly influenced by glucocorticoid hormones (Varmus et al., 1979b). On the other hand, we do not mean to exclude the potential significance of flanking cellular DNA (or of chromatin conformation at the integration sites) as a determinant of the transcriptional activity of proviruses; the environs of a provirus could, for example, account for the complete absence of expression of certain MMTV proviruses (e.g. the provirus in line 34; see also Ringold et al., 1979). Use of the cloned line 8 provirus and fragments of it as templates for transcription in vivo or in vitro should be helpful in defining regulatory domains.

SIGNIFICANCE OF THE STRUCTURAL SIMILARITIES BETWEEN PROVIRUSES AND TRANSPOSABLE ELEMENTS

The structure of MMTV proviruses (and of retrovirus proviruses generally) has a striking similarity to the structure of certain transposable elements of bacteria (Bukhari et al., 1977), yeast (Cameron et al., 1979) and Drosophila (Strobel et al., 1979). The structural homology is illustrated dramatically by a comparison of a provirus with the bacterial transposon it most resembles, Tn9 (Fig. 2; MacHattie and Jackowski, 1977). In both elements, domains encoding gene products are encompassed by large direct repeats, the direct

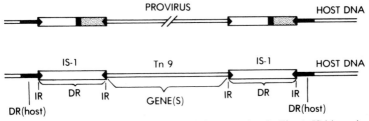

FIG. 2. Comparison of provirus with Tn9. Symbols are used as in Fig. 1. IS-l is an insertion element present as a direct repeat at the termini of the transposon, Tn9. Comparative properties are discussed in the text.

repeats conclude with short inverted repeats, and the entire units are flanked by short duplications of host DNA generated during insertion.

Despite the concordance of these physical symmetries, it would be premature to suggest that proviruses and transposons are functionally similar. For example, transposition in prokaryotes appears to depend upon the activity of transposases encoded by such elements (Heffron *et al.*, 1979), in conjunction with host enzymes used in DNA replication; movement of transposons is closely wedded to DNA synthesis, without the apparent generation of free forms. Proviruses can also be viewed as mobile genetic elements, but the transposition into new integration sites involves transcription into viral RNA by host RNA polymerase, transport of RNA in virus particles, DNA synthesis from RNA templates by virus-coded polymerase, and integration of free DNA into host chromosomes. It is entertaining in this context to consider reverse transcriptase as an analogue of transposases, but it is likely that the enzymatic functions are completely unrelated.

Although the mobility of transposons and proviruses probably differs mechanistically under usual circumstances, the strong structural relationship should prompt a concerted effort to seek behavior of proviral DNA akin to that exhibited by transposable elements. In addition to effecting transposition, transposons can occasionally generate deletions (precise excisions or deletions involving part of the element and/or flanking host DNA), inactivate genes by insertion, or exert polar or promotional influences upon "downstream" cellular DNA (Saedler *et al.*, 1974; Bukhari *et al.*, 1977). Strong selective pressures may well be required (as they are in prokaryotes) to detect such rare events. In several contexts in which such selective pressures exist for eukaryotic cells, we have identified proviruses affected by deletions involving the LTRs, proviruses acting as insertional mutagens, and proviruses acting as portable promoters of transcription of cellular sequences.

Insertion and Deletion Mutants in MuLV-infected ASV-transformed Rat Cell

We have developed an experimental system in which to investigate the mutagenic properties of a retrovirus which does not transform cultured cells (Moloney strain of murine leukemia virus, Mo-MuLV), using as a genetic target a single ASV provirus responsible for transformation of a cloned rat cell line (B31 cells; Varmus *et al.*, 1981). (These cells are non-permissive for replication of ASV, but fully permissive for replication of Mo-MuLV.) The objective has been to select from Mo-MuLV-superinfected B31 cells phenotypic revertants which have sustained lesions (e.g. insertion of an Mo-MuLV provirus) within the transforming (*src*) gene of ASV or in surrounding regions involved in its expression.

The prevalence of revertants in the non-superinfected B31 cells is approximately 10^{-5}. These can be isolated by preferential killing of transformed cells with antimetabolites (Varmus et al., 1981). Such spontaneous revertants have either lost the entire ASV provirus, including both copies of the LTR (presumably by chromosomal loss), or sustained point mutations (or small deletions) in src (Varmus et al., 1981; Oppermann et al., 1981). There is no appreciable increase in the rate of reversion after Mo-MuLV superinfection, but of the c. 50 revertant lines analyzed, two bear insertions of Mo-MuLV DNA within the ASV provirus. Four others have sustained deletions which remove the left-hand LTR and variable internal regions of the ASV provirus without evidence of nearby insertions of MuLV DNA (H.E.V., N.Q. and S.O., manuscript in preparation).

Both of the insertional mutants bear Mo-MuLV proviruses in the same orientation as the ASV provirus, in regions coding for replication genes several kb to the 5′ side of src. (The gene order of the ASV genome is 5′-gag-pol-env-src-3′.) In one case, the insert is near the 5′ end of gag, approximately 0.1 kb to the 3′ side of the donating splice point for src mRNA. The ASV promoter in the left-hand LTR appears to maintain normal transcriptional activity in this revertant, judging from hybridization with cDNA specific for the U_5 sequence. However, the resulting transcripts are presumably ASV-MuLV hybrids which are processed aberrantly; we find no stable transcripts from src. The other insertional mutant bears a Mo-MuLV provirus in the ASV pol gene. Although RNA has yet to be examined in this cell, retransformants have been isolated and shown to have lost c. 8.2 kb of the original 8.8 kb insert. It is possible that retransformation results from homologous recombination between the 0.6 kb LTRs of the Mo-MuLV provirus, thereby excising most of the insert and leaving one resident LTR.

The deletion mutants isolated thus far have in common the loss of the LTR at the left end of the ASV provirus. In the two cases examined for RNA, elimination of the LTR is accompanied by a complete absence of detectable ASV transcripts, conforming to the suggestion that the LTR is normally important for the initiation of transcription. The smallest deletion is c. 1.5 kb but the others are larger than 5–10 kb. We do not know whether the deletions are induced by the Mo-MuLV superinfection, mediated by the ASV LTRs, or unrelated mechanistically to the presence of retroviral genomes in these cells.

Deletion Mutants Among Proviruses Endogenous to Normal Chickens

Proviruses related to the genomes of laboratory strains of retroviruses are commonly found in the germ lines of many animals (Aaronson and Stephenson, 1976). We have prepared physical maps of 13 proviral elements, all related to the genome of ASV, which were segregating in flocks of White Leghorn chickens (Hughes *et al.*, 1979, 1981; Astrin, 1978). All of these elements (called "*ev*"s) are similar in structure to proviruses acquired after experimental infection, but most display some anomaly which impairs their transcriptional activity or their capacity to direct the production of infectious virus. As a result, they are likely to be innocuous to their host, thereby favoring their survival in the germ line.

Several of the anomalous endogenous proviruses bear lesions affecting the LTR (Hughes *et al.*, 1981; Hayward *et al.*, 1980). In three cases (*ev*-4, *ev*-5 and *ev*-6) the left-hand LTR has been deleted along with various amounts of adjacent coding sequence; in one of these cases (*ev*-6), the defective provirus is still expressed, presumably under the influence of a fortuitously adjacent cellular promoter. In two cases (*ev*-15 and *ev*-16), the elements are detected only with reagents specific for the U_3 and U_5 regions of the LTR; these may have been generated by homologous recombination between two copies of the LTR unit at the ends of intact endogenous proviruses, but further mapping is required for a full determination of their structure.

Proviruses in Avian Leukosis Virus-induced Tumors Bear Deletions Affecting Viral Gene Expression and Promote Transcription of Adjacent Cellular DNA

Inoculation of newborn chicks with avian leukosis viruses (ALVs) is followed several months later by the appearance of B-cell lymphomas, originating in the Bursa of Fabricius and often metastasizing to other organs (Gross, 1970). There are several reasons for suspecting that a viral genetic locus is not responsible for the oncogenic properties of this class of viruses. In contrast to viruses such as ASV which carry a clearly defined transforming gene (e.g. *src*), the ALVs induce tumors with long latency, do not transform cultured cells, display no biochemical evidence for a transforming gene or gene product, and do not carry sequences homologous to conserved cellular genes (the presumed progenitors of transforming genes such as *src* (Bishop *et al.*, 1980)). Although ALV proviruses can be inserted in many sites in the genome of infected hosts, tumors arising in infected animals appear to be clonal or semi-clonal as judged by analysis of proviruses with restriction endonucleases (Neiman *et al.*, 1980). This means that the tumor cells are derived from one or few infected cell(s), as described previously for mouse mammary tumors (Cohen *et al.*, 1979) and murine leukemias (Steffen and Weinberg, 1978; Jahner *et al.*, 1980).

In our recent studies of tumors induced by RAV-2, a typical ALV, we identified several tumors which bear a single RAV-2 provirus (G.P., L. Crittenden, S. Courtneidge, J.M.B. and H.E.V., manuscript in preparation). The most striking feature of these single proviruses is their aberrant structure. In the four examples studied in greatest detail, the proviruses have sustained deletions encompassing regions likely to be required for normal expression of viral genes. In two cases, the deletions are confined to small regions involving the left-hand LTR; in one case, a small deletion removes the region containing the donating splice point for *env* mRNA; and in the fourth case, an extensive deletion removes the left-hand LTR and most of the coding region of the provirus.

Since these defective proviruses are presumably involved in the initiation and/or maintenance of the oncogenic state, despite their structural deficiencies, we have attempted to determine whether they might influence the host cell in some fashion other than production of viral gene products. One possibility suggested by the structure of the provirus is the initiation of transcription of flanking cellular DNA by the right-hand LTR. Precedents for such "downstream" promotion exist among transposable elements of bacteria (Saedler *et al.*, 1974); similar examples have been found in ASV-transformed mammalian cells which contain RNA species, 1–3 kb in length, capable of annealing only to cDNA specific for the U_5 region of the viral genome (Quintrell *et al.*, 1980). Tests for such transcripts in cell lines derived from avian leukoses or in a tumor bearing a single RAV-2 provirus lacking the left-hand LTR have invariably demonstrated one or more transcripts which anneal only to $cDNA_5$ (Payne *et al.*, in preparation; personal communication from B. Neel, S. Astrin and W. Hayward). It is plausible that the right-hand LTR activates a flanking cellular gene instrumental in the oncogenic process. If so, integration at one of a limited number of sites in the cellular genome might be a requirement for tumor induction, perhaps explaining the prolonged latency and the derivations of tumors from one or a few cells. Two of the defective RAV-2 proviruses we have mapped extensively may, in fact, be positioned at the same site in the host genome, but comparison of flanking sequences after molecular cloning will be required to validate the mapping studies.

We do not know why the ALV proviruses in these tumors have sustained deletions or whether ALV proviruses undergo deletion formation at an abnormally high rate. It is conceivable that the immune response of the host exerts selective pressure against infected cells capable of producing immunoreactive viral proteins (e.g. *env* gene products). Alternatively, the promotional activity of the right-hand LTR (and hence the tumorogenic mechanism) may be augmented by deletions which impair transcriptional activity from the "upstream" region within the ALV provirus.

CONCLUSIONS

The replication of retroviruses involves the use of direct and inverted repetitions at several stages; the RNA genome is diploid, and each subunit bears a short-terminal direct repeat; the linear double-stranded product of reverse transcription carries a long-terminal direct repeat (LTR) composed of sequences unique to both the 3' and 5' ends of viral RNA; the LTRs are concluded with a short inverted repeat; closed circular DNA contains either one or two copies of the LTR; integrated (proviral) DNA is colinear with unintegrated linear DNA, including both copies of the LTR unit; and the integration events duplicate a short sequence of host DNA which forms a direct repeat flanking the provirus. Although the functional implications of these designs are not fully understood, the provirus bears a striking structural relationship to transposable elements of bacteria and other organisms, and the LTRs seem likely to be instrumental in viral DNA synthesis, integration and the control of transcription and processing of viral RNA.

In this chapter we have reviewed several aspects of work from our laboratory which bear upon proviral structure and behavior. Cloning and sequencing of unintegrated and integrated forms of avian sarcoma virus (ASV) and mouse mammary tumor virus (MMTV) DNA have defined the terminal inverted repeats, bases lost at the ends of proviral DNA, cellular sequences duplicated during insertion, and sequences within the LTRs which are likely to influence the initiation and processing of viral RNA. We have also obtained evidence for behavior of proviruses analogous to that occasionally exhibited by transposable elements of bacteria—insertional mutagenesis, deletion formation, and promotion of transcription of flanking cellular DNA. Some of these features may have implications for mechanisms of oncogenesis by certain retroviruses.

ACKNOWLEDGEMENTS

We would like to thank many of our colleagues, both intra- and extramural, for helpful discussions and for communication of results prior to publication. The work described here was supported by grants from the National Institutes of Health and the American Cancer Society to H.E.V. and J.M.B. The paper is a modest revision of a manuscript to appear in the 1980 *Cold Spring Harbor Symposium of Quantitative Biology*.

REFERENCES

Aaronson, S. and Stephenson, J. R. (1976). *BBA Rev. Cancer* **458**, 323–357.
Astrin, S. M. (1978). *Proc. natn. Acad. Sci. U.S.A.* **75**, 5941–5945.
Bishop, J. M. (1978). *A. Rev. Biochem.* **46**, 35–88.

Bishop, J. M., Courtneidge, S. A., Levinson, A. R., Oppermann, H., Quintrell, N., Sheinss, D. K., Weiss, S. R. and Varmus, H. E. (1980). *Cold Spring Harb. Symp. quant. Biol.* **44**, 919–930.

Bukhari, A. I., Shapiro, J. A. and Adhya, S. L. (eds) (1977). "DNA Insertion Elements, Plasmids, and Episomes". Cold Spring Harbor Laboratory, New York.

Cameron, J. R., Loh, E. Y. and Davis, R. W. (1979). *Cell* **16**, 739–757.

Coffin, J. M. (1979). *J. gen. Virol.* **42**, 1–26.

Cohen, J. C. and Varmus, H. E. (1977). *Nature, Lond.* **278**, 418–423.

Cohen, J. C., Shank, P. R., Morris, V. W., Cardiff, R. and Varmus, H. E. (1979). *Cell* **16**, 333–345.

Czernilofsky, A. P., DeLorbe, W., Swanstrom, R., Varmus, H. E., Bishop, J. M., Tischer, E. and Goodman, H. M. (1980). *Nucl. Acids Res.* **8**, 2967–2984.

DeLorbe, W. J., Luciw, P. A., Goodman, H., Varmus, H. E. and Bishop, J. M. (1980). *J. Virol.* **36**, 500–511.

Dhar, R., McClements, W. L., Enquist, L. W. and Vande Woude, G. F. (1980). *Proc. natn. Acad. Sci. U.S.A.* **77**, 3937–3941.

Fine, D. L., Plowman, J. K., Kelley, S. P., Arthur, L. O. and Hillman, E. A. (1974). *J. natn. Cancer Inst.* **52**, 1881–1886.

Grindley, N. O. F. and Sherratt, D. J. (1978). *Cold Spring Harb. Symp. quant. Biol.* **43**, 1257–1261.

Gross, L. (1970). "Oncogenic Viruses". Pergamon Press, New York.

Hager, G. L. and Donehower, L. A. (1980). *In* "Animal Virus Genetics" ICN-UCLA Symposium on Molecular and Cellular Biology (B. Fields, R. Jaenisch and C. F. Fox, eds), pp. 255–264. Academic Press, New York and London.

Haseltine, W. A., Maxam, A. and Gilbert, W. (1977). *Proc. natn. Acad. Sci. U.S.A.* **74**, 989–993.

Hayward, W., Braverman, S. B. and Astrin, S. M. (1980). *Cold Spring Harb. Symp. quant. Biol.* **44**, 1111–1122.

Heffron, F., McCarthy, B. J., Ohtsubo, H. and Ohtsubo, E. (1979). *Cell* **18**, 1153–1163.

Highfield, P. E., Rafield, L. F., Gilmer, T. M. and Parsons, J. T. (1980). *J. Virol.* **36**, 271–280.

Hilgers, J. and Bentvelzen, P. (1978). *Adv. Cancer Res.* **26**, 143–195.

Hsu, T. W., Sabran, J. L., Mark, G. E., Guntaka, E. V. and Taylor, J. W. (1978). *J. Virol.* **28**, 810–818.

Hughes, S. H., Shank, P. R., Spector, D. H., Kung, H.-J., Bishop, J. M., Varmus, H. E., Vogt, P. K. and Breitman, M. L. (1978). *Cell* **15**, 1397–1410.

Hughes, S. H., Payvar, F., Spector, D. H., Schimke, R. T., Robinson, H. L., Bishop, J. M. and Varmus, H. E. (1979). *Cell* **18**, 347–359.

Hughes, S. H., Toyoshima, K., Bishop, J. M. and Varmus, H. E. (1981). *Virology* **108**, 189–207.

Jahner, D., Stuhlmann, H. and Jaenisch, R. (1980). *Virology* **101**, 111–123.

Ju, G., Boone, L. and Skalka, A. M. (1980). *J. Virol.* **33**, 1026–1033.

Kung, H. J., Fung, Y. K., Majors, J. E., Bishop, J. M. and Varmus, H. E. (1981). *J. Virol.* **37**, 127–138.

Landy, A. and Ross, W. (1977). *Science, N.Y.* **197**, 1147–1160.

MacHattie, L. A. and Jackowski, J. B. (1977). *In* "DNA Insertion Elements, Plasmids and Episomes" (A. I. Bukhari, J. A. Shapiro and S. L. Adhya, eds), pp. 219–228. Cold Spring Harbor Laboratory, New York.

Majors, J. E. and Varmus, H. E. (1980). *In* "Animal Virus Genetics" ICN-UCLA Symposium on Molecular and Cellular Biology (B. Field, R. Jaenisch and D. F. Fox, eds), pp. 241–253. Academic Press, New York and London.

Majors, J. E. and Varmus, H. E. (1981). *Nature, Lond.* **289**, 253–258.

Mitra, S. W., Goff, S., Gilboa, E. and Baltimore, D. (1979). *Proc. natn. Acad. Sci. U.S.A.* **76**, 4355–4359.

Neiman, P., Payne, L. N. and Weiss, R. A. (1980). *J. Virol.* **34**, 178–186.

Oppermann, H., Levinson, A. D. and Varmus, H. E. (1981). *Virology* **108**, 47–70.

Peters, G. and Glover, C. C. (1980). *J. Virol.* **35**, 31–40.

Proudfoot, N. J. and Brownlee, G. G. (1974). *Nature, Lond.* **252**, 359–362.

Quintrell, N., Hughes, S. H., Varmus, H. E. and Bishop, J. M. (1980). *J. molec. Biol.* **143**, 363–393.

Ringold, G. M., Shank, P. R., Varmus, H. E., Ring, J. and Yamamoto, K. R. (1979). *Proc. natn. Acad. Sci. U.S.A.* **76**, 665–669.

Sabran, J. L., Hsu, T. W., Yeater, C., Kaji, A. Mason, W. S. and Taylor, J. M. (1979). *J. Virol.* **29**, 170–178.

Saedler, H., Reif, H.-J., Hu, S. and Davidson, N. (1974). *Molec. gen. Genet.* **132**, 265–289.

Schwartz, D. E., Zamecnik, P. C. and Weith, H. L. (1977). *Proc. natn. Acad. Sci. U.S.A.* **74**, 994–998.

Shank, P. R. and Varmus, H. E. (1978). *J. Virol.* **25**, 104–114.

Shank, P. R., Cohen, J. C., Varmus, H. E., Yamamoto, K. R. and Ringold, G. M. (1978a). *Proc. natn. Acad. Sci. U.S.A.* **75**, 2112–2116.

Shank, P. R., Hughes, S. H., Kung, H.-J., Majors, J. E., Quintrell, N., Guntaka, R. V., Bishop, J. M. and Varmus, H. E. (1978b). *Cell* **15**, 1383–1395.

Shapiro, J. A. (1979). *Proc. natn. Acad. Sci. U.S.A.* **76**, 1933–1937.

Shimotohno, K., Mizutani, S. and Temin, H. M. (1980). *Nature, Lond.* **285**, 550–554.

Shine, J., Czernilofsky, A. P., Friedrich, R., Goodman, H. and Bishop, J. M. (1977). *Proc. natn. Acad. Sci. U.S.A.* **74**, 1473–1477.

Shoemaker, C., Goff, S., Gilboa, E., Paskind, M., Mitra, S. W. and Baltimore, D. (1980). *Proc. natn. Acad. Sci. U.S.A.* **77**, 3932–3936.

Steffen, D. and Weinberg, R. A. (1978). *Cell* **15**, 1003–1010.

Strobel, E., Dunsmuir, P. and Rubin, G. M. (1979). *Cell* **17**, 429–439.

Sutcliffe, J. G., Shinnock, T. M., Verma, I. M. and Lerner, R. A. (1980). *Proc. natn. Acad. Sci. U.S.A.* **77**, 3302–3306.

Swanstrom, R., DeLorbe, W., Bishop, J. M. and Varmus, H. E. (1981). *Proc. natn. Acad. Sci. U.S.A.* **78**, 124–128.

Taylor, J. M. (1977). *Biochim. biophys. Acta* **473**, 57–72.

Taylor, J. M. (1979). *Curr. Top. Microbiol. Immunol.* **87**, 23–41.

Van Beveren, C., Goddard, J. G., Berns, A. and Verma, I. M. (1980). *Proc. natn. Acad. Sci. U.S.A.* **77**, 3307–3311.

Varmus, H. E., Heasley, S., Kung, H. J., Oppermann, H., Smith, V. C., Bishop, J. M. and Shank, P. R. (1978a). *J. molec. Biol.* **120**, 55–82.

Varmus, H. E., Cohen, J. C., Shank, P. R., Ringold, G. M., Yamamoto, K. R., Cardiff, R. and Morris, V. L. (1978b). *In* "Persistent Viruses" (J. G. Stevens, G. J. Todaro and C. F. Fox, eds), pp. 161–179. ICN-UCLA Symposium on Molecular and Cellular Biology. Academic Press, New York and London.

Varmus, H. E., Shank, P. R., Hughes, S. E., Kung, H. J., Heasley, S., Majors, J., Vogt, P. K. and Bishop, J. M. (1979a). *Cold Spring Harb. Symp. quant. Biol.* **43**, 851–864.

Varmus, H. E., Ringold, G. and Yamamoto, K. R. (1979b). *In* "Glucocorticoid Hormone Action" (J. Baxter and G. Rousseau, eds), p. 254. Springer-Verlag, Berlin.
Varmus, H. E., Quintrell, N. and Wyke, J. (1981). *Virology* **108**, 28–46.
Yoshimura, F. K. and Weinberg, R. A. (1979). *Cell* **16**, 323–332.
Ziff, E. G. and Evans, R. M. (1978). *Cell* **15**, 1463–1475.

III

Structure and Expression of RNA Virus Genes

Expression of the Genome of Single-stranded Messenger-sense RNA Viruses

PAUL KAESBERG

Biophysics Laboratory and Biochemistry Department, University of Wisconsin, Madison, Wisconsin 53706, U.S.A.

INTRODUCTION

The simplest viruses from the standpoint of virion structure and genome information content are the single-stranded, messenger-sense RNA viruses. Many have only a single species of coat protein geometrically arranged into an icosahedral or helical structure. The genomes of many such viruses encode only 3 or 4 proteins. Nevertheless, in only a few instances have other than virion structural proteins been identified unambiguously; generally the encoded proteins have not even been enumerated. This failing comes not only from technical problems associated with *in vitro* protein synthesis, a principal method for identifying encoded proteins, but more fundamentally from the numerous variations that exist in cistron expression. That is, one finds examples of post-translational protein-processing, of read-through to give related proteins, of untranslated "silent" cistrons and even of pretranslational modification of the mRNA itself. From a technical standpoint, *in vitro* systems differ radically in the efficiency of translation of a particular message, in the effectiveness of chain completion and particularly in their ability to initiate translation.

Clearly if general features of the replication of small RNA viruses are to be found, encoded proteins must be identified and the nuances of translational expression must be sorted out for a variety of viruses. It is particularly useful to include in such studies viruses whose virion structure is well-characterized because ultimately virion structure and its dynamic counterpart, virion assembly, must be relatable to the functions written into the genetic message. With this in mind we have been studying translation of the RNAs of several well-known viruses and here discuss translation of two of these, EMC (encephalomyocarditis virus) and SBMV (southern bean mosaic virus) RNA in two cell-free systems, rabbit reticulocyte lysates and wheat-embryo extracts. In the case of EMC, expression of the genome features an elaborate set of *post-translational* cleavages of the products of translation while expression of the SBMV genome features *pretranslational* cleavage of the viral RNA. As you will see, translation of these RNAs in the two systems is superficially very different, although the information generated is quite similar.

EMC RNA TRANSLATION

Long-standing *in vivo* studies of picornaviruses have shown that products of translation are proteolytically cleaved to produce virion and other proteins needed for virus proliferation. More recently the translation and processing have been observed in cell-free systems and, in particular, in rabbit reticulocyte lysates EMC RNA has been translated and the products have been processed with exceptionally high efficiency and fidelity to the *in vivo* events (Shih *et al.*, 1979). In an hour each EMC RNA molecule is translated more than eight times on average and except for virion proteins δ and β, the appearance of each of the proteolytic cleavage intermediates and final products detected *in vivo* can be charted as a function of time *in vitro*. Moreover, no products suggestive of incomplete translation or aberrant processing are identifiable. It is thus possible to study, *in vitro*, precursor–product relationships, synthesis of viral protease and VPg, and viral v. cellular protease specificity, with confidence that the results will be relevant to the natural events.

By contrast, wheat-embryo extracts which translate many mRNAs vigorously, induce no detectable protein synthesis with EMC RNA as a messenger. Addition of reticulocyte factors allows translation.

Synthesis and Processing of EMC Proteins in Rabbit Reticulocyte Lysates

Time course of translation and processing—pulse-chase experiments

In vivo and *in vitro* studies show that most of the EMC proteins are synthesized by translation initiated at a single site followed by proteolytic processing of both nascent and released products by proteases existing in the reticulocyte extracts and by a protease encoded in the genomic RNA and synthesized in the extracts by translation and subsequent proteolysis. Figure 1 summarizes our conception of the steps in translation and processing.

Figure 2 shows a polyacrylamide gel pattern of appearance and disappearance of ^{35}S-methionine-labeled products synthesized and proteolytically

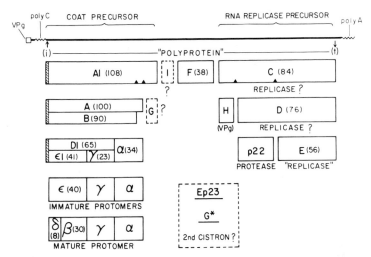

FIG. 1. Genome map of EMC showing positions of primary translation products, processing intermediates and final products. The diagram should be regarded as a working model. It was constructed from data compiled by Ann Palmenberg, Mark Pallansch and Roland Rueckert.

processed during a 3-hour period. Translation rate varies with the reticulocyte lysate. For Fig. 2, a lysate was chosen in which elongation was relatively slow in order to illustrate the earliest processing events. Note that A1 appears prior to A and B indicating that the latter are not produced by termination of translation.

Pulse-chase analyses, in which translation was carried out in the presence of ^{35}S methionine for specified times and continued for various times with the addition of excess cold methionine, identified what products were translated in various time periods.

Pulse-stop experiments

Although precursors of virion proteins are evident within 10 min in fast-translating lysates, their processing does not occur for another 10 min, suggesting that translation beyond some point is a prerequisite for their processing. This may be seen most readily in what we call "pulse-stop" analyses. In such experiments translation is terminated as specified times with sparsomycin, and processing is allowed to continue for various times. Thus processing can occur only with enzymes present (in active or precursor form) at the time of translation inhibition. Figure 3 shows that if translation continues beyond 20 min, the translation products are extensively processed following 3-hour incubation periods whereas if translation is stopped within 20 min, the extensive processing is absent.

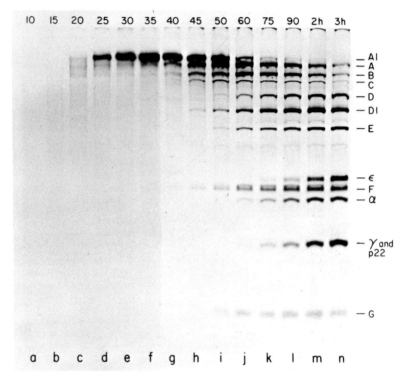

FIG. 2. Synthesis of EMC viral proteins. A lysate was used in which translation is relatively slow so that early products are delineated well. Incorporation was stopped at the indicated times by addition of EDTA and pancreatic ribonuclease. Analysis was on SDS polyacrylamide gels. Lanes (a) to (n) correspond to samples incubated for 10, 15, 20, 25, 30, 35, 40, 45, 50, 60, 75, 90, 120 and 180 min. Labeling was with ^{35}S methionine.

EMC genome-coded proteolytic activity

Figure 3 showed that termination of translation within 20 min, followed by long incubation to allow processing, resulted in a pattern comprised only of major bands A1, A and B. If, however, a second reaction mixture, translated for 60 min, is added, band A1 disappears, bands A and B are diminished, and strong bands ε, α and γ are evident (Fig. 4, lanes a to c). In this experiment translation proceeded in the presence of radioactive methionine for 20 min, sparsomycin was added, a non-radioactive translation mixture was added, and incubation was carried out for 2, 4 and 16 hours. Lane (d) shows the corresponding 4-hour experiment in which a non-radioactive reaction mixture devoid of EMC RNA was added. Lane (e) shows a corresponding experiment where the non-radioactive reaction

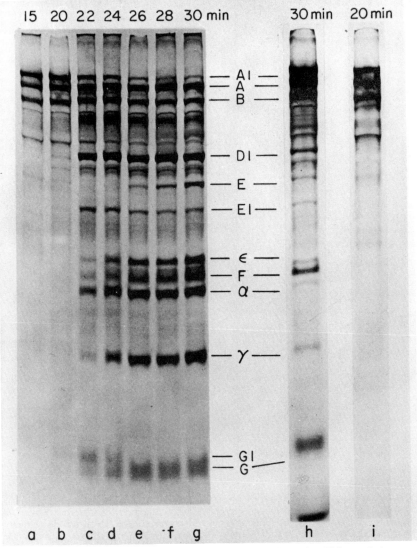

FIG. 3. Synthesis of EMC proteins. Pulse-stop analyses with a lysate in which translation was fast. Sparsomycin was added to the reaction mixtures at the indicated times so as to stop protein synthesis. Lanes (a) to (g): incubation was continued for a total of 3 hours during which time processing could continue. Lanes (h) and (i) were without further incubation.

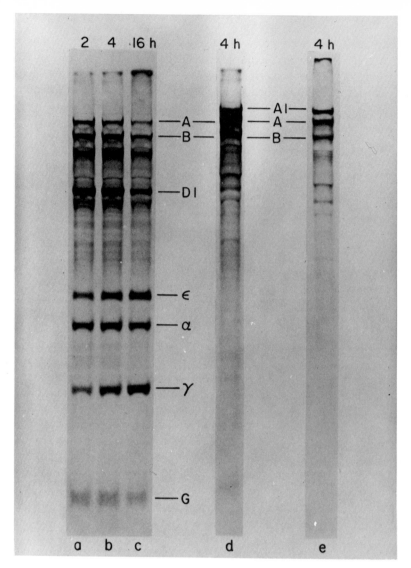

FIG. 4. Cleavage of the capsid protein precursors by a virion-coded protease. For preparing
[35]S-methionine-labeled capsid protein precursors, incorporation reaction mixtures were incubated for 20 min, followed by the addition of sparsomycin to stop further protein synthesis.
For preparing non-radioactive EMC or polio viral RNA translation products, ^{35}S methionine
was omitted from the normal reaction mixture and incubation was for 1 hour. For the mock-
reaction mixture, both the ^{35}S methionine and the viral RNA were omitted. Lanes (a) to (c): 1
volume of the radioactive precursor-containing reaction mixture was mixed with 1 volume of
the reaction mixture containing non-radioactive EMC viral RNA products and incubated for 2
hours (lane a), 4 hours (lane b), and 16 hours (lane c). Lanes (d) and (e): the radioactive
precursors were mixed with the mock-reaction mixture (lane d) or the polio viral protein-
containing reaction mixture (lane e) and incubated for 4 hours.

mixture contained polio RNA and its translation products. It may be seen that neither the mock-reaction mixture nor the polio products caused processing. A parallel experiment showed that the polio-reaction mixture had translated and processed polio proteins effectively. We conclude that completely translated and processed EMC extracts contain a specific protease that can generate EMC virion proteins from precursors.

The genome-coded protease

We have identified and extensively purified the protease responsible for the processing. Sedimentation in sucrose gradients provides a fractionation distinct from that of SDS polyacrylamide gels because some products (e.g., the virion proteins α, γ and ε) sediment as a unit. Such sedimentation analyses show that the protease activity migrates as does band P22, and have led us to conclude that P22 is the genomic protease (Palmenberg *et al.*, 1979). Sucrose fractions containing it, however, process only to yield D1 (the ε, γ precursor) and α. This could simply indicate dilution but could suggest loss of activity or loss of another protein. However, Ding and Lillian Shih have found ionic conditions in which the protease activity sediments with the reticulocyte ribosomes and can be eluted from them (unpublished work). Under these conditions P22 is recovered quantitatively and protein E is recovered in about 10% yield while proteolytic activity is recovered almost quantitatively. This preparation causes processing to yield ε and γ as well as α (Fig. 5).

Tryptic peptide analysis shows unequivocally that protein D is processed to yield E and P22 as indicated in Fig. 1. Furthermore, protein C is processed to yield D and H and VPg is found in H. Thus the remarkable fact emerges that the genome-attached VPg, the viral protease P22 and the proteins presumably implicated in replication, D and E, are all contiguous in the EMC map. This has led to persuasive models for interrelating the activities (Palmenberg *et al.*, 1979).

Processing of proteins A1, A and B

The evidence is strong that the virion proteins ε, α and γ come from action of the genome-coded protease on their precursors. But how are the precursors produced? Translation with "slow" lysates shows that A1 appears earlier than A and B even though the three are related and A1 is the largest. A and B appear before we can detect the genomic protease. These facts suggest that A1 arises from a reticulocyte protease event (or less likely by a termination of translation) and that A and B are derived from A1, again by action of a reticulocyte protease. However, that processing is discontinued if translation is stopped. It can be recommenced by addition of genome-coded protease. Thus it seems that A and B can be made in two

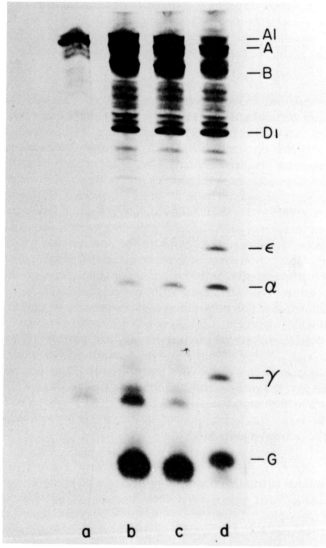

FIG. 5. Processing of capsid precursors by the second high-speed supernatant fraction. Synthesis products (consisting mostly of protein A1) were prepared and mixed with the fraction and were incubated for (a) 0 time; (b) 1 hour; (c) 3 hours; (d) 16 hours.

ways—by action of reticulocyte protease acting during translation or by genome protease acting even when translation has been aborted. In any case the further processing to yield ε, α and γ occurs by action of the genomic protease.

Processing to yield proteins C, D and E

The processing events involving these proteins are also curious. Given that P22 is an interior part of C one might conclude casually that production of C and D would require reticulocyte protease although E could be generated by either. However, an important observation by Ann Palmenberg and Roland Rueckert (unpublished work) suggests that these proteins arise from action of the genome protease as *monomolecular, self-cleavage* events. The critical observation is that some of the processing events continue nearly unabated when reaction mixtures are highly diluted.

Translation of EMC RNA in Wheat-embryo Extracts

Synthesis of EMC-specific proteins

Many mRNAs, deriving from plants, microorganisms and animals, are translated readily in wheat-embryo extracts. EMC RNA is not among them. It induces a low level of amino acid incorporation but we are unable to detect any viral proteins. However, if wheat-embryo extracts are supplemented with a ribosomal wash fraction from rabbit reticulocytes there is a large stimulation of amino acid incorporation and EMC-specific proteins are synthesized. Figure 6 shows the time course of synthesis of EMC proteins. Several distinct bands are evident. Bands designated WA1, WA and WB migrate like those of authentic EMC capsid precursors A1, A and B and a fourth WG migrates like EMC protein G. The tryptic maps of WA1, WA and WB are virtually identical to those of A1, A and B suggesting that proper translation and early processing has occurred.

However, most counterparts of authentic EMC proteins are absent, e.g., those of F and C, and the capsid proteins ε, α and γ. Product WD, which migrates roughly in the D position, turns out (by tryptic analysis) to be a subset of A1 and is unrelated to D. Clearly, only the early portion of the EMC genome has been translated.

WA1, WA and WB can be processed by protease derived from EMC RNA translation in reticulocyte lysates to yield ε, α and γ but the efficiency of cleavage is low, indicating that something is amiss. Possibly they have not folded correctly.

Translation of the second cistron

Among the bands evident when the gel of Fig. 1 is exposed for a longer period

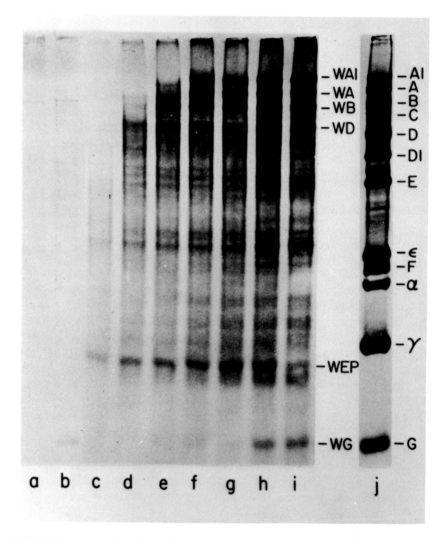

FIG. 6. Time course of synthesis of EMC proteins in wheat-embryo extracts to which has been added ribosome wash from a rabbit reticulocyte lysate. Lane (a), no wash, no EMC RNA, 60 min incubation; (b) no wash, with EMC RNA, 60 min incubation; (c) to (i), samples withdrawn at 10, 20, 30, 40, 50, 60 and 120 min from a reaction mixture containing ribosomal wash and EMC RNA; (j) EMC marker proteins.

is a band designated EP, an early product of translation (unpublished work by Ann Palmenberg). This protein has a tryptic map unlike that of A1 and may represent initiation of translation at a second cistron as has been reported for translation of polio-virus RNA. Correspondingly, translation of EMC RNA in wheat extracts yields, in addition to WA1, WA, WB, WD and WG, a product WEP which migrates identically to EP. WEP is made in considerable abundance in wheat extracts as compared to EP in reticulocyte lysates so that its firm characterization as a second cistron will be readily feasible.

SBMV RNA TRANSLATION

When we began attempts at translation of plant viral RNAs about a decade ago we wondered whether their mode of expression would resemble that of the RNA phages where the several cistrons are translated from the viral genome or picornaviruses where the functional proteins arise by proteolytic cleavage of polyprotein precursors. However, the results with brome mosaic virus in our laboratory, of Semliki Forest virus by Kääriäinen and his associates and with many other viral messengers show that a new mode of expression is generally operative—the creation of subgenomic RNAs which serve as messengers for translation of "silent" cistrons that exist in the genomic RNA. This mode, the existence of divided genomes, and proteolytic processing apparently make translation of internal cistrons, in the style of the RNA phages, unnecessary in the world of eukaryotic viruses. Nevertheless there are well-documented examples of the synthesis of more than one well-defined protein from an evidently homogeneous viral RNA to say nothing of numerous examples in which a multitude of sizes of translation products is displayed. SBMV RNA is an example where several products arise from a single messenger.

SBMV is one of a small number of viruses whose crystallography is much advanced so that there is reasonable expectation of relating the events of protein synthesis to virion protein assembly. The genomic RNA weighs 1.4 $\times$ 10^6 daltons. Its 5' terminus is blocked with a protein (SBMV VPg) and its 3' terminus lacks poly A (Ghosh et al., 1979).

Products of Translation of SBMV RNA in Reticulocyte Lysates and in Wheat-embryo Extracts

Translation of SBMV RNA in reticulocyte lysates or wheat-embryo extracts gives primarily four proteins designated P1, P2, P3 and P4 (Fig. 7) (Salerno-Rife et al., 1980). Although all four are detectable in both systems,

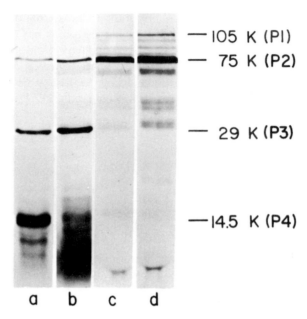

FIG. 7. Translation of SBMV virion RNA in wheat-embryo extracts and in reticulocyte lysates: (a) in wheat-embryo extracts with ³H leucine as the label; (b) same with ³⁵S methionine; (c) in rabbit reticulocyte lysates with ³H leucine; and (d) same with ³⁵S methionine.

P1 and P2 are much more prominent in reticulocyte lysates while P3 and P4 are much more prominent in wheat-embryo extracts. Tryptic peptide analysis shows that P1 and P2 are related. Evidence not described here suggests that P1 and P2 are initiated from the same site so that P1 may be a read-through of P2. P3 is virion coat protein. P4 is unrelated to P1, P2 and P3 and moreover does not seem related to SBMV RNA VPg.

SBMV RNA is not monodisperse. Both sucrose density sedimentation and polyacrylamide gel analyses show that many species of RNA exist shorter than full-length. Translation of fractionated RNA shows that P1 and P2 are induced by full-length RNA (Fig. 8). P3 is induced by a particular minor species, indeed a species that has a VPg. P4 is induced by full-length RNA and also by most of the minor species fractions when translation is induced in wheat-embryo extracts while one minor fraction, predominantly, induces its synthesis in reticulocyte lysates. This surprising result is explicable by the assumption that P4 mRNA exists as a minor species in the RNA but that, in addition, more P4 mRNA is created in the cell-free

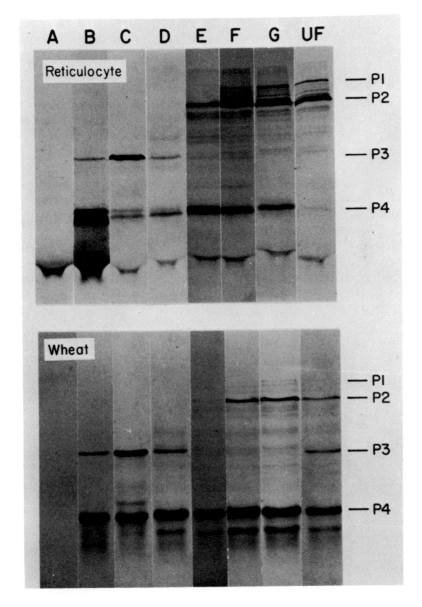

FIG. 8. SDS polyacrylamide gel electrophoresis of the ³H-leucine-labeled products synthesized by SBMV RNAs in the rabbit reticulocyte (top panel) and wheat-embryo systems (bottom panel). Lanes (A) to (G) represent the products made by RNAs from fractions (A) to (G) of the gradient, respectively. Lane UF represents the proteins synthesized by unfractionated SBMV RNA. In each case, the messenger concentration used was 66 μg/ml.

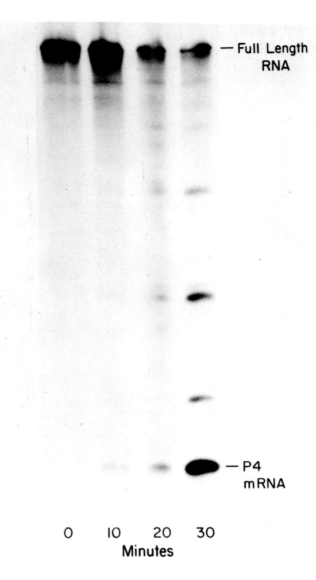

FIG. 9. Incubation of full-length SBMV RNA in reticulocyte lysates at 0 time; 10 min, 20 min and 30 min. The RNA was labeled at its 3' terminus by addition of ^{32}P-labeled pCp with the enzyme RNA ligase.

extracts. We examined the stability of SBMV RNA in reticulocyte lysates and in wheat-embryo extracts. In both extracts a fragment arising from the 3′ end of SBMV RNA is generated in a matter of minutes—a fragment we believe to be the P4 messenger (Fig. 9).

Clearly the biological importance of P4 and its mRNA is yet to be demonstrated. Correspondingly the significance of EMC early protein is not understood. However, what is obvious at this point is that results of cell-free translation must be interpreted with due regard for both the messenger and the cell-free system.

ACKNOWLEDGEMENTS

This research was supported by Public Health Service grants AI 1466 and AI 15342 and Career Award AI 21942 from the National Institutes of Health and by grant 7800002 from the Competitive Grants program of the Science and Education Administration of the U.S. Department of Agriculture. Ding Shih, Ann Palmenberg and Lillian Shih are largely responsible for the EMC work; the SBMV studies have been done by Tineke Rutgers, Terry Salerno-Rife, Amit Ghosh and Mang Keqiang. Throughout we have had Roland Rueckert's enthusiastic guidance.

REFERENCES

Ghosh, A., Dasgupta, R., Salerno-Rife, T., Rutgers, T. and Kaesberg, P. (1979). *Nucl. Acids Res.* **7**, 2137–2146.
Palmenberg, A. C., Pallansch, M. A. and Rueckert, R. R. (1979). *J. Virol.* **32**, 770–778.
Salerno-Rife, T., Rutgers, T. and Kaesberg, P. (1980). *J. Virol.* **34**, 51–58.
Shih, D. S., Shih, C. T., Zimmern, D., Rueckert, R. R. and Kaesberg, P. (1979). *J. Virol.* **30**, 472–480.

Antigenic Variation and Structure of Influenza Virus

G. G. BROWNLEE

Sir William Dunn School of Pathology, University of Oxford, South Parks Road, Oxford OX1 3RE, England

INTRODUCTION

Influenza has been known as a specific disease since the 1743 epidemic, the name deriving from the Italian *influenza* or literally an "influence". This originally meant* a "visitation" of *any* epidemic disease, but by application to the 1743 epidemic it became the specific English name. It has also long been known as a disease from which "many sicken and few die except the old and infirm". Epidemiologists of the last century knew the disease occurred in epidemics with relatively quiescent periods in between, but it was the 1918–1919 epidemic which marked the climax of this disease; it is estimated that 20 million people may have died directly or indirectly. Despite intensive efforts it took until 1933 before the first human virus was isolated and transmitted in ferrets.

Influenza is now classified within the *myxo* group of RNA viruses being sensitive to ether (because of the host-derived phospholipid membrane) releasing an internal antigen which allows classification into type A, B or C. Under the electron microscope the virus is a variably shaped (usually 80–120 nm diameter) particle, the only distinctive feature being the numerous external short projections known to be composed of hemagglutinin and neuraminidase.

Active vaccination on Man was attempted as early as 1937 but only transient immunity was achieved. Unlike other important viral diseases in Man such as poliomyelitis and measles, influenza has remained essentially uncontrolled. The reasons for this are in part due to its low mortality but are, of course, also directly related to the evolution of the surface hemagglutinin of the virus. Small, but epidemiologically significant changes (antigenic drift) occur in the current hemagglutinin subtype annually or

---

* "Shorter Oxford English Dictionary", 3rd edn, Oxford 1959.

biennially. Antigenically unrelated subtypes (antigenic shift) also appear, or reappear, in the population every decade or so. Table I shows the human subtypes of influenza A recognized since the first isolation of the virus. Using the new (1980) classification we see that only three subtypes have infected Man, two of which, H1 and H3, are cocirculating at present. The origin of "new" subtypes is uncertain, but Laver and Webster (1979) have shown that they could have arisen from animal or bird "reservoirs" by reassortment of the RNA genes during a mixed infection.

A modern biochemical approach to influenza has involved a study of its nine proteins and eight RNA segments. The eight RNA components are said to be "negatively stranded" as they are first transcribed in the cell nucleus by a specific influenza polymerase using the 5' end of host mRNA (see R. M. Krug, this volume) before "taking over" the host "translation" machinery and synthesizing specific viral proteins. The first structural work was directed at the hemagglutinin, as the important antigen, using protein chemical methods and a complete sequence of the HA1 subunit of Mem/102/72 (a 1972 strain) was achieved recently (Ward and Dopheide, 1980). But it was clear that recombinant DNA methods, which involved the *in vitro* synthesis of DNA from the RNA genes, followed by their "cloning" and DNA sequence analysis,was a much faster and easier approach. Using this methodology the hemagglutinin molecule of a number of different subtypes has now been completely sequenced.

TABLE I

Antigenic subtypes (H) of human influenza A

Subtype[a]	Year prevalent	Prototype strain
H1	1932–1957 and 1977–present	A/PR/8/34 A/FM/1/47 A/USSR/7/77
H2	1957–1968	A/Singapore/1/57
H3	1968–present	A/Hong Kong/1/68

[a] The nomenclature is based on the new 1980 classification. The 1918 human pandemic was probably the H1 subtype. This table omits nine other known H subtypes found in birds and animals, but not Man.

NUCLEOTIDE LENGTHS OF RNA SEGMENTS

Early estimates of the molecular weight of the eight RNA segments of influenza A strains (McGeoch *et al.*, 1976; Desselberger and Palaese, 1978) by electrophoresis on polyacrylamide gels were inaccurate. This was partly because conditions were not fully denaturing and partly because suitable RNA markers were not available. To overcome these difficulties we (Sleigh *et al.*, 1979a) have developed an alternative method. We first synthesized full-length complementary DNA copies of the mixture of the eight RNA genes using reverse transcriptase under optimal conditions. The mobility of the influenza cDNA on polyacrylamide gels under denaturing conditions was then compared with a marker mixture of single-stranded DNA derived from the fully sequenced *E. coli* plasmid pBR322. The results (Table II) for bands 1–6 were considerably lower than the earlier 1976 and 1978 estimates, although we observed slightly higher values for bands 7 and 8. The total number of residues is now estimated at 13 830 and is considerably less than previously thought. The accuracy of the new method has been vindicated (Table II) as the nucleotide lengths of bands 4, 7 and 8 became known subsequently from their full sequence determination (see below).

TABLE II

Comparison of various estimates of influenza A gene-lengths
(in numbers of nucleotides)

Gene	Protein	1976 estimate[a]	1978 estimate[b]	Our (1979 estimate[c]	From sequence
1	P3	5100	2800	2390	—
2	P1	3670	2800	2390	—
3	P2	4230	2700	2290	—
4	Hemagglutinin	2490	2080	1760	1765[d]
5	Nucleoprotein	2370	1760	1560	—
6	Neuraminidase	2400	1510	1480	—
7	Matrix and M2[g]	930	880	1060	1027[e]
8	NS1 and NS2	820	660	890	890[f]
Total	10	21 820	15 190	13 830	—

[a] McGeoch *et al.* (1976)
[b] Desselberger and Palaese (1978)
[c] Sleigh *et al.* (1979a)
[d] Sleigh *et al.* (1980)
[e] Winter and Fields (1980)
[f] Porter *et al.* (1980)
[g] R. A. Lamb (personal communication)

CLONING THE HEMAGGLUTININ GENE OF A HUMAN H3 SUBTYPE A/Mem/102/72

The virion RNA (negative strand) of band 4 of the Mem/102/72 strain (H3 subtype) was reasonably accessible to cDNA cloning methods developed previously for mRNA. Using standard methods we (Sleigh *et al.*, 1979b) obtained an almost full-length clone lacking only 30 residues from its 5′ end in the *E. coli* plasmid pBR322. This has been fully sequenced (Sleigh *et al.*,

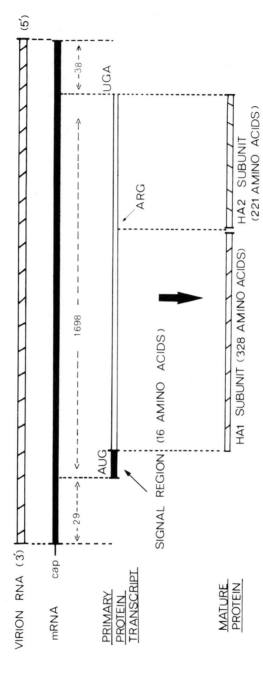

FIG. 1. Line diagram showing the organization of the HA1 and HA2 subunits of the mature hemagglutinin of A/Mem/102/72 with respect to its primary protein transcript and its gene.

1980) and from this and a knowledge of the N- and C-terminus of the HA1 subunit and the N-terminus of HA2, the complete amino acid sequence of the two hemagglutinin subunits has been deduced. Figure 1 shows the organization of the mature HA1 and HA2 subunits of the protein relative to its gene and mRNA. The gene is 1765 residues long (very close to our estimate of 1760 residues) and contains only 29 non-coding residues at one end and 38 non-coding residues at the other. The primary translation product is a precursor hemagglutinin. This is subsequently processed by the removal of the hydrophobic signal 16 amino acids and by the cleavage and loss of a single internal arginine residue to form the larger HA1 and the smaller HA2 subunit.

Figure 2 shows the primary amino acid sequence deduced for Mem/102/72 from the nucleic sequence numbering the mature HA1 and

HA1

```
         10          20          30          40          50          60
QDFFGNDNST ATLCLGHHAV PNGTLVKTIT NDQIEVTNAT ELVQSSSTGK ICNNPHRILD

         70          80          90         100         110         120
GIDCTLIDAL LGDPHCDGFQ NETWDLFVER SKAFSNCYPY DVPDYASLRL LVASSGTLEF

        130         140         150         160         170         180
INEGFTLTGV TQNGGSNACK RGPDSGFFSR LNWLYKSGST YPVLNVTMPN NDNFDKLYIW

        190         200         210         220         230         240
GVHHPSTDQE QTSLYVQASG RVTVSTKRSQ QTIIPNIGSR PWVRGQSSRI SIYWTIVKPG

        250         260         270         280         290         300
DILVINSNGN LIAPRGYFKM RTGKSSIMRS DAPIGTCISE CITPNGSIPN DKPFQNVNKI

        310         320
TYGACPKYVK QNTLKLATGM RNVPEKRT
```

HA2

```
         10          20          30          40          50          60
GLFGAIAGFI ENGWEGMIDG WYGFRHQNSE GTGQAADLKS TQAAIDQING KLNRVIEKTN

         70          80          90         100         110         120
EKFHQIEKEF SEVEGRIQDL EKYVEDTKID LWSYNAELLV ALGNQHTIDL TDSEMNKLFE

        130         140         150         160         170         180
KTRRQLRENA EDMGNGCFKI YHKCDNACIG SIRNGTYDHD VYRDEALNNR FQIKGVELKS

        190         200         210         220
GYKDWILWIS FAISCFLLCV VLLGFIMWAC QKGNIRCNIC I
```

FIG. 2. The amino acid sequence predicted from the nucleic acid sequence of A/Mem/102/72. HA1 and HA2 are numbered separately. An independent analysis of another subclone of A/Mem/102/72 using protein chemical methods (Ward and Dopheide, 1980) differs in four positions, viz: 110 (S), 127 (W), 226 (L) and 327 (Q) in HA1. The one-letter code for amino acids is: A (Ala), R (Arg), N (Asn), D (Asp), C (Cys), Q (Gln), E (Glu), G (Gly), H (His), I (Ile), L (Leu), K (Lys), M (Met), F (Phe), P (Pro), S (Ser), T (Thr), W (Trp), Y (Tyr) and V (Val).

HA2 separately. This sequence differs in only four places from a complete and independent study of the HA1 of the same strain using classical methods of protein chemistry (Ward and Dopheide, 1980). These differences appear to be real differences between substrains of the same strain.*

ANTIGENIC DRIFT IN THE HEMAGGLUTININ OF THE H3 SUBTYPE

Direct immunochemical analysis shows (Jackson et al., 1979) that the HA1 subunit carries the specific antigenic determinants for the individual variants within the H3 subtype. They appear to be most closely associated with the "CNBr 1" fragment (that is, the first 168 amino acids of HA1) although contributions from other parts of the HA1 molecule cannot be excluded. HA2 does not appear to contribute to antigenicity.

Field Strains

With the sequence determination, using recombinant DNA methods, of three distinct antigenic variants (field strains) within the H3 subtype (Sleigh et al., 1980; Both et al., 1980; Min Jou et al., 1980) we can compare antigenic variation with changes in amino acid sequence. Figure 3 shows that the following points can be made:

(a) There are 11 amino acid changes between the 1968 and 1972 field-strain variants.

(b) There are 14 changes between the 1972 and 1975 strains, 12 of them being "new" changes (not occurring between 1968 and 1972) and two being "reversions" to the amino acid of the 1968 strain. One of the "new" changes was the insertion of an S (Ser)-residue between positions 9 and 10.

(c) The 1975 strain still maintained 9 out of the 11 amino acid changes introduced in the 1968–1972 evolution of the virus. The remaining two positions reverted in the 1975 strain to the sequence observed in the 1968 strain.

(d) The amino acid changes in both comparisons (1968 v. 1972; 1972 v. 1975) were widely distributed over the primary sequence of the HA1 subunit.

Antigenic drift in influenza is widely believed to occur by the emergence of variants with sufficiently altered immunological properties in order to escape

---

* I have used the sequence of the substrain of A/Mem/102/72 recombined with A/Bellamy/42 (Ward and Dopheide, 1980) for the evolutionary comparisons because it had fewer changes than in the derivative substrain recombined with A/PR/8/34 (Sleigh et al., 1980) when compared to A/NT/60/68.

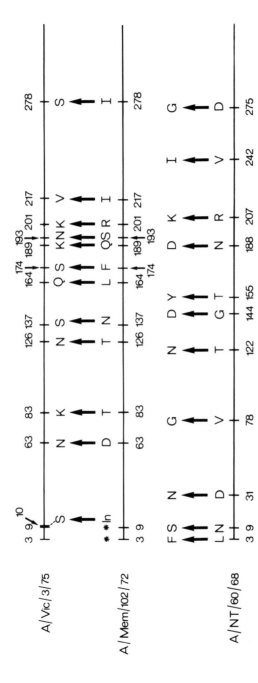

FIG. 3. Amino acid changes in the HA1 subunit of the hemagglutinin of the H3 subtype from a 1968 (A/NT/60/68) to a 1972 (A/Mem/102/72) to a 1975 (A/Victoria/3/75) strain. For A/Mem/102/72 the residue at positions 110, 127, 226 and 327 was from Ward and Dopheide (1980) (see text). All mutations introduced in the evolution A/NT/60/68 → A/Mem/102/72 were maintained in A/Victoria/3/75 with the exception of positions 3 and 9(*). The numbered residues, 63–278, for A/Victoria/3/75 are one residue displaced from the correct number (Min Jou et al., 1980) because of the insertion of a Ser residue between positions 9 and 10.

neutralization by antibodies already present in the population. For field strains with widespread epidemiological effects, such as the A/Victoria/3/75 strain and A/Mem/102/72 which effectively supplanted earlier variants, it is reasonable to assume that most of the amino acid changes from the parent strains are important in defining this altered antigenicity. Nevertheless this could be oversimplification and some changes may be less important than others, conferring only a marginal or neutral selective advantage. It is tempting to suggest that the changes at positions 201, 207 (both R → K) and those at 217 and 242, V → I or vice versa, which involve conservative amino acid changes, are antigenically unimportant. Also the two apparent reversions at positions 3 and 9 in the evolution from the 1972 to the 1975 strain to give amino acid residues characteristic of the 1968 strain strongly suggests that residues 3 and 9 are antigenically unimportant. This still leaves 7–10 "significant" amino acid changes in the HA1 subunit of each of these strains; while evolutionary theory may support the hypothesis that all changes are important, we cannot exclude that variation is determined by relatively few changes (say two to three) and that the remainder are neutral. But the results do exclude an early theory (Fazekas, 1975) that antigenic drift involves sequential changes of hydrophobic amino acids at one single amino acid residue. Specifically, none of the amino acid changes in A/Mem/102/72 (compared with A/NT/60/68) is subsequently altered in A/Victoria/3/75/ in a sequential fashion (we exclude the two reversions mentioned above).

Laboratory Mutants

Because field strains show fairly extensive differences from one another in their amino acid sequence, it has been instructive to isolate laboratory mutants with single or restricted mutations. Analysis of these mutations when correlated with altered antigenicity should pinpoint epidemiologically significant antigenic sites. Laboratory mutants have been isolated as variants escaping neutralization by either mixed antibodies (Fazekas, 1975) or by monoclonal antibodies. Moss et al. (1980) have analyzed a number of mutants by the former technique. They emphasize that the G → D (Gly → Asp) at position 144 varies in a number of mutants derived from A/NT/60/68. This change is particularly interesting as it is one of the first to appear in field strains. Monoclonally derived mutants have in general been disappointing in that mutants have failed to mimic changes occurring in the field. However, recently a mutant at this same residue 144 has been isolated, which had such altered antigenic properties that it could be distinguished by immunodiffusion (Laver et al., 1980).

In summary, the laboratory mutants provide the best opportunity for characterizing significant antigenic sites. Residue 144 is certainly one of these and undoubtedly it contributed to the early evolution of the H3 subtype. Whether one, two, or perhaps all of the amino acid changes in A/Mem/102/72 (see Fig. 3) are antigenically significant will probably only be known after a much more extensive analysis of mutants which have amino acid changes mimicking positions changing in the field.

ANTIGENIC SHIFT

Major changes in antigenicity (shift) occur much less frequently than the minor changes (drift) outlined above. These can now be correlated with the primary structure of the hemagglutinin as the complete amino acid sequence of an H2 and one avian strain H7 (previously called Hav1) is now available (Gething et al., 1980; Porter et al., 1979). Figure 4 shows a computer printout comparing the amino acid sequences numbering from the first methionine of the signal peptide (boxed) of the H3 subtype. The other two sequences were aligned with the H3 sequence by the occasional insertion of dashes in one or other of the sequences so as to maintain the position of the cysteine residues (dark shading) which are clearly well conserved (Brownlee, 1980). Boxes indicate the signal peptides and the peptide region between the HA1 and HA2 subunits. The signal peptides are out of alignment with respect to the initiator methionine residue because the N-terminal region of the mature HA1 of the H3 subtype is 10 residues longer than that of the other two subtypes. The HA1-HA2 interpeptide region also differs in the avian subtype (H7) in being longer (6 residues) than in the human subtypes (a single arginine residue). But in all cases there are arginine (R) or lysine (K) residues which would be susceptible to a trypsin-like enzyme, thus generating the HA1 and HA2 subunits.

Table III shows the extent of overall amino acid conservation expressed as percentage similarity in the three subtypes. The results are presented for

TABLE III
Amino acid conservation (%) of hemagglutinins of different subtypes

	H2 (A/Jap/57)	H3 (A/Mem/72)	H7 (Rostock)
H3 (A/Mem/72)	36, 49, 42[a]	—	—
H7 (Rostock)	35, 52, 42	39, 66, 49	—

[a] Values are as per cent (100 = identical). The first figure is for HA1 (residues 27–352 of Fig. 4), the second for HA2 (residues 358–580) and the third overall (residues 27–580). Deletions or insertions are counted as mismatches. (From Brownlee, 1980, with permission.)

HA1 and HA2 and for the whole hemagglutinin separately. The similarity of the HA2 subunits is striking, especially between H3 and H7 (66% identical). With HA2, it is well known that the N-terminal region (first 11 residues) is conserved and this region may interact with the cell "receptor". Another region of absolute identity is between residues 451 and 457. The HA1 subunit, by contrast, is less well conserved (35–39%) but considering that a fair proportion of the amino acid changes are conservative, e.g. V → I, R → K, a realistic assessment of similarity (not identity) would be somewhat higher than these figures.

Other regions of similarity have been noted in which "hydrophobicity" rather than any particular amino acid is conserved. Thus the signal peptide is poorly conserved between H3 and H7 even when realigned with respect to the initiator methionine. But both, as well as the H2 subtype, are hydrophobic. Similarly the C-terminal region of the HA2 subunit (residues 548–580 of Fig. 4) is generally hydrophobic in all three subtypes, consistent with its attachment to the cell membrane.

Overall the extent of similarity of the hemagglutinins of different subtypes is reasonably extensive and indicates that some of the sequence has a "structural" role not readily susceptible to variation in the evolution of the virus. Thus cysteine residues define the S–S bridges needed for the correct folding on the protein (Waterfield et al., 1981). Other regions must undoubtedly be conserved to allow the correct three-dimensional association of the hemagglutinin on the surface of the virus. "Packing" or association with a compatible neuraminidase molecule may also impose constraints. Presumably the location and possible function of the various conserved regions will be clearer when the three-dimensional structure is known.

CONCLUSION

We can now rapidly sequence the influenza hemagglutinin molecule and we know that it can function with widely differing amino acid sequences. There are clearly some constraints on this sequence but these are not so precise as to hinder extensive polymorphism from which the fittest virus strain is selected by the population immunity. It would seem likely that antigenicity, even within one subtype, is a function of many different parts of the HA1 subunit. To define these, in detail, promises to be a long task requiring many different experimental approaches, not least of these being a knowledge of the three-dimensional structure of the protein.

```
SIGNAL  PEPTIDE
   :
        10         20         30         40         50         60         70         80         90        100        110        120
MKTIIALSYI FCLVLGQDFP GNDNSTATLC LGHHAVPNGT LVKTITNDQI EVTNATELVQ SSSTGKICNN PHRI-LDGID CTLIDALLGD PHCDGFQNET WDLFV-ERSK A-FSNCYPYD
            MAIIYLILL FTAVRGDQIC IGYHANNSTE KVDTNLERNV TVTHAKDILE KTHNGKLCKL NGIPPLELGD CSIAGWLLGN PECDTLLSVP EWSYIMEKEN PRDGLCYPGS
        MN TQILVFALVA VIPTNADKIC LGHHAVSNGT KVNTLTERGV EVVNATETVE RTNIPKICSK GKRT-TDLGQ CGLLGTITGP PQCDQFLEFS ADLII-ERRE G-NDVCYPGK

       130        140        150        160        170        180        190        200        210        220        230        240
VPDYASLRLL VASSGTLEFI N--EGFTLTG VTQNGGSNAC KRGPDSGFFS RLNWL--YKS GSTYPVLNVT MPNNDNFDKL YIWGVHHPST DQEQTSLYVQ ASGRVTVSTK RSQQTIIPNI
FNDYEELKHL LSSVKHFEKV KILPKDRWTQ HTTTGGSRAC AVSGNPSFFR NMVWL--TKE GSDYPVAKGS YNNTSGEQML IIWGVHHPID ETEQRTLYQN VGTYVSVGTS TLNKRSTPGI
FVNEEALRQI LRGSGGIDKE T--MGFTYSG IRTNGTTSAC RRSGSS-FYA EMEWLLSNTD NASFPQMTKS YKNTRRESAL IVWGIHHSGS TTEQTKLYGS GNKLITVGSS KYHQSFVPSP

       250        260        270        280        290        300        310        320        330        340        350        360
GSRPWVRGQS SRISIYWTIV KPGDILVINS NGNLIAPRGY FKMRTGKSS- IMRSDAPIGT CISECITPNG SIPNDKPFQN VNKITYGACP KYVKQNTLKL ATGMRNVPEK RTR----GLF
ATRPKVNGQG GRMEFSWTLL DMWDTINFES TGNLIAPEYG FKISKRGSSG IMKTEGTLEN CETKCQTPLG AINTTLPFHN VHPLTIGECP KYVKSEKLVL ATGLRNVPQI ESR----GLF
GTRPQINGQS GRIDFHWLIL DPNDTVTFSF NGAFIAPNRA SFLR-GKSMG IQSDVQVDAN CEGECYHSGG TITSRLPFQN INSRAVGKCP RYVKQESLLL ATGMKNVPEP SKKREKRGLF

       370        380        390        400        410        420        430        440        450        460        470        480
GAIAGFIENG WEGMIDGWYG FRHQNSEGTG QAADLKSTQA AIDQINGKLN RVIEKTNEKF HQIEKEFSEV EGRIQDLEKY VEDTKIDLWS YNAELLVALG NQHTIDLTDS EMNKLFEKTR
GAIAGFIEGG WQGMVDGWYG YHHSNDQGSG YAADKESTQK AFDGITNKVN SVIEKMNTQF EAVGKEFGNL ERRLENLNKR MEDGFLDVWT YNAELLVLME NERTLDFHDS NVKNLYDKVR
GAIAGFIENG WEGLVDGWYG FRHQNAQGEG TAADYKSTQS AIDQITGKLN RLIEKTNQQF ELIDNEFTEV EKQIGNLINW TKDFITEVWS YNAELLVAME NQHTIDLADS EMNKLYERVR

       490        500        510        520        530        540        550        560        570        580
RQLRENAEDM GNGCFKIYHK CDNACIGSIR NGTYDHDVYR DEALNNRFQI KGVELKSGYK DWILWISFAI SCFLLC-VVL LGFIMWACQK -GNIRCNICI.........H3
MQLRDNVKEL GNGCFEFYHK CDDECMNSVK NGTYDYPKYE EESKLNRNEI KGVKLSSMGV YQILAIYATV AGSLSLAIMM AGISFWMCS- NGSLQCRICI.........H2
KQLRENAEED GTGCFEIFHK CDDDCMASIR NNTYDHSKYR EEAMQNRIQI DPVKLSSGYK DVILWFSFGA SCFLLL-AIA VGLVF-ICVK NGNMRCTICI.........H7
```

FIG. 4. Comparison of the amino acid sequence of: *top line*, H3 (A/Mem/102/72); *middle line*, H2 (A/Jap/307/57); *bottom line*, H7 (formerly Hav1), (Rostock). Cysteine (C) residues, on which alignment was made, are dark-shaded. Boxed regions mark the signal region (top left) and HA1-HA2 interpeptide region (middle right). (Modified from Brownlee, 1980, with permission.)

REFERENCES

Both, G. W., Sleigh, M. J., Bender, V. J. and Moss, B. A. (1980). *In* "Structure and Variation in Influenza Virus" (G. Laver and G. Air, eds) pp. 81–89. Elsevier/North Holland, New York.

Brownlee, G. G. (1980). *In* "Structure and Variation in Influenza Virus" (G. Laver and G. Air, eds) pp. 385–390. Elsevier/North Holland, New York.

Desselberger, U. and Palaese, P. (1978). *Virology* **88**, 394–399.

Fazekas de St. Groth, S. (1975). *In* "Negative Strand Viruses" (B. W. J. Mahy and R. D. Barry, eds) Vol. 2, pp. 741–754. Academic Press, London and New York.

Gething, M.-J., Bye, J., Skehel, J. and Waterfield, M. (1980). *In* "Structure and Variation in Influenza Virus" (G. Laver and G. Air, eds) pp. 1–10. Elsevier/North Holland, New York.

Jackson, D. C., Dopheide, T. A., Russell, R. J., White, D. O. and Ward, C. W. (1979). *Virology* **93**, 458–465.

Laver, W. G. and Webster, R. G. (1979). *Brit. med. Bull.* **35**, 29–33.

Laver, W. G., Air, G. M., Webster, R. G., Gerhard, W., Ward, C. W. and Dopheide, T. A. (1980). *In* "Structure and Variation in Influenza Virus" (G. Laver and G. Air, eds) pp. 295–306. Elsevier/North Holland, New York.

McGeoch, D., Fellner, P. and Newton, C. (1976). *Proc. natn. Acad. Sci. U.S.A.* **73**, 3045–3049.

Min Jou, W., Verhoeyen, M., Devos, R., Saman, R., Huylebroeck, D., Rompuy, L., Fang, R.-X. and Fiers, W. (1980). *In* "Structure and Variation in Influenza Virus" (G. Laver and G. Air, eds) pp. 63–68. Elsevier/North Holland, New York.

Moss, B. A., Underwood, P. A., Bender, V. J. and Whittaker, R. G. (1980). *In* "Structure and Variation in Influenza Virus" (G. Laver and G. Air, eds) pp. 329–338. Elsevier/North Holland, New York.

Porter, A. G., Barber, C., Carey, N. H., Hallewell, R. A., Threlfall, G. and Emtage, J. S. (1979). *Nature, Lond.* **282**, 471–477.

Porter, A. G., Smith, J. C. and Emtage, J. S. (1980). *Proc. natn. Acad. Sci. U.S.A.* **77**, 5074–5078.

Sleigh, M. J., Both, G. W. and Brownlee, G. G. (1979a). *Nucl. Acids Res.* **6**, 1309–1321.

Sleigh, M. J., Both, G. W. and Brownlee, G. G. (1979b). *Nucl. Acids Res.* **7**, 879–893.

Sleigh, M. J., Both, G. W., Brownlee, G. G., Bender, V. J. and Moss, B. A. (1980). *In* "Structure and Variation in Influenza Virus" (G. Laver and G. Air, eds) pp. 69–78. Elsevier/North Holland, New York.

Ward, C. W. and Dopheide, T. A. (1980). *In* "Structure and Variation in Influenza Virus" (G. Laver and G. Air, eds) pp. 27–38. Elsevier/North Holland, New York.

Waterfield, M., Scrace, G. and Skehel, J. J. (1981). *Nature, Lond.* **289**, 422–424.

Wiley, D. C., Wilson, I. A. and Skehel, J. J. (1981). *Nature, Lond.* **289**, 373–378.

Wilson, I. A., Skehel, J. J. and Wiley, D. C. (1981). *Nature, Lond.* **289**, 366–373.

Winter, G. and Fields, S. (1980). *Nucl. Acids Res.* **8**, 1965–1974.

NOTE ADDED IN PROOF

Since writing this chapter (April 1980) the three-dimensional structure of the hemagglutinin has been published (Wilson *et al.*, 1981); using this information, four distinct antigenic sites have been proposed (Wiley *et al.*, 1981). Further, sequence analysis (Sleigh, Both and Moss, personal communication) of A/Texas/77 and A/Bangkok/79—late strains of the H3 subtype—support the concept of at least three separate antigenic sites. These late strains also show amino acid changes in positions common to a number of monoclonally derived variants, which had not been previously correlated with changes in the field.

The 5' Ends of Influenza Viral Messenger RNAs are Donated by Capped Cellular RNAs

ROBERT M. KRUG, MICHELE BOULOY and
STEPHEN J. PLOTCH

Memorial Sloan-Kettering Cancer Center, New York, New York 10021, U.S.A.

INTRODUCTION

One feature distinguishing influenza virus from other non-oncogenic RNA viruses is that the functioning of the host nuclear RNA polymerase II is required for virus replication, specifically for viral RNA transcription (Lamb and Choppin, 1977; Spooner and Parry, 1977; Mark *et al.*, 1979). Our recent studies of viral RNA transcription have provided an explanation for this requirement for RNA polymerase II. We have shown that capped eukaryotic mRNAs strongly stimulate viral RNA transcription *in vitro* and donate their 5'-terminal methylated cap and a short stretch of internal nucleotides to the 5' end of the viral RNA transcripts (Bouloy *et al.*, 1978, 1979; Plotch *et al.*, 1979; Robertson *et al.*, 1980). A similar mechanism appears to operate *in vivo*, because we have found that the viral messenger RNAs (mRNAs) synthesized in the infected cell contain a short stretch of nucleotides at their 5' end, including the cap, that are not viral-coded (Krug *et al.*, 1979). The need for newly synthesized host mRNA primers to stimulate viral RNA transcription would explain the requirement for the proper functioning of host RNA polymerase II in the infected cell. In addition, our results may provide an insight into cellular functions; in particular, they suggest that the fully methylated cap (cap 1) structure, $m^7GpppNm$, which is found at the 5' end of all mammalian cellular mRNAs (Shatkin, 1979), may have an important role in cellular mRNA synthesis. Here we will summarize some of the results of these recent studies.

CELLULAR mRNAs AS PRIMERS FOR INFLUENZA VIRAL RNA TRANSCRIPTION

We had proposed a few years ago (Plotch and Krug, 1977, 1978; Plotch *et* al., 1978) that viral RNA transcription *in vivo* requires initiation by primer RNAs synthesized by RNA polymerase II and that the 5'-terminal cap found on *in vivo* viral mRNA (Krug et al., 1976) is derived from these primer RNAs. These proposals were based on two sets of data: (i) the strong stimulation by a primer dinucleotide, ApG or GpG, of viral RNA transcription *in vitro* catalyzed by the virion-associated transcriptase (McGeoch and Kitron, 1975; Plotch and Krug, 1977, 1978); and (ii) the absence of detectable capping and methylating enzymes in virions (Plotch et al., 1978).

Our later experiments essentially proved these proposals. We first identified primer RNAs in rabbit reticulocyte extracts, where they were shown to be globin mRNAs (Bouloy et al., 1978). We showed that β-globin mRNA, purified by polyacrylamide gel electrophoresis, stimulated viral RNA transcription about 80-fold and, on a molar basis, was about 1000 times more effective as a primer than ApG. Other capped eukaryotic mRNAs were also found to be extremely effective primers (Bouloy et al., 1978; Plotch et al., 1979; Bouloy et al., 1979). The viral RNA transcripts primed by these eukaryotic mRNAs were effectively translated into virus-specific proteins in cell-free systems (Bouloy et al., 1978).

To demonstrate that the cap of the primer mRNA was transferred to the viral mRNA during *in vitro* transcription, we used as primer globin mRNA containing ^{32}P only in its cap (prepared by enzymatically recapping β-eliminated globin mRNA in the presence of (α-^{32}P) GTP) (Plotch et al., 1979). After transcription in the presence of unlabeled nucleoside triphosphates, the resulting viral mRNA segments were shown to contain ^{32}P-labeled cap structures. As the initial approach for determining whether sequences in addition to the cap were transferred, we compared the size of the globin mRNA-primed viral mRNA segments to that of the ApG-primed segments (Plotch et al., 1979). The latter segments initiated exactly at the 3' end of the virion RNA (vRNA) templates (Skehel and Hay, 1978; Robertson, 1979). Gel electrophoretic analysis indicated that the globin mRNA-primed segments were 10–15 nucleotides larger than the ApG-primed segments (Plotch et al., 1979). The segments primed by other mRNAs were also about 10–15 nucleotides larger (Bouloy et al., 1979). Thus, approximately the same number of nucleotides, about 10–15, plus the cap, were transferred from globin and other mRNA primers.

CHARACTERIZATION OF THE TRANSFERRED HOST NUCLEOTIDE SEQUENCE

These studies did not allow us to identify which bases of a primer mRNA were transferred. To accomplish this, we labeled globin mRNA *in vitro* with

^{125}I to high specific activity, thereby labeling the C residues, and used this globin mRNA as a primer in the presence of unlabeled nucleoside triphosphates (Robertson *et al.*, 1980). The ^{125}I-labeled region transferred to each of the viral mRNA segments was sequenced. Each segment had the same T1 ribonuclease and pancreatic ribonuclease fingerprints (Fig. 1), indicating that the same region of the globin mRNA primer was transferred to each of the segments. The two major T1-oligonucleotides, Nos. 1 and 2, containing 75% of the total radioactivity, together comprise a sequence identical to the first 13 nucleotides (plus the cap) at the 5′ terminus of β-globin mRNA, which has the sequence (Lockard and RajBhandary, 1976):

m^{7}Gpppm^{6}AmpC(m)pApCpUpUpGpCpUpUpUpUpGpApCp . . .

1   8   13   15

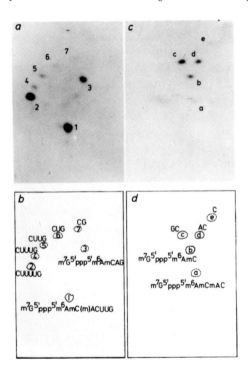

FIG. 1. RNase T1 and pancreatic RNase fingerprints of ^{125}I-labeled viral mRNAs. ^{125}I-labeled globin mRNA was used as a primer for influenza viral RNA transcription in the presence of unlabeled nucleoside triphosphates, and the viral mRNA segments after enzymatic deadenylation were separated by electrophoresis on 3% acrylamide gels containing 6M urea (5–8, 11). The bands were individually eluted and fingerprinted as described (Robertson *et al.*, 1980). Each band yielded the same fingerprint. (a) RNase T1 fingerprint. (b) Schematic drawing of panel (a). (c) Pancreatic RNase fingerprint. (d) Schematic drawing of panel (c). From Robertson *et al.* (1980).

Because only the C residues were labeled with ^{125}I, we could not conclude with certainty that all 13 5'-terminal nucleotides were transferred from β-globin mRNA. The results, however, certainly indicate that the β-globin mRNA donated at least the first 8, but no more than the first 14, 5'-terminal bases to the viral mRNAs. It was almost certain that the ... UUUU ... sequence (residues 9–12) originated from β-globin mRNA since there are no A residues in the 12-base 3'-terminal sequence of the vRNA templates (3' UCGUUUUCGUCC ...; Skehel and Hay, 1978; Robertson, 1979). On the other hand, the G residue at position 13 of the viral mRNA could have arisen by transfer from the β-globin mRNA or as a result of transcription. This data did not establish whether or not the A residue at position 14 of the β-globin mRNA was transferred to the viral mRNA. Thus, this data indicated that the predominant sequence transferred from β-globin mRNA includes the first 12, 13 or 14 5'-terminal nucleotides.

The minor ^{125}I-labeled RNase T1-resistant oligonucleotides found in the viral mRNAs (Fig. 1) indicated that shorter 5'-terminal pieces of β-globin mRNA were at times transferred and also strongly suggested that the transferred sequences were linked to G as the first base transcribed. For example, 14% of the RNase T1 products were the sequences CUUUGp, CUUGp, CUGp and CGp (T1 spots Nos. 4 → 7), which can be presumed to derive their C and U residues from β-globin mRNA (residues 8–11) and which must obtain their G residue by transcription. Similarly, m^{7}Gpppm^{6}AmC(m)AGp(T1 spot No. 3)—containing 11% of the ^{125}I label— clearly must derive its m^{7}Gpppm^{6}AmC(m)p from β-globin mRNA, could derive its next A residue either from the β-globin mRNA or as a result of transcription, and must obtain its G residue as a result of transcription. Thus, 14% and 11% of the time, 5'-terminal, cap-containing pieces of β-globin mRNA—8–11 bases and 2–3 bases in length, respectively—were transferred and linked to G as the first base transcribed.

MECHANISM FOR TRANSFER OF THE HOST SEQUENCE

The predominant sequence transferred from β-globin mRNA is not complementary at its 3' end to the 3'-terminal common sequence of the vRNA templates, favoring a mechanism of priming that does not involve hydrogen-bonding between the primer mRNA and the template vRNA. Other data even more strongly eliminated the possibility that hydrogen-bonding is needed for the stimulation of transcription. We have found that capped, 5'-terminal fragments of mRNAs are effective primers, in fact 4–8-fold more effective than the mRNA from which they were derived (Bouloy and Krug, submitted for publication). Fragments of globin mRNA and A1MV (alfalfa mosaic virus) RNA 4 as short as 14–23 nucleotides in length were good primers, but these fragments were too short to contain any sequence complementary to the 3' end of the vRNA segments (Lockard and RajBhandary, 1976; Koper-Zwarthoff et al., 1977).

In addition, in collaboration with Aaron Shatkin's laboratory at the Roche Institute of Molecular Biology, we have found that capped ribopolymers which lack a sequence complementary to the 3' end of vRNA are effective primers (Krug et al., 1980). As shown in Table I, capped poly A and capped poly AU, neither of which contains a complementary sequence, were about as effective as primers as globin mRNA. Capped poly U was also active, even though the 3'-terminal 12-nucleotide sequence common to the 8 influenza vRNA segments contains no A residues (Skehel and Hay, 1978; Robertson, 1979). Of the various polymers tested as primers, capped poly C was the least effective. Thus, while there may be some effect of sequence on the efficiency of transcriptase-priming by an exogenous capped ribopolymer, there is no requirement for the presence of a sequence complementary to the 3' end of the vRNA template.

In the absence of hydrogen-bonding between the primer mRNA and the template vRNA, the most likely mechanism for priming is that shown in

TABLE I

Priming activity of capped synthetic ribopolymers

Experiment[a]	Primer added (pmol)	GMP incorp (pmol)
1	None	1.9
	Globin mRNA (6)	84.4
	Capped poly U (3)	29.1
	Capped poly A (3)	59.1
	Capped poly AU (3)	62.7
2	None	1.1
	Capped poly A (3)	33.0
	Capped poly AG (3)	16.3
3	None	1.2
	Globin mRNA (6)	31.6
	Capped poly AU (3)	27.0
	Capped poly C (3)	5.5

[a] In the three experiments shown, different virus preparations with different specific transcriptase activities (ranging from 14 to 32 nmol of GMP incorporated/mg viral protein/hour in a globin mRNA-primed reaction) were used. From Krug et al. (1980).

Fig. 2. A 5′-terminal fragment of β-globin mRNA is cleaved by a virion-associated nuclease, and this fragment is then linked to the first base transcribed, a G residue. The stimulation of the initiation of transcription would result from a specific interaction of the capped RNA and/or its 5′-terminal fragments with one or more virion-associated transcriptase proteins. One of the consequences of G, rather than A, being the first base transcribed is that the viral RNA transcripts would not necessarily contain an A complementary to the 3′-terminal U of the template vRNA. It is most likely that this G would be directed by the 3′-penultimate C of the vRNA, though it could also be directed by the 3′-terminal U since the dinucleotide GpG, like ApG, has been shown to initiate transcription exactly at the 3′-terminal UC of the vRNA (Hay and Skehel, 1979). The only way that an A could be found opposite the 3′-terminal U of the vRNA template would be if the 5′-terminal fragment cleaved from the primer mRNA had a 3′-terminal A.

This mechanism predicts the existence of an intermediate in the priming reaction: 5′-terminal fragment(s) cleaved from the mRNA primer. In addition, in a reaction in which the only triphosphate present is GTP, it should be possible to demonstrate the presence of the cleaved 5′-terminal

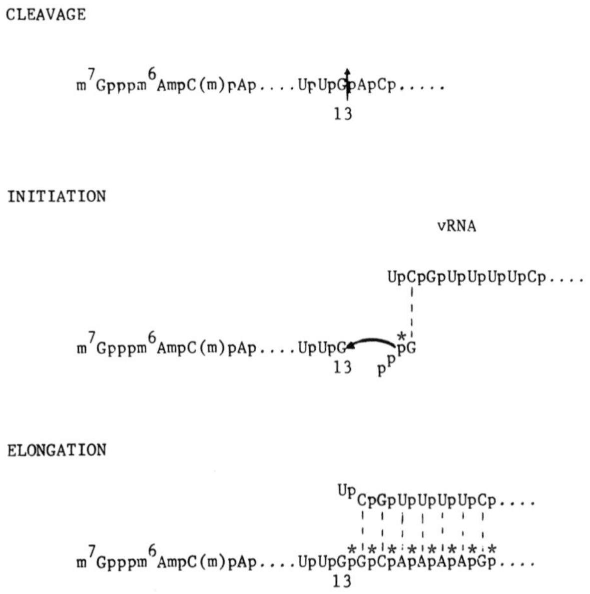

CLEAVAGE

$$m^7Gpppm^6AmpC(m)pAp\ldots.UpUpGpApCp\ldots.$$
13

INITIATION

vRNA

$$UpCpGpUpUpUpUpCp\ldots.$$

$$m^7Gpppm^6AmpC(m)pAp\ldots.UpUpG\quad pG$$
13   pP

ELONGATION

$$^{Up}CpGpUpUpUpUpCp\ldots.$$

$$m^7Gpppm^6AmpC(m)pAp\ldots.UpUpGpGpCpApApApApGp\ldots.$$
13

FIG. 2. Mechanism for the priming of influenza viral RNA transcription by β-globin mRNA and other capped RNAs.

fragment to which one or more Gs were added. We have identified both these species. These results will be published elsewhere (Plotch, Bouloy and Krug, unpublished experiments), and we will only present the conclusions here. To identify the second species (i.e., the fragment with Gs added), globin mRNA was incubated with detergent-treated virus in the presence of $(\alpha\text{-}^{32}P)$ GTP as the only ribonucleoside triphosphate precursor. Two terminally labeled species predominated: a fragment resulting from the cleavage of β-globin mRNA at G_{13} with either one or two Gs added at its 3' end, indicating that the endonucleolytic cleavage occurred mainly at this position (as shown in Fig. 2). In addition, some ($\sim 10\%$ of the time) cleavage at U_{12} was seen, with two Gs added. When the same experiment was done with either labeled ATP or labeled UTP as the only triphosphate, no terminally labeled fragments of β-globin mRNA were found. With labeled $(\alpha\text{-}^{32}P)$ CTP, however, the G_{13} fragment was terminally labeled with a single C residue. This C incorporation was presumably directed by the G residue at the third position at the 3' end of vRNA and occurred only because the major 5' fragment cleaved from β-globin mRNA contained a 3'-terminal G. With A1MV RNA 4 (containing no G residues in the first 37 nucleotides at its 5' end (Koper-Zwarthoff et al., 1977)), capped poly A, and capped poly AU as primers, the only base incorporated in single triphosphate reactions was G (and not C). We were able to show the cleavage products of A1MV RNA 4 directly, without G-addition, using A1MV RNA 4 labeled with ^{32}P in its cap. The 5'-terminal sequence of A1MV RNA 4 is capGUUUUUAUUUUUAAUUU . . . (Koper-Zwarthoff et al., 1977). The predominant cleavage occurred after the A 13 nucleotides from the cap. We also showed that this cleavage yielded a 3'-terminal hydroxyl group in the 5' fragment. In the presence of GTP, one, two or three Gs were added to the 3' end of this cleavage product. By analyzing several different capped primers, we have found that the cleavage is predominantly at a purine (A or G) at a position 10–13 nucleotides from the 5' end.

The specific interaction between the primer mRNA and one or more transcriptase proteins, which leads to the stimulation of the initiation of transcription, most probably involves primarily the 5'-terminal methylated cap. Only capped mRNAs are active as primers (Bouloy et al., 1978, 1979; Plotch et al., 1979). Removal of the m^7G of the cap of an mRNA by chemical (β-elimination) or enzymatic treatment eliminates all priming activity, and most of this activity can be restored by enzymatically recapping the β-eliminated mRNA (Plotch et al., 1979). The cap must contain methyl groups, since reovirus mRNAs with 5' GpppG ends are not active as primers (Bouloy et al., 1979).

Indeed, we have recently shown that each of the two methyl groups in the cap, the 7-methyl on the terminal G and the 2'-O-methyl on the penultimate base, strongly influences the priming activity of a mRNA (Bouloy et al., 1980). Of particular interest is the effect of the 2'-O-methyl group. To demonstrate the importance of this group, we used several plant viral RNAs containing the monomethylated cap O structure, m⁷GpppG. As shown in Table II, brome mosaic virus (BMV) RNA 4 stimulated influenza viral RNA transcription only about 10–15% as effectively as globin mRNA. After treatment with the enzyme mRNA-(nucleoside-2'-)methyltransferase (mRNA-N2'-MTase) in the presence of AdoMet (S-adenosylmethionine), the priming activity of BMV RNA 4 was greatly increased, about 14-fold. Analysis of the cap structure established that the enzyme methylated only the 2'-O-group of the penultimate base of the cap. When methylation was blocked by substituting AdoHcy (S-adenosylhomocysteine) for AdoMet during the enzyme treatment, no significant increase in the priming activity of BMV RNA 4 occurred. Qualitatively similar results were obtained with other plant viral RNAs: priming activity increased 3- to 20-fold following 2'-O-methylation. This is the first instance in which the 2'-O-methyl group of the cap has been shown to have a strong and clear-cut effect on a specific function of an mRNA. Consequently, the full methylated cap 1 structure, m⁷GpppNm, which is found in all mammalian cellular mRNAs and most animal virial mRNAs (Shatkin, 1976), is more stringently required for priming influenza viral RNA transcription than for translation in cell-free systems.

The cap 1 structure is presumably recognized at the initiation step of viral RNA transcription (see Fig. 2), thereby mediating the stimulation of initiation. Recent experiments indicate that the cap is actually recognized at the initial step of the reaction by the virion-associated endonuclease: the specific cleavage of an mRNA described above was found to occur only when

TABLE II
Requirement of the 2'-O-methyl group in the penultimate base of the cap structure for the priming activity of BMV RNA 4[a]

Primer (pmol)	Treatment of primer	pmol GMP incorp.
None	—	0.9
ApG (20 000)	None	101.8
Globin mRNA (15)	None	37.2
BMV RNA 4 (15)	None	5.1
BMV RNA 4 (15)	mRNA-N2'-MTase + AdoMet	72.3
BMV RNA 4 (15)	mRNA-N2'-MTase + AdoHcy	6.1

[a] Transcriptase assays were carried out for 1 hour at 31°C with (α-³²P) GTP as labeled precursor (5–7, 10, 11). BMV RNA 4 (60 pmol) was treated with mRNA-N2'-MTase in the presence of either 3 μM AdoMet or 200 μM AdoHcy. From Bouloy et al. (1980).

the mRNA contained a fully methylated cap structure (Bouloy, Plotch and Krug, unpublished experiments). Clearly, it will be of great interest to identify the virus-associated protein(s) which recognize the cap 1 structure and to compare it to the cap-recognizing protein associated with eukaryotic ribosomes (Sonenberg *et al.*, 1979).

Although no specific nucleotide sequence in a capped RNA is required for its priming activity, our results indicate that reducing the secondary structure of an RNA enhanced its activity as a primer (Krug *et al.*, 1980). Two different modifications of reovirus mRNA (substitution of inosine for guanosine and bisulfite treatment) which abolished G-C pairing and thus reduced secondary structure were shown to increase priming activity by at least 3- to 5-fold. Diminished secondary structure in an RNA, particularly in its 5'-terminal region, probably facilitates cleavage of the RNA by the virion-associated nuclease. This is probably the reason why 2'-O-methylated AlMV RNA 4, in which 14 of the first 18 5'-terminal nucleotides are Us (Koper-Zwarthoff *et al.*, 1977), is the best mRNA primer so far examined. A similar explanation may hold for the observation that capped 5'-terminal, T1-ribonuclease-derived fragments of mRNAs are 4- to 8-fold better primers on a molar basis than the intact mRNAs from which they were derived (Bouloy and Krug, submitted for publication).

INHIBITION OF INFLUENZA VIRUS TRANSCRIPTASE

We found that uncapped RNAs with reduced secondary structure inhibited the transcriptase reaction primed by either ApG or globin mRNA (Krug *et al.*, 1980). Inosine substitution or bisulfite treatment of the uncapped form of reovirus mRNAs converted them from essentially inactive species to potent inhibitors of the transcriptase. In addition, certain uncapped ribopolymers with low secondary structure, e.g. poly S^4U (4-thiouridylic acid) and poly U, were strong inhibitors, whereas those with ordered structures, e.g. poly AU and poly CG, did not inhibit. Of these ribopolymers, poly S^4U was the most effective, 1 µg (in a 25 µl assay) reducing transcription primed by either globin mRNA or ApG by 95 %. Inhibition of ApG-primed transcription by poly U and poly S^4U has also been reported recently by others (Smith *et al.*, 1980). However, the absence of hydrogen-bonding in a ribopolymer was apparently by itself not sufficient for conferring inhibitory activity. In the transcriptase assay mixture, poly A and poly C should be free of most hydrogen-bonding, yet these two homopolymers did not inhibit. This suggests that nucleotide specificity may also be involved in inhibition. Thus, it was found that poly AG (which should also be free of hydrogen-bonding) was an effective inhibitor in contrast to poly A, consistent with the localized

G-containing regions in poly AG conferring inhibitory activity. Our results suggest that U, S⁴U, G and I (but not A and C) contribute to the inhibitory activity of a ribopolymer when these nucleotides are not base-paired in the polymer. We have found that the ribopolymers with the highest inhibitory activity, poly S⁴U and poly AG, inhibit the first step in mRNA-primed transcription, the endonucleolytic cleavage of the mRNA primer. It is conceivable that further studies of this type could lead to the development of compounds which may be potent and clinically useful anti-influenza agents.

TRANSFER OF THE 5′ END FROM HOST TO VIRAL mRNA ALSO TAKES PLACE *IN VIVO*

We have obtained strong evidence that priming of viral RNA transcription by capped host RNAs occurs in the infected cell (Krug et al., 1979). If such priming occurs *in vivo*, then the viral mRNA synthesized in the infected cell should contain 10–15 nucleotides at its 5′ end, including the cap, which are not viral-coded. This is exactly what we found. First, gel electrophoretic analysis indicated that the segments of *in vivo* viral mRNA, like the segments synthesized *in vitro* with mRNA primers, were 10–15 nucleotides longer at their 5′ end than the ApG-primed *in vitro* segments. In addition, when ^{3}H-methyl-labeled *in vivo* viral mRNA was hybridized to vRNA, the 5′-terminal cap structure of the mRNA was not protected against pancreatic or T1 RNase digestion (Krug et al., 1979) (Fig. 3). All of the cap (the -4.4 to 5.2 charge species) was released from the double strands by the nuclease digestion. Only internal m^6A residues (charge -2) remained in the double-strands, although approximately a third of these residues were also released. As each molecule of *in vivo* viral mRNA contains an average of three m^6A residues (Krug et al., 1976), these results indicate that one of these m^6As is in the 5′-terminal sequence that is not viral-coded. Thus, our results strongly suggest that host-cell mRNAs and/or their precursors serve as primers for viral RNA transcription in the infected cell, and that they donate their cap and 10–15 internal nucleotides, one of which is m^6A, to the resulting viral mRNA molecules (Krug et al., 1979). Because our *in vitro* results show that most capped RNAs are effective primers, it is extremely likely that many different host capped RNAs serve as primers for transcription *in vivo*. Consequently, the initial sequence at the 5′ end of *in vivo* viral mRNA would be heterogenous. Consistent with such heterogeneity, two different bases, Am or Gm, are found at the 5′-penultimate position (Krug et al., 1976).

The synthesis of these host-cell mRNA primers can be presumed to constitute the α-amanitin-sensitive step (RNA polymerase II function)

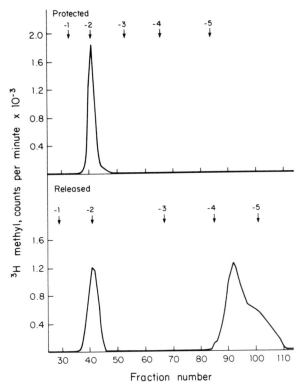

FIG. 3. ³H-methyl-labeled species retained in (protected) or released from vRNA: *in vivo* viral mRNA hybrids by pancreatic RNase digestion. The *in vivo* viral mRNA contained the ³H-methyl label. The hybrids were digested with pancreatic RNase, and the RNase-resistant double-strands and RNase-released material were collected. The double-strands (after heating and fast-cooling) and the released material were digested with RNase T2, and the hydrolyzates were analyzed by DEAE-Sephadex chromatography. From Krug *et al.* (1979).

required for viral RNA transcription. A critical, unanswered question is why new and continuous synthesis of these host mRNA primers is required. It may be that pre-existing capped RNAs are tied up in ribonucleoprotein structures (including polyribosomes) and cannot be used by the virion transcriptase. Also, the requirement for new synthesis may be due at least in part to the site of viral RNA transcription in the infected cell. The presence of m⁶A residues in the viral mRNA (Krug *et al.*, 1976) suggests that viral RNA transcription may have a nuclear phase, and other experiments measuring the steady-state level of viral mRNA (Barrett *et al.*, 1979; Mark *et al.*, 1979) suggest that the nucleus is the site of at least primary transcription. If viral RNA transcription occurs in the nucleus, then the need for continued host-cell mRNA synthesis may reflect the fact that the amount of available cellular mRNA and/or its precursors in the nucleus is limited and rapidly depleted. Clearly, it will be of great interest to establish definitively where both primary and amplified transcription occurs in the cell.

POSSIBLE IMPLICATIONS FOR CELLULAR TRANSCRIPTION

Another important question is whether these results with influenza viral RNA transcription shed any light on cellular functions. For example, is the recognition of methylated cap (cap 1) structures unique to influenza viral RNA transcription, or does it also occur during cellular transcription-related processes? One possibility is that the methylated cap structure is required for the proper processing of cellular heterogeneous nuclear RNA. An initial step in this processing, cleavage of heterogeneous nuclear RNA near its 5' end (Klessig, 1977; Berget *et al.*, 1977; Darnell, 1978), is comparable to the cleavage of the primer RNA near its 5' end which occurs in the influenza-virion transcriptase reaction (see Fig. 2). The cellular enzyme(s), like the virion enzyme, may require the presence of the cap 1 structure. A later step in the processing of heterogeneous nuclear RNA, the ligation of the 5'-terminal region on to distal regions of the RNA (Klessig, 1977; Berget *et al.*, 1977; Darnell, 1978), could also involve cap recognition. This hypothesis can be tested when the appropriate cellular enzyme(s) are identified and characterized. At the present time, the cap-recognizing protein(s) and cleavage enzyme(s) in influenza virions may serve as a useful model system.

ACKNOWLEDGEMENTS

We would like to thank Barbara Broni for expert technical assistance. These investigations were supported in part by U.S. Public Health Service Grants CA 08748 and AI 11772, and by U.S. Public Health Service International Fellowship TW 02590-01 to M.B.

REFERENCES

Barrett, T., Wolstenholme, A. J. and Mahy, B. W. J. (1979). *Virology* **98**, 211–225.
Berget, S. M., Moorce, C. and Sharp, P. A. (1977). *Proc. natn. Acad. Sci. U.S.A.* **74**, 3171–3175.

Bouloy, M., Plotch, S. J. and Krug, R. M. (1978). *Proc. natn. Acad. Sci. U.S.A.* **75**, 4886–4890.
Bouloy, M., Morgan, M. A., Shatkin, A. J. and Krug, R. M. (1979). *J. Virol.* **32**, 895–904.
Bouloy, M., Plotch, S. J. and Krug, R. M. (1980). *Proc. natn. Acad. Sci. U.S.A.* **77**, 3952–3956.
Darnell, J. E., Jr. (1978). *Science, N.Y.* **202**, 1257–1260.
Hay, A. J. and Skehel, J. J. (1979). *J. gen. Virol.* **44**, 599–608.
Klessig, D. F. (1977). *Cell* **12**, 9–21.
Koper-Zwarthoff, E. C., Lockard, R. E., Alzner-DeWeerd, B., RajBhandary, U. L. and Bol, J. T. (1977). *Proc. natn. Acad. Sci. U.S.A.* **74**, 5504–5508.
Krug, R. M., Morgan, M. M. and Shatkin, A. J. (1976). *J. Virol.* **20**, 45–53.
Krug, R. M., Broni, B. A. and Bouloy, M. (1979). *Cell* **18**, 329–334.
Krug, R. M., Broni, B. A., LaFiandra, A., Morgan, M. A. and Shatkin, A. J. (1980). *Proc. natn. Acad. Sci. U.S.A.* **77**, 5874–5878.
Lamb, R. A. and Choppin, P. W. (1977). *J. Virol.* **23**, 816–819.
Lockard, R. E. and RajBhandary, U. L. (1976). *Cell* **9**, 747–760.
Mark, G. E., Taylor, J. M., Broni, B. and Krug, R. M. (1979). *J. Virol.* **29**, 744–752.
McGeoch, D. and Kitron, N. (1975). *J. Virol.* **15**, 686–695.
Plotch, S. J. and Krug, R. M. (1977). *J. Virol.* **21**, 24–34.
Plotch, S. J. and Krug, R. M. (1978). *J. Virol.* **25**, 579–586.
Plotch, S. J., Tomasz, J. and Krug, R. M. (1978). *J. Virol.* **28**, 75–83.
Plotch, S. J., Bouloy, M. and Krug, R. M. (1979). *Proc. natn. Acad. Sci. U.S.A.* **76**, 1618–1622.
Robertson, H. D., Dickson, E., Plotch, S. J. and Krug, R. M. (1980). *Nucl. Acids Res.* **8**, 925–942.
Robertson, J. S. (1979). *Nucl. Acids Res.* **6**, 3745–3757.
Shatkin, A. J. (1976). *Cell* **9**, 645–653.
Skehel, J. J. and Hay, A. J. (1978). *Nucl. Acids Res.* **4**, 1207–1219.
Smith, J. C., Raper, R. H., Bell, L. D., Stebbing, N. and McGeoch, D. (1980). *Virology* **103**, 245–249.
Sonenberg, N., Rupprecht, K., Hecht, S. and Shatkin, A. J. (1979). *Proc. natn. Acad. Sci. U.S.A.* **76**, 4345–4349.
Spooner, L. L. R. and Barry, R. D. (1977). *Nature, Lond.* **268**, 650–652.

Drift and Shift of Influenza A-type Virus: Comparison of the Hemagglutinin Gene in H3 Subtypes

WILLY MIN JOU, MARTINE VERHOEYEN,
RONGXIANG FANG and WALTER FIERS

*Laboratory of Molecular Biology, State University of Ghent, Ledeganckstraat 35,
B-9000 Ghent, Belgium*

INTRODUCTION

The ability of A-type influenza viruses to change their antigenic properties is a major obstacle to controlling the viral infection by vaccination, and the virus continues to cause serious respiratory illness in man. These changes may be drastic (antigenic shift), resulting in the appearance of a new subtype, or they may be smaller, progressive changes (antigenic drift) within a particular subtype. In order to relate changes in the antigenic character to changes in the structure of the hemagglutinin (the major surface antigen), we have determined the primary structure of two hemagglutinin genes from viruses belonging to the same subtype (H3N2) and separated from each other by a seven-year drift period: A/Aichi/2/68 (Verhoeyen *et al.*, 1980) and A/Victoria/3/75 (Min Jou *et al.*, 1980). These sequences were compared with the hemagglutinin gene from an intermediate H3N2 isolate, A/Memphis/102/72 (Sleigh *et al.*, 1980), with the sequence of the H2N2 strain A/Japan/305/57 (Gething *et al.*, 1980) and with the H7 fowl plague hemagglutinin gene (Porter *et al.*, 1979).

The influenza A virus has a divided genome consisting of eight negative-stranded RNA segments; the hemagglutinin protein is coded for by gene 4 (Palese and Schulman, 1976; Scholtissek *et al.*, 1976). Our approach was to study the hemagglutinin information as cloned DNA, as this procedure is often faster than protein or RNA sequencing.

CLONING AND SEQUENCING OF HEMAGGLUTININ GENES FROM INFLUENZA TYPE A

Between 10 and 70 ng of gene 4 poly(dT)-tailed dsDNA, synthesized as described (Min Jou *et al.*, 1980), was annealed to a 1.5 to 3 times molar excess of poly(dA)-tailed pBR322 DNA and used to transform competent *E. coli* cells. The colonies obtained were screened for the presence of full-size hemagglutinin information by colony hybridization and by restriction enzyme analysis. In both cases a plasmid containing a presumed full-size hemagglutinin gene (approximately 1860 base-pairs including the tails) was chosen for sequence determination according to the method of Maxam and Gilbert (1977).

The total length of the Aichi hemagglutinin gene (Verhoeyen *et al.*, 1980) and also of the Memphis hemagglutinin RNA (Sleigh *et al.*, 1980) is 1765 nucleotides; Victoria has an extra insertion of three nucleotides (specifying an asparagine residue at position 8' of the HA1 part of the hemagglutinin protein) (Min Jou *et al.*, 1980). Apart from this difference the detailed organization of the hemagglutinin RNAs belonging to the H3 subtype (Aichi, Memphis and Victoria) seems to be identical: 29 nucleotides precede the signal for the AUG translation initiation codon; the translated region is 1698 (1701) nucleotides long; and the 3'-terminal non-coding stretch comprises 38 nucleotides. Considering the deduced amino acid sequences, the signal sequence (Blobel and Dobberstein, 1975) is 16 amino acids long, the HA1 portion of the hemagglutinin protein consists of 328 (329) amino acids, the region excised from between HA1 and HA2 by post-translational processing is a single arginine residue, and the HA2 part comprises 221 amino acids. In total there are five apparent reversions from Memphis to Victoria, apparent because there is no reason to believe that prototypes of successive epidemics as isolated from nature form a direct geneological

TABLE I
Antigenic drift in influenza virus

Comparison	Number of amino acid changes in			
	Signal	HA1	HA2	Total
Aichi–Memphis	1	15	4	19
Memphis–Victoria	2	17	2	21
Aichi–Victoria	3	22	4	29
Total variable sites between the three strains	3	27	5	35

lineage. There is not a single case of two successive changes (to a different amino acid) between these three H3 strains.

As expected (Table I), most of the changes are found in the HA1 part of the molecule, and there is some clustering within the middle third of this portion (amino acids 110–226) in the three strains examined. The amino acid changes in HA1 of Victoria with respect to Aichi (Fig. 1) are at positions 8′ (Asn), 31 (Asn), 63 (Asn), 78 (Gly), 83 (Lys), 122 (Asn), 126 (Asn), 137 (Ser), 144 (Asp), 155 (Tyr), 164 (Gln), 174 (Ser), 183 (Val), 188 (Asp), 189 (Lys), 193 (Asn), 201 (Lys), 207 (Lys), 217 (Val), 242 (Ile), 275 (Gly) and 278 (Ser). Although a not unimportant fraction of these changes must have influenced the antigenic character, it is too early to understand the basis of drift from the sequence information presently available.

The nucleotide changes in the different hemagglutinin genes are summarized in Table II. The single-base mutations which do lead to an amino acid change are more frequent in the HA1 part compared to the HA2 part, as discussed above at the protein sequence level. But, very surprisingly, the number of silent mutations is also substantially higher in the HA1 region. This can perhaps be explained on the basis of a specific secondary and/or tertiary structure of the viral RNA. Now, silent mutations may be selectively retained in the HA1 region due to immunological selection, but their negative effect on the specific folding of the RNA may be counteracted by complementary but silent mutations.

Comparison between different subtypes reveals considerable variability, especially in the HA1 region. The HA2 region shows about 50% conservation when human H3 is compared to human H2, about 52% for H2 compared to fowl plague virus (FPV), but there is 66% homology between

TABLE II
Mutations in the HA1 and HA2 regions of the H3 hemagglutinin gene

Comparison	HA1		HA2	
	Causing amino acid change	Silent	Causing amino acid change	Silent
Aichi–Memphis	16	14	4	9
Memphis–Victoria	16	13	2	3
Aichi–Victoria	22	23	4	10
Total variable sites	27	24	5	11
%	2.7	2.4	0.8	1.7

H3: A/Aichi/2/68
H2: A/Jap/305/57
H7: A/FPV/Rostock/34

Signal peptide (−15 to −1 / 1)

```
          -15            -10             -5        -1  1         5             10
H3:  Met-Lys-Thr-Ile-Ile-Ala-Leu-Ser-Tyr-Ile-Phe-Cys-Leu-Ala-Leu-Gly↓Gln-Asp-Leu-Pro-Gly-Asn-Asp-Asn-Ser-Thr-
H2:  Met-Ala-Ile-Ile-Tyr-Leu-Ile-Leu-Leu-Phe-Thr-Ala-Val-Arg-Gly↓
H7:  Met-Asn-Thr-Gln-Ile-Leu-Val-Phe-Ala-Leu-Val-Ala-Val-Ile-Pro-Thr-Asn-Ala↓
```

```
      15            20            25            30            35            40            45
H3:  Ala-Thr-Leu-Cys-Leu-Gly-His-His-Ala-Val-Pro-Asn-Gly-Thr-Leu-Val-Lys-Thr-Ile-Thr-Asp-Asp-Gln-Ile-Glu-Val-Thr-Asn-Ala-Thr-Glu-Leu-Val-Gln-Ser-
H2:  Asp-Gly-Ile-Cys-Ile-Gly-Tyr-His-Ala-Asn-Asn-Ser-Thr-Glu-Lys-Val-Asp-Thr-Ile-Leu-Glu-Arg-Asn-Val-Thr-Val-Thr-His-Ala-Lys-Asp-Ile-Leu-Glu-Lys-
H7:  Asp-Lys-Ile-Cys-Leu-Gly-His-His-Ala-Val-Ser-Asn-Gly-Thr-Lys-Val-Asn-Thr-Leu-Thr-Glu-Arg-Gly-Val-Glu-Val-Val-Asn-Ala-Thr-Glu-Thr-Val-Glu-Arg-
```

```
      50            55            60            65            70            75            80
H3:  Ser-Ser-Thr-Gly-Lys-Ile-Cys-Asn-Asn-Pro-His-Arg-Ile-Leu-Asp-Gly-Ile-Asp-Cys-Thr-Leu-Ile-Asp-Ala-Leu-Leu-Gly-Asp-Pro-His-Cys-Asp-Val-Phe-Gln-
H2:  Thr-His-Asn-Gly-Lys-Leu-Cys-Lys-Leu-Asn-Gly-Ile-Pro-Pro-Leu-Glu-Leu-Gly-Asp-Cys-Ser-Ile-Ala-Gly-Trp-Leu-Leu-Gly-Asn-Pro-Glu-Cys-Asp-Arg-Leu-
H7:  Thr-Asn-Ile-Pro-Leu-Ile-Cys-Ser-Lys-Gly-Lys-Arg-Thr-Val-Asp-Leu-Gly-Gln-Cys-Gly-Leu-Leu-Gly-Thr-Ile-Thr-Gly-Pro-Pro-Gln-Cys-Asp-Gln-Phe-Leu-
```

```
      85            90            95           100           105           110
H3:  Asn-Glu-Thr-Trp-Asp-Leu-Phe-Val-Glu-Arg-Ser-Lys-Ala-Phe-Ser-Asn-Cys-Tyr-Pro-Tyr-Asp-Val-Pro-Asp-Tyr-Ala-Ser-Leu-Arg-Ser-Leu-Val-Ala-Ser-
H2:  Ser-Val-His-Glu-Trp-Ser-Tyr-Ile-Val-Glu-Lys-Ala-Asn-Pro-Ala-Asn-Asp-Leu-Cys-Tyr-Pro-Gly-Asn-Phe-Asn-Asp-Tyr-Glu-Glu-Leu-Lys-His-Leu-Leu-Ser-Ser-
H7:  Glu-Phe-Ser-Ala-Asp-Leu-Ile-Ile-Glu-Arg-Arg-Glu-Gly-Ser-Asp-Val-Cys-Tyr-Pro-Gly-Lys-Phe-Val-Asn-Glu-Glu-Ala-Leu-Arg-Gln-Ile-Leu-Arg-Glu-Ser-Gly-Gly-
```

```
     115           120           125           130           135           140           145
H3:  Ser-Gly-Thr-Leu-Glu-Phe-Ile-Thr-Glu-Gly-Phe-Thr-Trp-Thr-Gly-Val-Thr-Gln-Asn-Gly-Gly-Ser-Asn-Ala-Cys-Lys-Arg-Gly-Pro-Gly-Ser-Gly-Phe-Phe-Ser-
H2:  Val-Lys-His-Phe-Glu-Lys-Val-Lys-Ile-Leu-Pro-Lys-Asp-Arg-Trp-Thr-Gln-His-Thr-Thr-Thr-Gly-Gly-Ser-Arg-Ala-Cys-Ala-Val-Ser-Gly-Asn-Pro-Ser-Phe-Phe-
H7:  Ser-Gly-Ile-Glu-Asp-Lys-Thr-Met-Gly-Phe-Thr-Tyr-Ser-Gly-Ile-Arg-Thr-Asn-Gly-Thr-Thr-Ser-Ala-Cys-Arg-Arg-Ser-Gly-Ser-Ser-Phe-Tyr-
```

```
     150           155           160           165           170           175           180
H3:  Ser-Arg-Leu-Asn-Trp-Leu-Thr-Lys-Ser-Gly-Ser-Thr-Tyr-Pro-Val-Leu-Asn-Val-Thr-Met-Pro-Asn-Asn-Asp-Asn-Phe-Asp-Lys-Leu-Tyr-Ile-Trp-Gly-Ile-
H2:  Arg-Asn-Met-Val-Trp-Leu-Thr-Lys-Lys-Gly-Ser-Asp-Tyr-Pro-Val-Ala-Lys-Gly-Ser-Tyr-Asn-Asn-Thr-Ser-Gly-Glu-Gln-Met-Leu-Ile-Ile-Trp-Gly-Val-
H7:  Ala-Glu-Met-Lys-Trp-Leu-Leu-Ser-Asn-Thr-Asp-Asn-Ala-Ala-Phe-Pro-Gln-Met-Thr-Lys-Ser-Tyr-Lys-Asn-Thr-Arg-Arg-Glu-Ser-Ala-Leu-Ile-Val-Trp-Gly-Ile-
```

```
     185           190           195           200           205           210           215
H3:  His-His-Pro-Ser-Thr-Asn-Gln-Glu-Gln-Thr-Ser-Leu-Tyr-Val-Gln-Ala-Ser-Gly-Arg-Val-Thr-Val-Ser-Thr-Arg-Arg-Ser-Gln-Gln-Thr-Ile-Ile-Pro-Asn-Ile-Gly-
H2:  His-His-Pro-Ile-Asp-Glu-Thr-Glu-Gln-Arg-Thr-Leu-Tyr-Gln-Asn-Val-Gly-Thr-Tyr-Val-Ser-Val-Gly-Thr-Ser-Thr-Leu-Asn-Lys-Arg-Ser-Thr-Pro-Glu-Ile-Ala-
H7:  His-His-Ser-Val-Glu-Ala-Ser-Gln-Arg-Thr-Leu-Tyr-Gln-Asn-Val-Gly-Thr-Tyr-Val-Ser-Val-Gly-Thr-Ser-Thr-Leu-Asn-Lys-Arg-Ser-Thr-Pro-Glu-Ile-Ala-Thr-Arg-Pro-
```

```
     220           225           230           235           240           245           250
H3:  Ser-Arg-Pro-Trp-Val-Arg-Gly-Leu-Ser-Ser-Arg-Ile-Ser-Ile-Tyr-Trp-Thr-Ile-Val-Lys-Pro-Gly-Asp-Val-Leu-Val-Ile-Asn-Ser-Asn-Gly-Asn-Leu-Ile-Ala-Pro-
H2:  Thr-Arg-Pro-Lys-Val-Asn-Gly-Gln-Gly-Gly-Arg-Met-Glu-Phe-Ser-Trp-Thr-Leu-Leu-Asp-Met-Trp-Asp-Thr-Ile-Asn-Phe-Glu-Ser-Thr-Gly-Asn-Leu-Ile-Ala-Pro-
H7:  Gln-Ile-Asn-Gly-Gln-Ser-Gly-Arg-Ile-Asp-Phe-His-Trp-Leu-Ile-Leu-Asp-Pro-Asn-Asp-Thr-Val-Thr-Phe-Ser-Phe-Asn-Gly-Ala-Phe-Ile-Ala-Pro-
```

```
     255           260           265           270           275           280           285
H3:  Arg-Gly-Tyr-Phe-Lys-Met-Arg-Thr-Gly-Lys-Ser-Ser-Ile-Met-Arg-Ser-Asp-Ala-Pro-Ile-Asp-Thr-Cys-Ile-Ser-Glu-Cys-Ile-Thr-Pro-Asn-Gly-Ser-Ile-Pro-
H2:  Glu-Tyr-Gly-Tyr-Lys-Ile-Ser-Lys-Arg-Gly-Ser-Ser-Gly-Ile-Met-Lys-Thr-Glu-Gly-Thr-Leu-Glu-Asn-Cys-Glu-Thr-Lys-Cys-Gln-Thr-Pro-Leu-Gly-Ala-Ile-Asn-
H7:  Asn-Arg-Ala-Ser-Phe-Leu-Arg-Gly-Lys-Ser-Met-Gly-Ile-Gln-Ser-Asp-Val-Gln-Val-Asp-Ala-Asn-Cys-Glu-Gly-Glu-Cys-Tyr-His-Ser-Gly-Gly-Thr-Ile-Thr-
```

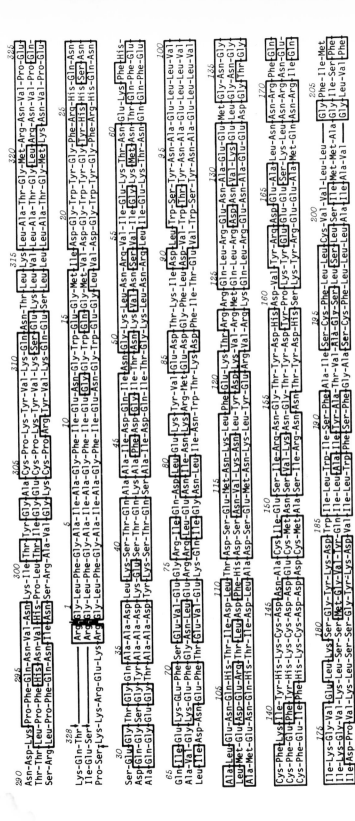

FIG. 1. Antigenic shift in influenza virus.

FPV and human H3. This greater similarity between a human and an avian subtype than between two human subtypes is in agreement with the hypothesis that shifts arise by integration into a human influenza viral genome of a hemagglutinin subtype derived from an animal reservoir.

SUMMARY

DNA copies of the gene-coding for the hemagglutinin from two human influenza A strains of the H3N2 subtype, A/Aichi/2/68 and A/Victoria/3/75, have been cloned in the plasmid pBR322 and their complete nucleotide sequence has been determined. Comparison of the deduced amino acid sequences reveals at the molecular level the result of seven years of antigenic drift. Comparison with the hemagglutinin from a human strain of the previous (H2N2) subtype, A/Japan/305/57, documents the second independent mode of antigenic variation: antigenic shift. The structure is also compared with a third (H7) hemagglutinin, that of the avian strain A/FPV/Rostock/34.

ACKNOWLEDGEMENT

This work was supported by grants from the Fonds voor Kollektief Fundamenteel Onderzoek and from the Gekoncerteerde Akties of the Belgian Ministry of Science. M.V. is the recipient of an N.F.W.O. fellowship.

REFERENCES

Blobel, G. and Dobberstein, B. (1975). *J. Cell Biol.* **67**, 835–851.
Gething, M.-J., Bye, J., Skehel, J. J. and Waterfield, M. (1980). *In* "Structure and Variation in Influenza Virus" (W. G. Laver and G. M. Air, eds), pp. 1–10. Elsevier, Amsterdam.
Maxam, A. and Gilbert, W. (1977). *Proc. natn. Acad. Sci. U.S.A.* **74**, 560–564.
Min Jou, W., Verhoeyen, M., Devos, R., Saman, E., Fang, R.-X., Huylebroeck, D., Fiers, W., Threlfall, G., Barber, C., Carey, N. and Emtage, S. (1980). *Cell* **19**, 683–696.
Palese, P. and Schulman, J. L. (1976). *Proc. natn. Acad. Sci. U.S.A.* **73**, 2142–2146.
Porter, A. G., Barber, C., Carey, N., Hallewell, R. A., Threlfall, G. and Emtage, S. (1979). *Nature, Lond.* **282**, 471–477.
Scholtissek, C., Harms, E., Rohde, W., Orlich, M. and Rott, R. (1976). *Virology* **74**, 332–344.

Sleigh, M. J., Both, G. W., Brownlee, G. G., Bender, V. J. and Moss, B. A. (1980). *In* "Structure and Variation in Influenza Virus" (W. G. Laver and G. M. Air, eds), pp. 69–79. Elsevier, Amsterdam.
Verhoeyen, M., Fang, R.-X., Min Jou, W., Devos, R., Huylebroeck, D., Saman, E. and Fiers, W. (1980). *Nature, Lond.* **286**, 771–776.

Phosphoproteins of Vesicular Stomatitis Virus and of the Host Cell

GAIL M. CLINTON and ALICE S. HUANG

Department of Microbiology and Molecular Genetics, Harvard Medical School
and
Division of Infectious Diseases, Children's Hospital Medical Center, 300 Longwood Avenue, Boston, Massachusetts 02115, U.S.A.

INTRODUCTION

Phosphorylation of proteins is a post-translational modification which can regulate the ability of proteins to recognize other proteins (Taborsky, 1974; Rubin and Rosen, 1975; Uy and Wold, 1977; Krebs and Beavo, 1979) and, possibly, to bind to specific nucleic acid sequences. To define some of these regulatory events during macromolecular synthesis, we initiated studies on the phosphoproteins of vesicular stomatitis virus (VSV).

STRUCTURAL PHOSPHOPROTEINS OF VSV

Among the five structural proteins coded for by VSV (Fig. 1), two are phosphorylated (Imblum and Wagner, 1974; Moyer and Summers, 1974; Wagner, 1975; Clinton et al., 1978a). They are the nucleocapsid-associated NS protein and the matrix or membrane-associated M protein. Both of these phosphoproteins can be separated by charge into multiple sub-species (unpublished observations). However, when the NS and the M proteins are separated, based largely on their overall conformation, two major forms of each protein are seen (Clinton et al., 1978a). There appears to be a 10% difference in the phosphate content between the less phos-phorylated NS1 species and the more phosphorylated NS2 species. The M1 form contains 10-fold less phosphate than the M2 form.

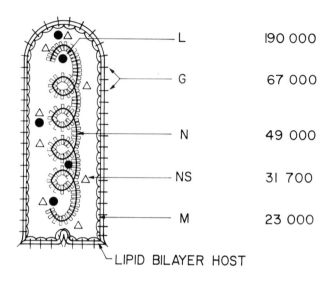

FIG. 1. Diagrammatical representation of the vesicular stomatitis virions with the molecular weights of the structural proteins (see Wagner, 1975).

REGULATORY ROLES OF VSV PHOSPHOPROTEINS

Although both phosphoproteins are structural elements of the virions, each plays an important role during viral multiplication. The NS1 protein is found tightly bound to the core of the virion (Clinton *et al.*, 1978a). Intracellularly, the amount of NS1 bound to VSV cores varies depending on the pH of the medium. When the medium is at pH 7.4, more NS1 is bound and more infectious virus is made. At pH 6.8, less NS1 is bound to cores and the amount of virus progeny is greatly reduced (Fiszman *et al.*, 1974; Clinton *et al.*, 1978a). The matrix or membrane-associated M protein, not only defines the bullet shape of VSV (McSharry *et al.*, 1971), but also affects VSV transcription in the infected cell (Clinton *et al.*, 1978b; Martinet *et al.*, 1979). This type of negative control appears to be mediated by the binding of the less phosphorylated M1 species to cores (Clinton *et al.*, 1978a). With a temperature-sensitive (ts) mutant of VSV, where the lesion has been mapped to the M protein (Knipe *et al.*, 1977), VSV transcription is increased. Moreover, the genes that are more distal to the promotor are transcribed to a greater extent than those proximal to the promotor (Clinton *et al.*, 1978b). This suggests that M protein inhibits transcription by attenuating the progressive synthesis of each of the VSV messenger RNAs.

EFFECTS OF HOST PHOSPHATASES AND KINASES

In vitro binding assays or cell-fractionation studies are complicated by the presence of kinases and phosphatases. Cellular phosphatases affect only NS2, whereas the native NS1 is stable to dephosphorylation (Clinton *et al.*, 1979). The effects of cellular phosphatases on M protein are variable because cell extracts also contain protease activity which partially degrades M protein (unpublished observations). The dephosphorylation of NS2 to NS1 is inhibited by the addition of pyrophosphate (Clinton *et al.*, 1979).

Kinase activity associated with VSV particles is extremely active (Silberstein and August, 1973; Imblum and Wagner, 1974; Moyer and Summers, 1974). The bulk of the activity purifies away from VSV structural proteins, suggesting that it is a cellular constituent (Imblum and Wagner, 1974; Clinton, unpublished observations). Hopefully, by defining and controlling these enzymatic functions during *in vitro* reconstitution experiments or during cell fractionations, a more exact role of phosphorylation during VSV multiplication can be defined.

IDENTITY OF CYTOPLASMIC AND VIRION PROTEINS

As another approach to studying the VSV phosphoproteins, chemical analyses were made on the NS and M proteins. Both conformational forms of NS and M proteins are found in virions and virions provide relatively pure VSV proteins. For these reasons, the starting material for biochemical studies was purified extracellular virus. To ensure that these virion proteins are comparable to their regulatory counterparts found in cells during VSV growth, intracellular as well as virion-associated NS and M proteins were compared by partial chymotryptic digests following the method described by Cleveland *et al.* (1977). Identity appears to exist for each of the NS1, NS2 and M forms irrespective of their source (Clinton *et al.*, 1978a, 1979 and Fig. 2).

DISTRIBUTION OF PHOSPHOSERINE, PHOSPHOTHREONINE AND PHOSPHOTYROSINE IN VSV PROTEINS

Recently, tyrosine has been added to serine and threonine as potential amino acids in proteins which accept phosphate (Eckhart *et al.*, 1979; Witte *et al.*, 1980; Hunter and Sefton, 1980). To analyze these three phosphoamino acids from the NS and M proteins the ^{32}P-labeled proteins were hydrolyzed at 110°C in 6N HCl for 1.5 hours. In our hands, approximately 40–60% of the phosphate label in VSV phosphoproteins was found in

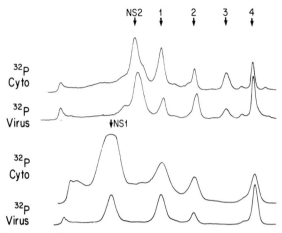

FIG. 2. Partial protease digestion of the intracellular and virion forms of NS1 and NS2 proteins. The proteins were obtained from the cytoplasm of Chinese hamster ovary cells infected with VSV at a multiplicity of 20 and exposed to 50 μCi/ml of carrier-free ^{32}P in phosphate-free medium between 2 and 4 hours post infection (Clinton et al., 1979). Cytoplasmic extracts were prepared (Manders and Huang, 1972) in conjunction with similarly labeled virions obtained from the extracellular medium. The NS1 and NS2 forms were first separated by polyacrylamide gel electrophoresis containing SDS and urea (Clinton et al., 1979). Individual forms were eluted and digested with α-chymotrypsin (50 μg/ml); the peptides were separated by electrophoresis in SDS-15% polyacrylamide gels containing urea (Cleveland et al., 1977; Clinton et al., 1979).

phosphoamino acids. The remainder is distributed among free phosphate, phosphopeptides and, under certain circumstances, uridine and cytidine monophosphate (Clinton and Huang, 1981; Sefton et al., 1980). Detection of these various phosphate-containing molecules is usually done by two-dimensional, high-voltage electrophoresis on Whatmann paper (Clinton and Huang, 1981). The two dimensions involve pH 3.5 which separates phosphothreonine from phosphotyrosine and pH 1.9 which separates uridine monophosphate from phosphotyrosine. When ribonucleic acid is not a contaminant or has been removed by prior ribonuclease treatment, quantitation of phosphoamino acids can be done after separation in one dimension at pH 3.5.

The distribution of radioactivity in each of the three phosphoamino acids of NS1 and NS2 forms is 60% for phosphoserine, 40% for phosphothreonine and <0.05% for phosphotyrosine (Fig. 3, lanes F and G). In contrast, M protein contains about 82% phosphoserine, 10% phosphothreonine and 8% phosphotyrosine (Fig. 3, lane H). These results indicate that NS1 and NS2, although they differ in their overall content of phosphate, have an identical distribution of phosphorylated amino acids.

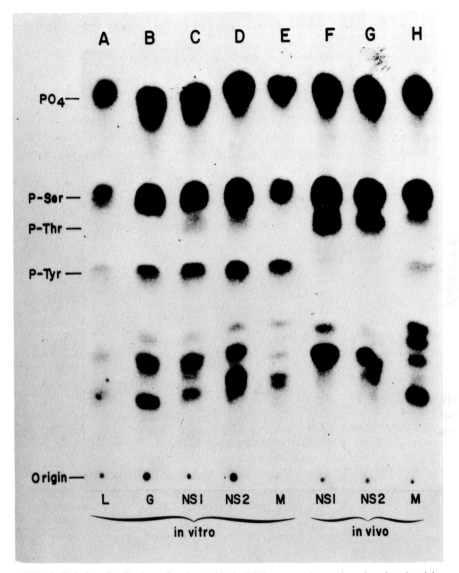

FIG. 3. Relative distribution of amino acids in VSV structural proteins phosphorylated in infected BHK21 cells or *in vitro* by the virion-associated kinase(s). (Lanes F–H) Purified ³²P-labeled virions from infected cell cultures were disrupted with NP40 and treated with ribonuclease prior to separation of the individual proteins. (Lanes A–E) For *in vitro*-labeled proteins, the kinase assay was for 1 hour at 31°C (Clinton *et al.*, 1979). After separation of the individual proteins by electrophoresis on polyacrylamide gels, the proteins were eluted and hydrolyzed for 1.5 hours in 6N HCl. High-voltage paper electrophoresis of the products was at pH 3.5 (Clinton and Huang, 1981).

M protein, however, differs from the NS forms by having a different distribution of the three phosphoamino acids, and particularly by its enrichment in phosphotyrosine.

Because of the apparent selectivity of the kinases in cells for different phosphoamino acids between the NS and M proteins, the products of the *in vitro* virion-associated kinase activity were also examined for selective phosphorylation. Figure 4 shows the kinetics of the *in vitro* transfer of ^{32}P to VSV structural proteins by purified virions disrupted with nonidet P40 and incubated in the presence of Mg^{2+} and γ^{32}P-ATP. Initially, NS and M proteins, those normally phosphorylated in cells, are the major phosphate acceptors. As the assay time progresses, four out of the five structural proteins are phosphorylated.

When the distribution of phosphoamino acids labeled *in vitro* with ^{32}P are

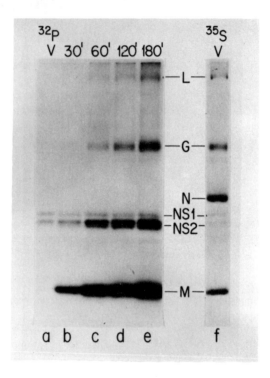

FIG. 4. *In vitro* phosphorylation of the VSV structural proteins. Whole, unlabeled purified virions were incubated under the conditions of the kinase activity with γ^{32}P-ATP (Clinton *et al.*, 1979). The proteins were analyzed on a 10% polyacrylamine gel containing SDS and 5M urea. (a) ^{32}P-VSV labeled *in vivo*. Kinase reaction incubated at 31°C for 30 min (b); 60 min (c); 120 min (d); and 180 min (e). (f) Marker ^{35}S-methionine-labeled VSV.

separated, there is phosphotyrosine in the L, G, NS and M proteins of VSV (Fig. 3, lanes A–E). This amount of tyrosine-specific activity is high and rather indiscriminating in relation to acceptor proteins. However, some specificity is seen *in vitro* for threonine on the NS protein (Fig. 3, lanes C and D).

PHOSPHOAMINO ACIDS IN INFECTED AND UNINFECTED BHK21 CELLS

When such analyses are extended to baby hamster kidney (BHK21) cells grown in suspension cultures, and the pattern of phosphorylated amino acids compared, it becomes apparent that VSV concentrates tyrosine-specific kinase activity and is more highly phosphorylated in its tyrosine residues than overall host proteins (Table I). Whether the tyrosine-specific kinase associated with virions is related to that associated with some viral oncogene products (Collet and Erickson, 1978; Eckhart *et al.*, 1979; Witte *et al.*, 1980; Hunter and Sefton, 1980) remains to be determined.

Table I also shows that the pattern of distribution of phosphoamino acids for BHK21 cells remains unchanged after infection by VSV. However, there is an overall inhibition of ^{32}P-incorporation into host phosphoamino acids by 75% after infection.

TABLE I

The relative amounts of phosphoserine, phosphothreonine and phosphotyrosine in proteins from BHK21 cells and from vesicular stomatitis virus

Source of protein	Relative %[a]		
	P-tyr	P-thr	P-ser
BHK21 cells[b]	0.5	9.5	89.8
VSV-infected BHK21 cells[b]	0.3	13.4	86.3
VSV phosphorylated in BHK21 cells	3.7	14.6	81.7
VSV phosphorylated *in vitro*	21.2	11.4	68.4

[a] About 10^6 ^{32}P cpm of cellular proteins or 10^4 ^{32}P cpm of viral proteins were acid-hydrolyzed and subjected to high-voltage, two-dimensional paper electrophoresis. The ^{32}P label which comigrated with the phosphoserine, phosphothreonine and phosphotyrosine markers is expressed as a percentage of the radioactivity in each amino acid divided by the total radioactivity in all three amino acids.

[b] When equal aliquots of proteins from infected and uninfected BHK21 cells were analyzed, the total amount of radioactivity in all three phosphoamino acids was 1:4, respectively, indicating an inhibition of phosphorylation in infected cells. To compare the relative percentage of each of the three phosphoamino acids, equal amounts of radioactivity were applied to the paper before high-voltage, two-dimensional electrophoresis.

CONCLUSIONS

These observations support the concept of regulation via a mechanism of phosphorylation and dephosphorylation of host and viral proteins after infection by VSV. The VSV system has the potential for studying two different mechanisms of control via phosphorylation: (1) change in the overall extent of phosphorylation of species of the same NS protein without affecting the distribution of the three amino acids that are phosphorylated; and (2) change in the pattern of amino acids that are phosphorylated among groups of proteins with similar functions or locations in cells. Conceivably, the M protein may be representative of a class of membrane-associated proteins which are involved in nucleic acid regulation and are enriched in phosphotyrosine.

ACKNOWLEDGEMENTS

This work was supported by research grants VC-63 and MV-54 from the American Center Society and AI 16625 from the National Institutes of Health. G.M.C. was supported in succession by an American Cancer Society postdoctoral fellowship and an institutional National Research Service Award T32 CA 09031 from the National Cancer Institute. This manuscript was prepared by Suzanne Ress utilizing the UNIX-NROFF Text Processing Program of the Health Sciences Center of the Harvard School of Public Health.

REFERENCES

Cleveland, D. W., Fischer, S. G., Kirschner, M. W. and Laemmli, U. K. (1977). *J. biol. Chem.* **252**, 1102–1106.
Clinton, G. M. and Huang, A. S. (1981). *Virology* **108**, 510–514.
Clinton, G. M., Burge, B. W. and Huang, A. S. (1978a). *J. Virol.* **27**, 340–346.
Clinton, G. M., Little, S. P., Hagen, F. S. and Huang, A. S. (1978b). *Cell* **15**, 1455–1462.
Clinton, G. M., Burge, B. W. and Huang, A. S. (1979). *Virology* **99**, 84–94.
Collett, M. S. and Erikson, R. L. (1978). *Proc. natn. Acad. Sci. U.S.A.* **75**, 2021–2024.
Eckhart, W., Hutchinson, M. A. and Hunter, T. (1979). *Cell* **18**, 925–933.
Fiszman, M., Leaute, J. B., Chany, C. and Girard, M. (1974). *J. Virol.* **13**, 801–808.
Huang, A. S. and Manders, E. K. (1972). *J. Virol.* **9**, 909–916.
Hunter, T. and Sefton, B. M. (1980). *Proc. natn. Acad. Sci. U.S.A.* **77**, 1311–1315.
Imblum, R. L. and Wagner, R. R. (1974). *J. Virol.* **13**, 113–124.
Knipe, D. M., Baltimore, D. and Lodish, H. F. (1977). *J. Virol.* **21**, 1149–1158.
Krebs, E. G. and Beavo, J. A. (1979). *A. Rev. Biochem.* **48**, 923–959.

Martinet, C., Combard, A., Printz-Ane, C. and Printz, P. (1979). *J. Virol.* **29**, 123–134.

McSharry, J. J., Compans, R. W. and Choppin, P. W. (1971). *J. Virol.* **8**, 722–729.

Moyer, S. A. and Summers, D. F. (1974). *J. Virol.* **13**, 455–465.

Rubin, C. S. and Rosen, O. M. (1975). *A. Rev. Biochem.* **44**, 831–887.

Sefton, B. M., Hunter, T., Beemon, K. and Eckhart, W. (1980). *Cell* **20**, 807–816.

Silberstein, H. M. and August, J. T. (1973). *J. Virol.* **12**, 511–522.

Taborsky, G. (1974). *In* "Advances in Protein Chemistry" (C. B. Anfinsen and J. Edsall, eds) Vol. 28, pp. 1–210. Academic Press, New York and London.

Uy, R. and Wold, F. (1977). *Science, N.Y.* **198**, 890–895.

Wagner, R. R. (1975). *In* "Comprehensive virology" (H. Frankel-Conrat and R. R. Wagner, eds) Vol. 4, pp. 1–94. Plenum Press, New York.

Witte, O. N., Dasgupta, A. and Baltimore, D. (1980). *Nature, Lond.* **283**, 826–831.

Interactions of the mRNA 5'-terminal Region During Initiation of Eukaryotic Protein Synthesis

AARON J. SHATKIN, EDWARD DARZYNKIEWICZ,[1]
KUNIO NAKASHIMA,[2] NAHUM SONENBERG[3] and
STANLEY M. TAHARA

Roche Institute of Molecular Biology, Nutley, New Jersey 07110, U.S.A.

STRUCTURE, SYNTHESIS AND FUNCTIONAL EFFECTS OF 5' CAPS

A distinguishing feature of eukaryotic cellular as well as most viral mRNAs is the 5'-terminal cap structure, $m^7G(5')ppp(5')N$ (Shatkin, 1976). Caps are synthesized at an early stage of mRNA formation. They are present on nascent genome transcripts in the nuclei of mammalian cells and apparently are conserved during processing of heterogeneous nuclear RNA to cytoplasmic messengers (Rottman, 1978; Darnell, 1979). Caps are also found on 5'-terminal oligonucleotide precursors of viral mRNAs synthesized *in vitro*, consistent with capping occurring as part of the initiation step of transcription. A mechanism of initiation-related cap synthesis has been elucidated by taking advantage of the fact that many different purified animal viruses contain, in addition to RNA polymerases that copy the virion genome nucleic acids, capping enzymes that modify the transcripts. Thus, enzyme activities in purified human reovirus type 3 catalyze the synthesis of capped oligonucleotides by a series of reactions in which the initiating nucleotide of the nascent chain retains its β-phosphate in the middle position of the triphosphate bridge of the cap (Furuichi *et al.*, 1976). Capping of cellular RNA probably occurs by a similar series of reactions since comparable cap-forming activities have been purified and characterized from HeLa cell nuclei (Venkatesan *et al.*, 1980).

Present addresses
[1] Department of Biophysics, University of Warsaw, Poland.
[2] Department of Biochemistry, Yamagata University, Yamagata, Japan.
[3] Department of Biochemistry, McGill University, Montreal, Canada.

The presence of 5'-terminal caps affects mRNAs in several important ways (Banerjee, 1980). (i) Stability is enhanced; e.g. reovirus mRNAs with 5'-terminal (p)ppG were degraded more rapidly than the corresponding capped mRNAs when microinjected into *Xenopus* oocytes or incubated in cell-free translating extracts (Furuichi *et al.,*, 1977). (ii) Transcription initiation is influenced by caps or cap-related processes in some viral systems. For example, influenza virus transcriptase is markedly stimulated by capped reovirus mRNAs but not by the corresponding uncapped RNAs (Bouloy *et al.*, 1979). The stimulatory RNAs act as primers of the virion enzyme complex, and the cap and adjacent ~ 10–15 nucleotides are transferred to form the 5' termini of the influenza transcripts both *in vitro* (Robertson *et al.*, 1980) and *in vivo* (Krug *et al.*, 1979). The RNA polymerase activity associated with purified insect cytoplasmic polyhedrosis virus is stimulated more than 50-fold by S-adenosylmethionine (SAM) (Furuichi, 1974). In this case the methyl donor is an allosteric activator of the RNA polymerase and specifically promotes initiation by lowering the K_m for the initiating nucleotide (Furuichi and Shatkin, 1979). (iii) Translation initiation is facilitated by the presence of caps, and ribosome binding of capped viral and cellular mRNAs is diminished in the absence of a 5'-terminal m^7G (Filipowicz, 1978; Shatkin *et al.*, 1979). Various other kinds of experiments lend strong support to the hypothesis that caps promote translation initiation by facilitating the stable binding of mRNA to ribosomes. In reovirus mRNAs, caps form part of the 40S ribosome binding sites as defined by protection against RNase digestion (Kozak, 1977; Kozak and Shatkin, 1979). The binding sites also include the 5'-proximal AUG. Similarly, in most other eukaryotic mRNAs examined, the 5'-proximal AUG functions as the initiator codon. Thus eukaryotic ribosomes may initiate correctly by first binding to the capped 5'-end of mRNA and then repositioning at the closest AUG to begin polypeptide synthesis (Kozak, 1978).

Recognition of the capped end of mRNA during initiation is consistent with the observation that ribosome binding is very effectively inhibited *in vitro* by the presence of cap analogs such as $m^7G^{5'}p$ and $m^7G^{5'}pp$ as well as by the 7-ethyl and 7-benzyl derivatives of GDP (Adams *et al.*, 1978). These results emphasize the functional importance of 7-substitution, not necessarily by a methyl group, for cap recognition. Since 7-alkylated G contains an extra positive charge that could interact with negatively charged phosphate groups, it has been suggested that caps assume a conformation that is preferred for ribosome binding (Hickey *et al.*, 1977). Recently we observed that methyl esterification of the phosphate group of $m^7G^{5'}p$ reversibly and almost completely eliminated its activity as a cap analog. However, comparative analyses of the $m^7G^{5'}p \cdot$methyl ester and $m^7G^{5'}p$

revealed no significant differences in conformation about the C-4'—C-5' bond. While these preliminary results do not support the preferred conformation model of cap function, they do imply that specific structural features of the 5'-terminal cap of mRNA are recognized during initiation.

We have isolated from eukaryotic initiation factor preparations by m^7GDP-Sepharose affinity chromatography cap-binding proteins that differentially stimulate capped mRNA translation *in vitro* (Sonenberg et al., 1979 and 1980). We have also looked for other kinds of initiation-dependent interactions of the 5'-region of mRNA. By using photoreaction with a bifunctional pyrimidine cross-linking agent, 4'-substituted psoralen, we observed cross-linking between ^3H-methyl-labeled 5' fragments of mRNA and 18S rRNA in eukaryotic initiation complexes.

CAP-BINDING PROTEINS THAT STIMULATE TRANSLATION OF CAPPED mRNAs *IN VITRO*

Because the presence of the cap promotes ribosome binding during initiation of translation, we tested ribosomal high-salt-wash fractions for protein factor(s) that can recognize and bind to the cap. For this purpose the cross-linking scheme shown in Fig. 1 was used. ^3H-methyl-labeled reovirus mRNA was oxidized to convert the $5'$-m^7G *cis*-diol to the dialdehyde form. Factors with free amino groups that bound at or near the cap thus formed Schiff bases with the oxidized mRNA 5' end and

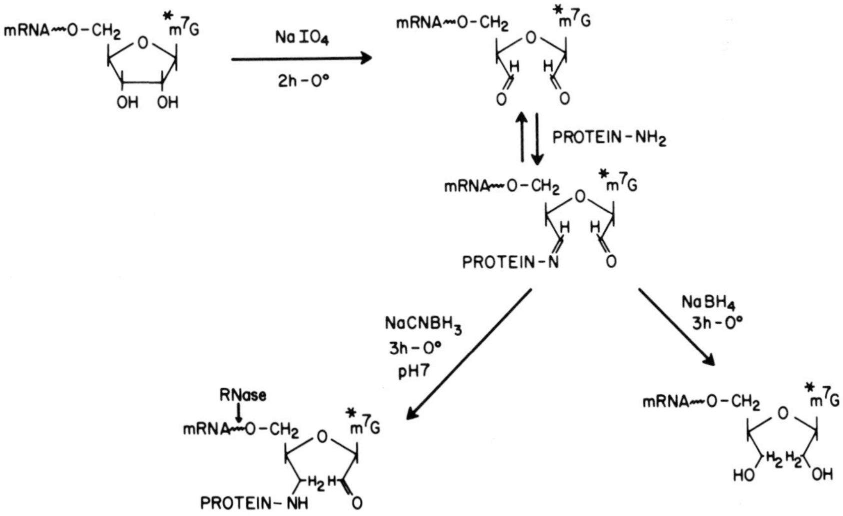

FIG. 1. Scheme for cross-linking mRNA 5' end to protein.

cyanoborohydride reduction then stabilized the covalent linkages. After digestion with RNase, the "cap-binding" proteins containing nuclease-resistant radiolabeled caps were resolved by gel electrophoresis and detected by fluorography. In a mixture of reticulocyte initiation factors analyzed by this method one cap-specific polypeptide (mol. wt. $\sim$ 24 000) was detected (Fig. 2, left). Cross-linking of this component to mRNA was inhibited by m⁷GDP but not by GDP. (The additional radiolabeled bands at $\sim$ 50 000 daltons correspond to EF-1; GDP but not the cap analog inhibited their cross-linking, consistent with EF-1 affinity for GTP.) Among the many partially purified factors tested, the 24 000 cap-binding protein (CBP) was detected in eIF-3 and eIF-4B but in less than stoichiometric amounts (Fig. 2). Consistent with the presence of CBP, these two factors differentially stimulated the translation of capped Sindbis virus mRNA relative to the naturally uncapped encephalomyocarditis (EMC) virus RNA in HeLa cell-free protein synthesizing extracts (Table I). Furthermore, treatment of eIF-3 with 0.5 M KCl which removes the 24 000-dalton CBP (Trachsel *et al.*, 1980) eliminated its discriminatory activity for capped mRNA translation (Table I).

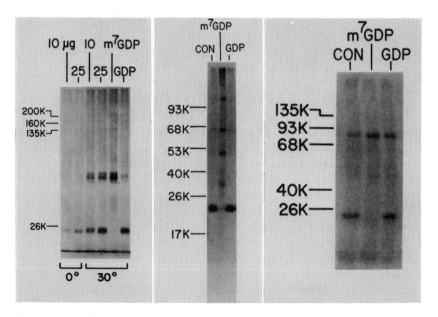

FIG. 2. Cross-linking of reticulocyte initiation factors to cap in reovirus mRNA. (Left) DEAE-cellulose purified factors; 15 µg protein in samples with 1 mM m⁷GDP. (Center) 15 µg eIF-3, 0.4 mM m⁷GDP or GDP. (Right) 2 µg eIF-4B, 0.4 mM m⁷GDP or GDP. CON = control without G nucleotide (from Sonenberg *et al.*, 1978).

TABLE I

Effect of protein synthesis factors on translation of capped and uncapped RNA

Factor added	Fold-stimulation		Ratio of stimulation
	Sindbis RNA	EMC RNA	Sindbis RNA/ EMC RNA
Expt. 1			
eIF-2 (3.5 µG)	2.2	1.1	2.0
eIF-3|(5 µg of 0.5M KCl washed)	0.7	1.4	0.5
eIF-4A (6 µg)	5.6	2.2	2.5
eIF-4B (1.5 µg)	13.6	1.5	9.1
eIF-4C + D (2 µg)	0.8	0.5	1.6
eIF-5 (0.5 µg)	0.6	1.6	0.4
EF-2 (2 µg)	1.0	1.8	0.6
Expt. 2			
EF-1 (1 µg)	1.2	0.9	1.3
eIF-3 (4 µg of 0.1M KCl washed)	12.2	0.9	12.3

RNAs (0.5 µg 25 µl^{-1}) were incubated in HeLa cell-free extracts for 60 min with ^{35}S methionine (24 µCi, specific activity 893 Ci mmol^{-1}) as radioactive precursor. For the extracts in experiments 1 and 2 the levels of synthesis without added factors were 1754 and 7880 cpm with Sindbis RNA and 8830 and 87 652 cpm with EMC RNA, respectively (from Sonenberg *et al.*, 1980).

In order to obtain purified cap-binding protein(s) for functional studies, m^{7}GDP-Sepharose was prepared for affinity chromatography by carbodiimide coupling of levulinic acid-(7-methyl guanosine 0$^{5'}$-diphosphate-0$^{2',3'}$-acetal) to AH-Sepharose-4B (Seela and Waldek, 1975; Sonenberg *et al.*, 1979). Highly purified 24 000-dalton CBP was obtained by adsorption to the resin and elution with excess m^{7}GDP. This polypeptide cross-linked specifically to oxidized mRNA and also differentially stimulated the translation of capped Sindbis mRNA as compared to uncapped EMC RNA in HeLa extracts (Fig. 3).

Poliovirus, like EMC virus, contains an uncapped RNA that is translated by cap-independent mechanisms (Nomoto *et al.*, 1976; Hewlett *et al.*, 1976; Frisby *et al.*, 1976). In HeLa cells infected with poliovirus, initiation of capped mRNA translation is effectively shut off (Leibowitz and Penman, 1971). Addition of highly purified 24 000-dalton CBP to extracts of poliovirus-infected HeLa cells restored their ability to translate capped mRNAs (Trachsel *et al.*, 1980). However, the restoring activity of the purified CBP was highly unstable as compared to cruder preparations,

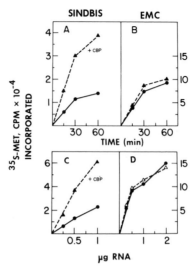

FIG. 3. Effect of CBP on translation of Sindbis virus mRNA and EMC virus RNA in extracts of uninfected HeLa cells. Experimental conditions were as described (from Sonenberg *et al.*, 1980).

suggesting that maintenance of the restoring activity may depend upon the presence of additional component(s) that could have been degraded or separated from the 24 000-dalton CBP during purification. To test this possibility, protein factors were obtained in the presence of proteolysis inhibitors, and two different pooled fractions from the sucrose gradient purification step were analyzed on m^7GDP-Sepharose. The material sedimenting more slowly than 5S, designated CBP I, included mainly the cap cross-linking 24 000-dalton component while a heavier fraction, CBP II, pooled from the 10–16S region of the gradient contained in addition nearly equal amounts of proteins of molecular weights $\sim$47 000 daltons and $\sim$210 000 daltons (Fig. 4). Both CBP I and II, like 24 000-dalton CBP, stimulated Sindbis virus mRNA translation in extracts of uninfected HeLa cells (Fig. 5). However, only CBP II had stable restoring activity for capped mRNA translation in extracts of poliovirus-infected cells (Fig. 5). Consistent with the importance of higher molecular weight components for maintaining the restoring function of CBP, Trachsel *et al.* have found that reticulocytes contain 210 000- and 50 000-dalton polypeptides that share tryptic and chymotryptic peptides with the 24 000-dalton CBP. The shut-off of capped mRNA translation in poliovirus-infected cells thus may be due to destabilization of the CBP activity by proteolytic conversion of the high molecular

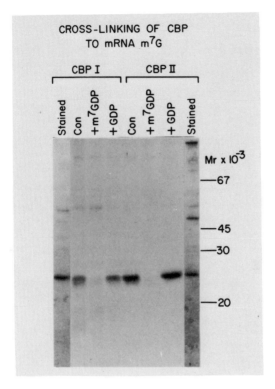

FIG. 4. Gel analysis of purified cap-binding proteins. Rabbit reticulocytes were lysed in the presence of proteolysis inhibitors including 0.5 mM phenylmethylsulfonyl fluoride, 50 units/ml of Trasylol, 0.1 mg/ml of soybean trypsin inhibitor and 1 mM EGTA, and initiation factors were prepared from the 0.5M KCl ribosomal wash (Crystal et al., 1974). A 0–70% ammonium sulfate precipitate of the wash was dialyzed against buffer containing 0.5M KCl and analyzed by sedimentation in a 15–30% linear sucrose gradient in 0.5M KCl with marker proteins. Fractions corresponding to <5S were pooled as CBP I and those between 10 and 16S as CBP II. The pools were purified by affinity chromatography on m^7GDP-Sepharose, and the m^7GDP-eluted samples were cross-linked to ^{3}H-methyl-labeled, oxidized reovirus mRNA and analyzed by polyacrylamide gel electrophoresis (Sonenberg et al., 1979). For the Coomassie blue-stained profiles, 1.8 μg and 0.7 μg of CBP I and II were used, respectively; for cross-linking, the amounts were 0.9 μg and 0.5 μg for CBP I and II.

weight polypeptides to a functionally less stable form of the CBP. Differential inhibition of capped mRNA translation by functional inactivation of CBP may be a regulatory phenomenon not restricted to poliovirus-infected HeLa cells and may also occur in reovirus-infected L cells (Skup and Millward, 1980; Zarbl et al., 1980) and in other situations involving mRNA discrimination.

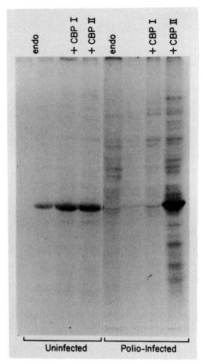

FIG. 5. Differential stimulation of Sindbis virus mRNA translation by CBP I v. II. Translating extracts were prepared from uninfected HeLa cells and from cells 3 hours after infection with poliovirus type 1 at a m.o.i. of 50 (Rose *et al.*, 1978; Sonenberg *et al.*, 1980). ^{35}S-methionine incorporation into viral capsid protein in 25 μl incubation mixtures containing 0.4 μg CBP was assayed by gel electrophoresis.

PROXIMITY OF THE mRNA 5′ REGION AND 18S RIBOSOMAL RNA IN INITIATION COMPLEXES AS DETECTED BY PSORALEN CROSS-LINKING

Eukaryotic mRNAs are in most cases functionally monocistronic. Ribosome binding apparently occurs only at or near the mRNA 5′ end, perhaps by a mechanism involving threading (Kozak, 1979) and scanning (Kozak, 1978). Cap-dependent enhancement of initiation may be mediated by CBP interacting with the cap and opening base-paired 5′ sequences in mRNA, thus facilitating ribosome threading and/or scanning. In prokaryotes ribosomes bind to mRNA at multiple internal sites, and initiation complexes are stabilized by base-pairing between the 3′ end of 16S rRNA and a purine-rich sequence on the 5′ side of the AUG initiator codon (Steitz, 1979). A similar situation appears not to apply in eukaryotes since a nucleotide sequence complementary to the highly conserved 3′ sequence of 18S rRNA is not a common feature of the 5′-terminal leaders of eukaryotic mRNAs (Hägenbuchle et al., 1978). However, some viral (Ziff and Evans, 1978) and cellular (Schroeder et al., 1979) mRNAs do contain such a sequence which could influence initiation complex formation.

To explore the possibility that the 5′ region of mRNA interacts with rRNA during initiation, we photoreacted initiation complexes with 4′-aminomethyl-4,5′,8-trimethylpsoralen (AMT), a bifunctional reagent that cross-links pyrimidines in base-paired regions of polynucleotides (Pathak et al., 1974; Isaacs et al., 1977). Wheat-germ 80S complexes were formed with reovirus mRNA fragments that were radiolabeled at the 5′ end with ^{3}H methyl and internally with ^{32}P UMP. The isolated complexes were irradiated with long wavelength UV light in the presence or absence of AMT, and the extracted RNA was analyzed by gradient sedimentation. As shown in Fig. 6A, complexes irradiated in the absence of psoralen yielded radioactive mRNA fragments that sedimented near the top of the gradient, well resolved from the two species of rRNA. Irradiation in the presence of AMT resulted in cross-linking of a large fraction of the mRNA fragments to 18S rRNA (Fig. 6B). Although there was little cross-linking to the larger rRNA species, both rRNAs in the complexes were accessible to psoralen photoreaction under these conditions as shown by the formation of similar levels of adducts with ^{3}H-labeled AMT (Fig. 6C).

Results essentially the same as those in Fig. 6 were obtained with other mRNAs and with wheat-germ 40S complexes and rabbit reticulocyte 80S initiation complexes (Nakashima et al., 1980). Cross-linking to 18S rRNA was dependent on initiation complex formation, and no hybrid molecules were obtained when a mixture of mRNA and purified rRNAs was photoreacted with AMT or when extracts in which initiation was prevented by ATP depletion were photoreacted.

To test whether the sites in 18S rRNA that interacted with mRNA included 3′ termini, AMT cross-linked hybrids consisting of 5′-^{3}H-methyl-labeled reovirus mRNA fragments and wheat-germ 18S rRNA were radiolabeled with ^{32}pCp. Since the mRNA was fragmented by treatment with alkali, only the rRNA 3′ ends should contain free 3′ hydroxyls necessary for RNA ligase catalyzed ^{32}P-labeling (England and Uhlenbeck, 1978). The radiolabeled sample, which included an excess of 18S rRNA from ribosomes that did not bind mRNA in addition to rRNA–mRNA cross-linked complexes, was digested with RNase T1 and loaded on to a column of acetylated dihydroxyboryl (DBAE) cellulose to bind selectively those oligonucleotides that contained 2′-3′-cis-diols (Rosenberg, 1974; Furuichi et al., 1975). Since the ^{32}P was present as ^{32}pCp at the 3′ end of rRNA and T1

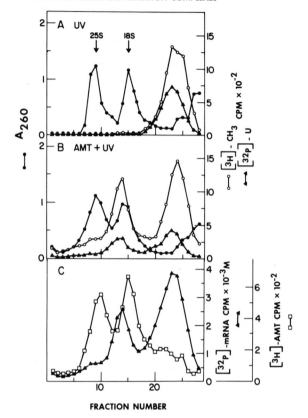

CROSSLINKING OF REOVIRUS mRNA TO 18S rRNA
IN WHEAT GERM 80S INITIATION COMPLEXES

FIG. 6. Cross-linking of reovirus mRNA to 18S rRNA and covalent binding of ^{3}H-AMT to
rRNAs in wheat-germ 80S initiation complexes. For panels A and B, complexes containing
alkali-fragmented reovirus mRNA labeled in the 5' cap with ^{3}H methyl and internally with [32]P
UMP (average chain-length ~200) were isolated by glycerol gradient centrifugation and
irradiated for 20 min at 0°C at 352nm in the absence (A) or presence (B) of 2.8×10^{-4}M
AMT. For panel C, uniformly ^{32}P-labeled capped mRNA fragments were used and the
incubation mixture, rather than isolated initiation complexes, was irradiated for 20 min in 1.4
$\times 10^{-4}$M ^{3}H-labeled AMT (specific activity = 1.4×10^5 cpm/μg, kindly provided by Dr J.
Hearst, University of California, Berkeley). RNA was extracted and analyzed by glycerol
gradient centrifugation (from Nakashima *et al.*, 1980).

digestion yields products with a 3' phosphate, 99% of the ^{32}P-labeled
fragments did not adsorb to the resin. The bulk of the [3]H (78%) was
retained by the resin as expected for capped fragments containing a 2,3'-*cis*-
diol in the 5'-linked m[7]G. The bound material was eluted, concentrated and
chromatographed a second time, the ^{3}H-labeled material rebound quantitatively together with ~70% of the [32]P (Fig. 7). Presumably the [32]P-

binding was due to cross-linking of 3'-end-labeled rRNA fragments to capped mRNA fragments since in the absence of cross-linking there was little rebinding of the ^{32}P-labeled T1-oligonucleotides. The DBAE-bound material was concentrated, irradiated at 254 nm to photoreverse the AMT cross-links (Rabin and Crothers, 1979), and analyzed in a polyacrylamide sequencing gel in parallel with partial enzymatic digests of 3'-^{32}P-labeled wheat-germ 18S rRNA. Although some of the photoreversed oligo-nucleotide sample remained near the gel origin, the released radioactivity migrated in the position of the T1 limit oligonucleotide of 18S rRNA. The partial digestion patterns also were consistent with the cross-linked, DBAE-selected ^{32}P-radioactivity being derived at least in part from the ^{32}P-pCp-labeled 3' termini of 18S rRNA.

The sites of cross-linking in the radiolabeled mRNA 5'-terminal fragments were similarly examined by gel analyses. Capped reovirus mRNA labeled internally with ^{32}P UMP was bound to wheat-germ ribosomes in the presence of sparsomycin, and the complexes were treated with RNase T1 before separation by glycerol gradient sedimentation (Kozak and Shatkin, 1979). The isolated initiation complexes containing the RNase T1-protected mRNA fragments were photoreacted with AMT, and the radioactivity that

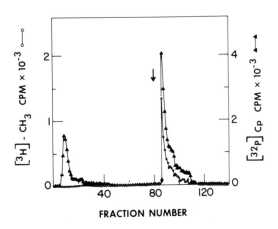

FIG. 7. Chromatography of cross-linked T1 fragments of 5'-^{3}H-methyl-labeled reovirus mRNA and 3'-^{32}P-labeled 18S rRNA. Wheat-germ 80S initiation complexes were isolated, photoreacted with AMT and the cross-linked complexes of mRNA fragments with 18S rRNA were gradient-purified. rRNA in the complexes was radiolabeled at the 3' end by incubation with [32]P pCp and T$_4$ RNA ligase (England and Uhlenbeck, 1978). After digestion with RNase T1, the digest was applied to a column of acetylated dihydroxyboryl-substituted cellulose (DBAE-cellulose) (Furuichi et al., 1975). Bound material was eluted with 1M sorbitol (arrow), concentrated and rechromatographed on DBAE-cellulose. Aliquots of fractions were counted in Aquasol (from Nakashima et al., 1980).

was cross-linked to rRNA was purified by gradient sedimentation and analyzed by polyacrylamide gel electrophoresis. Most of the ^{32}P-labeled mRNA fragments remained at the origin consistent with cross-linking to the high molecular weight rRNA (Fig. 8A, lane 1). After photoreversal of the

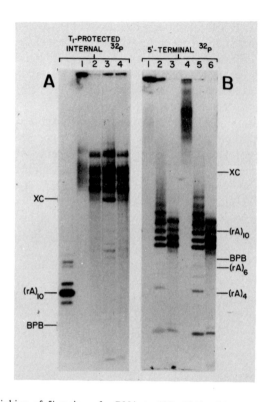

FIG. 8. Cross-linking of 5′ region of mRNA to 18S rRNA. (A) Reovirus capped mRNA labeled internally with ^{32}P UMP (specific activity $\sim 10^{6}$ cpm/μg) was incubated in wheat-germ extract (20 μg mRNA/0.4 ml incubation mixture) under conditions of initiation in the presence of 0.2 mM sparsomycin (Kozak and Shatkin, 1979). After 10 min at 23°C, 150 units of RNase T1 were added and the incubation continued for 15 min. Initiation complexes were then separated by gradient sedimentation, and the pooled fractions were photoreacted ± AMT as in Fig. 6. The extracted RNA was purified in another gradient and radioactivity cosedimenting with rRNA was collected and analyzed in a 20% polyacrylamide gel before (lanes 1 and 3) and after (lanes 2 and 4) photoreversal of the AMT crosslinks (Rabin and Crothers, 1979; Darzynkiewicz *et al.*, 1980). (B) Reovirus mRNA ^{32}P-labeled in the 5′-terminal cap was bound to wheat-germ ribosomes, isolated without RNase digestion, and photoreacted as in panel (A). After gradient purification of the radioactive material cross-linked to 18S rRNA, the samples were analyzed by gel electrophoresis before (lane 1) and after digestion with 1200 units/ml of RNase T1 for 1 hour at 37°C (lane 2). For lane 3 the T1 digest was also photoreversed. Lanes 4–6 contain mRNA that was photoreacted in the absence of AMT and analyzed before (4) or after RNase T1 treatment (5) plus photoreversal (6).

AMT cross-links, essentially all of the ^{32}P-labeled fragments migrated near the xylene cyanol marker dye (lane 2). This is the region expected for T1-protected reovirus mRNA fragments previously determined to be in the size ranges 26–32 and 30–54 nucleotides for 80S and 40S protected species, respectively (Kozak, 1977). Another sample of T1-protected radioactivity that was treated identically, except that the photoreaction step was done in the absence of AMT, was not cross-linked to rRNA. It yielded gel patterns similar to that of the cross-linked and subsequently photoreversed sample (lanes 3 and 4). The results suggest that in wheat-germ initiation complexes the 5'-proximal region of reovirus mRNA that is protected against RNase T1 digestion by ribosomes, i.e. the initiation site, is in close proximity to, and possibly base-paired with, 18S rRNA.

From these experiments with internally ^{32}P-labeled mRNA it was not possible to determine if cross-linking to 18S rRNA also occurs at nucleotides very close to the 5' cap, e.g. within the 5'-terminal T1 nucleotide. Photoreaction of nucleic acids with AMT yields pyrimidine adducts, and the 5'-end sequence of the multiple species of reovirus mRNA is m^{7}GpppG^m-C-U-(N)$_{3-6}$-G .... Thus, if AMT cross-linking of 5'-labeled mRNA in initiation complexes occurs at cap-adjacent pyrimidines, digestion with RNase T1 should yield larger mRNA–rRNA hybrid oligonucleotides. Reovirus mRNA with 5'-terminal ppG (Furuichi and Shatkin, 1976) was capped with α-^{32}P-GTP post-transcriptionally to form ^{32}P-end-labeled molecules containing m^{7}G*pppG^m. Wheat-germ initiation complexes made with this mRNA were photoreacted with AMT and the 18S rRNA-cross-linked radioactivity was analyzed in comparison to oligo(A) markers kindly provided by Y. Furuichi. As shown in Fig. 8B, the high molecular weight, cross-linked ^{32}P-labeled material (lane 1) was converted by RNase T1 to oligonucleotides migrating in the positions of ~8–12 nucleotides (lane 2). After nuclease digestion only a relatively small fraction of the radioactivity remained at the origin and no distinct intermediate-sized fragments were apparent (lane 2). Upon photoreversal of the AMT cross-links, the T1 digested sample was essentially completely converted to short oligonucleotides (lane 3). Similarly treated, ^{32}P-end-labeled mRNA that was irradiated without AMT migrated into the gel suggesting that some nicking had occurred during the various experimental manipulations (lane 4). RNase T1 digestion alone (lane 5) and enzyme treatment followed by photoreversal (lane 6) both yielded the expected series of 5'-terminal T1 oligonucleotides. Thus, although reovirus mRNA in initiation complexes may be cross-linked to 18S rRNA at the cap-adjacent pyrimidines in a small fraction of the mRNA molecules, there appears to be considerably greater interaction in the region near the initiator AUG that is protected by 80S ribosomes. Further studies of 40S and 80S complexes with radioactive AMT should make it possible to localize the exact sites of interaction that occur between mRNA and rRNA during initiation of protein synthesis.

It should be emphasized that although initiation-dependent interactions, perhaps base-pairing, between mRNA 5′ regions and 18S rRNA can be detected by AMT cross-linking, their functional role in protein synthesis remains to be assessed. Nevertheless, the possibility should be considered that eukaryotic translation initiation may be modulated not only by factors such as cap-binding proteins but also by interactions of the ribosome binding site with the rRNA in the small ribosomal subunit.

ACKNOWLEDGEMENTS

We would like to thank A. J. LaFiandra and M. A. Morgan for assistance and S. Hecht, W. C. Merrick, K. Rupprecht and H. Trachsel for fruitful collaborations on various aspects of these studies.

REFERENCES

Adams, B. L., Morgan, M., Muthukrishnan, S., Hecht, S. M. and Shatkin, A. J. (1978). *J. biol. Chem.* **253**, 2589–2595.

Banerjee, A. K. (1980). *Microbiol. Revs.* **44**, 174–205.

Bouloy, M., Morgan, M., Shatkin, A. J. and Krug, R. M. (1979). *J. Virol.* **32**, 895–904.

Crystal, R. G., Elson, N. A. and Anderson, W. F. (1974). *Meth. Enzymol.* **30**, 101–136.

Darnell, Jr., J. E. (1979). *Progr. nucl. Acid Res. molec. Biol.* **22**, 327–353.

Darzynkiewicz, E., Nakashima, K. and Shatkin, A. J. (1980). *J. biol. Chem.* **255**, 4973–4975.

England, T. E. and Uhlenbeck, O. C. (1978). *Nature, Lond.* **275**, 560–561.

Filipowicz, W. (1978). *FEBS Lett.* **96**, 1–11.

Frisby, D., Eaton, M. and Fellner, P. (1976). *Nucl. Acids Res.* **3**, 2771–2779.

Furuichi, Y. (1974). *Nucl. Acids Res.* **1**, 802–809.

Furuichi, Y. and Shatkin, A. J. (1976). *Proc. natn. Acad. Sci. U.S.A.* **73**, 3448–3452.

Furuichi, Y. and Shatkin, A. J. (1979). *In* "Transmethylation" (E. Usdin, R. T. Borchardt and C. R. Creveling, eds), pp. 351–360. Elsevier/North Holland, New York.

Furuichi, Y., Shatkin, A. J., Stavnezer, E. and Bishop, J. M. (1975). *Nature, Lond.* **275**, 618–620.

Furuichi, Y., Muthukrishnan, S., Tomasz, J. and Shatkin, A. J. (1976). *J. biol. Chem.* **251**, 5043–5053.

Furuichi, Y., LaFiandra, A. and Shatkin, A. J. (1977). *Nature, Lond.* **266**, 235–239.

Hagenbüchle, O., Santer, M., Steitz, J. A. and Mans, R. J. (1978). *Cell* **18**, 551–563.

Hewlett, M. J., Rose, J. K. and Baltimore, D. (1976). *Proc. nat. Acad. Sci. U.S.A.* **73**, 327–330.

Hickey, E. D., Weber, L. A., Baglioni, C., Kim, C. H. and Sarma, R. H. (1977). *J. molec. Biol.* **109**, 173–183.
Isaacs, S. T., Shen, C. J., Hearst, J. E. and Rapoport, H. (1977). *Biochemistry* **16**, 1058–1064.
Kozak, M. (1977). *Nature, Lond.* **269**, 390–394.
Kozak, M. (1978). *Cell* **15**, 1109–1123.
Kozak, M. (1979). *Nature, Lond.* **280**, 82–85.
Kozak, M. and Shatkin, A. J. (1979). *Meth. Enzymol.* **60**, 360–375.
Krug, R. M., Broni, B. A. and Bouloy, M. (1979). *Cell* **18**, 329–334.
Leibowitz, R. and Penman, S. (1971). *J. Virol.* **8**, 661–668.
Nakashima, K., Darzynkiewicz, E. and Shatkin, A. J. (1980). *Nature, Lond.* **286**, 226–230.
Nomoto, A., Lee, Y. F. and Wimmer, E. (1976). *Proc. natn. Acad. Sci. U.S.A.* **73**, 375–380.
Pathak, M. A., Kramer, D. M. and Fitzpatrick, T. B. (1974). *In* "Sunlight and Man—Normal and Abnormal Photobiologic Responses" (M. A. Pathak, L. C. Harber, M. Seiji and A. Kukita, eds), pp. 335–387. University of Tokyo Press, Tokyo.
Rabin, D. and Crothers, D. M. (1979). *Nucl. Acids Res.* **7**, 689–703.
Robertson, H. D., Dickson, E., Plotch, S. J. and Krug, R. M. (1980). *Nucl. Acids Res.* **8**, 925–942.
Rose, J. K., Trachsel, H., Leong, K. and Baltimore, D. (1978). *Proc. natn. Acad. Sci. U.S.A.* **75**, 2732–2736.
Rosenberg, M. (1974). *Nucl. Acids Res.* **1**, 653–671.
Rottman, F. M. (1978). *In* "Biochemistry of Nucleic Acids II" (B. F. C. Clark, ed.) Vol. 17, pp. 45–73. University Park Press, Baltimore.
Schroeder, H. W. Jr., Liarakos, C. D., Gupta, R. C., Randerath, K. and O'Malley, B. W. (1979). *Biochemistry* **18**, 5798–5808.
Seela, F. and Waldek, S. (1975). *Nucl. Acids Res.* **2**, 2343–2354.
Shatkin, A. J. (1976). *Cell* **9**, 645–653.
Shatkin, A. J., Furuichi, Y., Kozak, M. and Sonenberg, N. (1979). *In* "12th FEBS Meeting—Gene Functions" (S. Rosenthal *et al.*, eds) Vol. 51, pp. 297–306. Pergamon Press, New York.
Skup, D. and Millward, S. (1980). *Proc. natn. Acad. Sci. U.S.A.* **77**, 152–156.
Sonenberg, N., Morgan, M. A., Merrick, W. C. and Shatkin, A. J. (1978). *Proc. natn. Acad. Sci. U.S.A.* **75**, 4843–4847.
Sonenberg, N., Rupprecht, K. M., Hecht, S. M. and Shatkin, A. J. (1979). *Proc. natn. Acad. Sci. U.S.A.* **76**, 4345–4349.
Sonenberg, N., Trachsel, H., Hecht, S. and Shatkin, A. J. (1980). *Nature, Lond.* **285**, 331–333.
Steitz, J. A. (1979). *In* "Biological Regulation and Development" (R. F. Goldberger, ed.), pp. 349–399. Plenum Press, New York.
Trachsel, H., Sonenberg, N., Shatkin, A. J., Rose, J. K., Leong, K., Bergmann, J. E., Gordon, J. and Baltimore, D. (1980). *Proc. natn. Acad. Sci. U.S.A.* **77**, 770–774.
Venkatesan, S., Gershowitz, A. and Moss, B. (1980). *J. biol. Chem.* **255**, 2829–2834.
Zarbl, H., Skup, D. and Millward, S. (1980). *J. Virol.* **34**, 497–505.
Ziff, E. B. and Evans, R. M. (1978). *Cell* **15**, 1463–1475.

Synthesis and Functions of Alphavirus Gene Products

LEEVI KÄÄRIÄINEN,[1] KATSUYUKI HASHIMOTO,[2]
NISSE KALKKINEN,[3] SIRKKA KERÄNEN,[1] PÄIVI
LEHTOVAARA,[3] MARJUT RANKI,[1] JAAKO SARASTE,[1]
DOROTHEA SAWICKI,[4] STANLEY SAWICKI,[4] ISMO
ULMANEN[1] and PERTTI VÄÄNÄNEN[1]

[1] *Department of Virology, University of Helsinki, 00290 Helsinki 29, Finland*
[2] *Department of Virology and Rickettsiology, National Institute of Health,
Tokyo, Japan*
[3] *Department of Biochemistry, University of Helsinki, 00290 Helsinki 29, Finland*
[4] *Department of Microbiology, Medical College of Ohio, Ohio 43699, U.S.A.*

INTRODUCTION

The alphaviruses consist of a nucleocapsid which is surrounded by an envelope membrane. About 240 capsid protein molecules (M_r 30 000) and one positive-stranded RNA molecule (42S RNA M_r 4.3×10^6) are the components of the nucleocapsid (for review see Kääriäinen and Söderlund, 1978). The lipoprotein envelope of Semliki Forest virus (SFV) consists of 16 000 to 17 000 phospholipid-cholesterol pairs and about 240 trimers consisting of envelope proteins E1 (M_r 49 000), E2 (M_r 52 000) and E3 (M_r 10 000) (Kääriäinen and Söderlund, 1978).

The 5' end of the 42S RNA genome has a cap structure and the 3' end has a poly(A) 60–70 nucleotides long (Dubin and Stollar, 1977; Sawicki and Gomatos, 1976; Frey and Strauss, 1978; Pettersson *et al.*, 1980). These typical features of eukaryotic messenger RNA have been confirmed by *in vitro* translation of the 42S RNA (Simmons and Strauss, 1974; Glanville *et al.*, 1976, 1978; Glanville and Lachmi, 1977; Lehtovaara *et al.*, 1980).

The replication of alphaviruses has been studied fairly well (for review see Pfefferkorn and Shapiro, 1974; Strauss and Strauss, 1977; Kääriäinen and Söderlund, 1978) not least as an example of membrane biogenesis (for reviews see Kääriäinen and Renkonen, 1977; Simons *et al.*, 1978).

Here we summarize some of our recent studies on the replication of SFV and Sindbis virus, which allow us to assign tentative functions for different gene products.

ATTACHMENT AND PENETRATION

Comparison of N-terminal amino acid sequences of SFV (Kalkkinen *et al.*, 1980) and Sindbis virus (Bell *et al.*, 1978) envelope proteins E1 and E2 reveal that E1 proteins, which are responsible for the hemagglutination of these viruses (Dalrymple *et al.*, 1976; Helenius *et al.*, 1976), are closely related. The E2 proteins, according to the analyzed amino acids at the N-terminus, show less homology (Table I). These results are in good agreement with previous findings that hemagglutination of SFV and Sindbis virus is inhibited by antisera against other alphaviruses, whereas antiserum against Sindbis virus E2 protein neutralizes only Sindbis virus but not SFV (Taylor, 1967; Dalrymple *et al.*, 1976).

TABLE I

Comparison of N-terminal amino acid sequences of Semliki Forest and Sindbis virus envelope proteins E1 and E2[a]

Envelope protein	Number of analyzed amino acids	Identical amino acids	Amino acid replacements	
			Compatible with one base change	Others
E1	24	14	7	3
E2	20	4	6	10

[a] Data from Kalkkinen *et al.* (1980), Bell *et al.* (1978).

We have recently studied interaction of SFV and Sindbis virus with different red cells to illustrate the functions of E1 protein. As has been known for a long time, hemagglutination and attachment of virus to red cells takes place only in slightly acid medium (pH 6.2 or lower) (Clarke and Casals, 1958, Väänänen and Kääriäinen, 1979, 1980). Attachment and hemagglutination occur already at 0°C. If the temperature is thereafter raised to 20–37°C, fusion of virus envelope with the red-cell membrane takes place even at neutral pH. In this process nucleocapsid is released to the cytoplasm and the virus glycoproteins remain on the cell surface (Väänänen *et al.*, to be published). Extensive red-cell fusion can be obtained at 42°C provided that medium has pH 5.8 (Väänänen and Kääriäinen, 1980).

Helenius and coworkers have studied the interaction of SFV with host cells in some detail (Helenius *et al.*, 1978a, 1980a,b; Marsh and Helenius, 1980; White and Helenius, 1980). Adsorption of the virus takes place in neutral medium and at least one of the virus receptors is the histocompatibility antigen (HL-A, H-2). Thus the conditions of adsorption differ from those of hemagglutination. The adsorbed virus is rapidly internalized by endocytosis and the whole virus is taken into coated vesicles (Helenius *et al.*, 1980a,b). The coated vesicles fuse with lysosomes resulting finally in the drop of pH. At this phase the virus membrane apparently fuses with the lysomal membrane whereby the nucleocapsid is released to the cytoplasmic side of the vesicle. This *penetration stage* resembles greatly that of virus fusion with the erythrocyte membrane and is thus most probably mediated by E1 protein (Fig. 1). SFV can be readily fused with protein-free liposomes consisting of cholesterol and phospholipids provided that the medium has a pH 6.0 or lower (White and Helenius, 1980). Since no fusion of the virus can be accomplished in neutral or alkaline media, it is probable that the drop of pH elicits a configurational change in the E1 protein. This presumably exposes a hydrophobic amino acid sequence capable of interacting directly with lipids. The adsorption of virus to host-cell receptors is probably mediated by E2 protein, which carries the type-specific antigenic determinants (Dalrymple *et al.*, 1976).

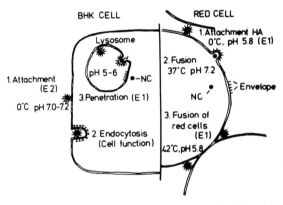

FIG. 1. Proposed roles of Semliki Forest virus envelope proteins E1 and E2 in interactions with baby hamster kidney (BHK) and red cells.

TRANSLATION OF SFV-SPECIFIC NON-STRUCTURAL PROTEINS

One of our temperature-sensitive mutants of SFV, ts-1 (Keränen and Kääriäinen, 1974), turned out to direct the synthesis of non-structural proteins in excess (Keränen and Kääriäinen, 1975; Lachmi *et al.*, 1975; Kääriäinen *et al.*, 1976). When the initiation of protein synthesis was synchronized after release of a hypertonic block of initiation, a sequential synthesis of several different proteins was observed (Lachmi and

Kääriäinen, 1976). From these results we postulated that the non-structural proteins were synthesized as a large polyprotein, which was subsequently cleaved to yield four non-structural proteins: ns70, ns86, ns72, ns60. We found two short-lived proteins nsp155 and nsp135 which were postulated to be precursors of ns70 plus ns86 and ns72 plus ns60, respectively (Fig. 2).

Larger proteins nsp220 (Kääriäinen *et al.*, 1978) and nsp250 (Keränen and Kääriäinen, 1979) have been found in cells infected with RNA-negative temperature-sensitive mutants. The proteins were synthesized in infected cultures which were first incubated at the permissive temperature (28°C) to start the infection. At 6 hours post-infection the cultures were shifted to the restrictive temperature (39°C) and labeled with ^{35}S methionine (Keränen and Kääriäinen, 1979).

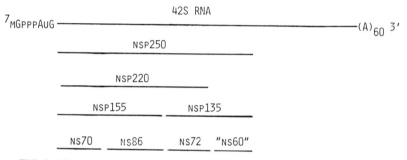

FIG. 2. Cleavage scheme of Semliki Forest virus-specific non-structural proteins.

Comparison of the products after limited proteolysis with *Staphylococcus aureus* V8 protease was carried out with the postulated non-structural precursor proteins (nsp250, nsp220, nsp155 and nsp135) and the cleavage products (ns70, ns86 and ns72) as shown in Table II. First, nsp155 and nsp135 are unique polypeptides as previously shown by tryptic peptide mapping (Glanville *et al.*, 1978). Secondly, all their V8 peptides are found in nsp250, suggesting that this large protein is indeed the non-structural polyprotein. Thirdly, nsp220 has the peptides of nsp155 and many but not all of nsp135, supporting our previous results that nsp220 is a product from the N-terminal part of the polyprotein (Kääriäinen *et al.*, 1978). Fourthly, the peptides present in nsp250 and in nsp135 but absent from nsp220 are most probably those from "ns60", a protein which we have not been able to isolate so far (Lehtovaara *et al.*, 1980).

The translational products of SFV 42S RNA in wheat-germ cell-free extract yielded tryptic peptides of nsp155 as well as some of those of nsp135 (ns72), confirming their virus-specificity (Glanville and Lachmi, 1977). More recently, we have obtained translation of nsp220, nsp155, ns70, ns86 and

TABLE II

Relationship of SFV-specific non-structural proteins to their precursors according to limited proteolysis[a]

V8-peptides derived from	Presence of V8-peptides in			
	nsp250	nsp220	nsp155	nsp135
nsp250	—	most	partly	partly
nsp220	all	—	most	partly
nsp155	all	all	—	none
nsp135	all	partly	none	—
ns70	all	all	all	—
ns86	all	all	all	none
ns72	all	all	none	all

[a] ^{35}S-methionine-labeled proteins were isolated from cells infected with ts-6 (nsp250, nsp220) and ts-1 mutants (nsp155, nsp135, ns70, ns86 and ns72). The isolation of proteins and digestion with V8-protease from *Staphylococcus aureus* as well as the polyacrylamide gel electrophoresis has been described elsewhere (Lehtovaara *et al.*, 1980).

ns72 in rabbit reticulocyte lysate, in response to added SF virion 42S RNA (Lehtovaara *et al.*, 1980). This is an indication of processing of the non-structural polyprotein *in vitro*.

Translation of 42S RNA in the presence of formyl ^{35}S methionyl-tRNA gave only one initiation dipeptide; F-met-ala, after pronase treatment of the translational product (Glanville *et al.*, 1976). This result is in agreement with our other findings that all SFV-specific non-structural proteins are indeed translated as a polyprotein. Since a 250 000-dalton protein has also been found in Sindbis virus-infected cells (Keränen and Kääriäinen, 1979) the results obtained with SFV are probably true for the whole alphavirus group (Brzesky and Kennedy, 1977).

From the non-structural proteins only ns86 has been shown to be associated with ribosomes (Ranki *et al.*, 1979).

REGULATION OF MINUS-STRAND RNA SYNTHESIS

At 37°C the synthesis of 42S RNA negative strands starts about 1 hour after infection. The maximum rate of synthesis is attained at about 2.5 hours followed by complete shut-off at 4 hours after infection (Bruton and Kennedy, 1975; Sawicki and Sawicki, 1980). During the minus-strand synthesis about 5-fold excess of positive strands are synthesized (Sawicki and Sawicki, 1980). It is thus difficult to study the specific role of non-structural proteins in the synthesis of minus strands with biochemical methods. We have approached the problem using temperature-sensitive mutants of Sindbis virus, which fall into four different complementation groups (A, B, F and G).

To start the infection the mutants were grown first at the permissive temperature (28°C) for 3 to 4 hours followed by shift to the restrictive temperature (39°C). The cells were labeled at 39°C for 15 to 30 min with ^{3}H uridine at different times after shift up. Double-stranded RNA was isolated and the amount of ^{3}H label annealing to virion 42S RNA positive strands was taken as the measure for the synthesis of minus strands.

Screening of representatives of the four complementation groups revealed that two mutants ts-6 (Group F) and ts-11 (Group B) stopped the synthesis of minus strands soon after shift-up (Fig. 3). As shown earlier, ts-6 has a temperature-sensitive defect in the polymerase function also late in infection when minus-strand synthesis has been shut off (Keränen and Kääriäinen, 1979). Thus it seems that one gene product "B" is specifically required in the synthesis of negative-stranded RNA. Since the minus-strand synthesis stops also in ts-6 infected cells at 39°C it seems probable that the polymerization itself is carried out by the gene product "F" suggesting that "B" may be important, say in the initiation of the minus-strand synthesis.

In SFV- (Sawicki and Sawicki, 1980) and Sindbis virus-infected cells (Sawicki *et al.*, to be published) the minus-strand RNA synthesis is shut off rapidly if protein synthesis is inhibited. This would suggest that the factor controlling the synthesis (gene product "B") is a short-lived protein, the synthesis of which is continuously required.

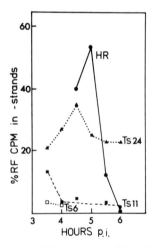

FIG. 3. Time course of minus-strand RNA synthesis of Sindbis virus (HR) and temperature-sensitive mutants ts-24 (Group A), ts-11 (Group B) and ts-6 (Group F) after shift to the restrictive temperature after 3 hours incubation at the permissive temperature.

One of the Sindbis-virus mutants, ts-24 (Group A) allowed the synthesis of minus strands to continue for many hours after shift to the restrictive temperature (Fig. 3) whereas at 28°C normal shut-off took place as with the wild type. Interestingly, cycloheximide failed to shut off the minus-strand synthesis at 39°C in contrast to the wild-type infected cells. These results suggest that once the gene product "A" becomes inactive at 39°C, the function of gene product "B" (ts-11 function) is prolonged. Apparently no *de novo* synthesis of "B" is required when "A" (ts-24 function) is thermally inactivated, suggesting that "A" in normal infection somehow inactivates the function of "B" (Sawicki *et al.*, to be published).

COMPONENTS REQUIRED FOR THE SYNTHESIS OF 42S AND 26S RNA POSITIVE STRANDS

The synthesis of 42S and 26S RNA positive strands continues throughout the infection. As templates for both, RNA species serve 42S RNA negative strands which are synthesized early in infection. The synthesis of the positive-stranded RNA continues even when protein synthesis is inhibited, indicating that the RNA polymerase is a stable one (for review see Kääriäinen and Söderlund, 1978).

For the identification of polymerase components we have used genetic and biochemical methods. In the genetic studies of the late RNA synthesis temperature-sensitive RNA-negative mutants were grown at 28°C for 6 hours to allow the polymerase components to be synthesized and assembled normally. The cultures were then transferred to the restrictive temperature in the presence of cycloheximide in order to find temperature-sensitive lesions in the function of the polymerase, which was made during the low-temperature incubation. Labeling of RNA with ^{3}H uridine was carried out at different times after shift to 39°C and the labeled RNA was analyzed in most cases by sucrose gradient centrifugation.

Only one mutant, Sindbis ts-6 (Group F) showed a temperature-sensitive polymerase function (Table III). The defect was reversible and the synthesis of both 42S and 26S RNA was equally affected (Keränen and Kääriäinen, 1979). The cessation of RNA synthesis after shift to 39°C as well as the resumption after shift back to 28°C occurred also in the presence of cycloheximide. From these results and those described above we conclude that gene product "F" is required in the synthesis of 42S RNA negative and positive strands as well as in the synthesis of 26S RNA.

Simmons and Strauss (1972a, b) have shown that the template for 26S RNA is the 42S RNA negative strand. We have recently shown that the 5' end of the 42S RNA positive strand has the structure [7]mGpppAUG whereas

TABLE III

Defects of Sindbis virus temperature-sensitive RNA-negative mutants manifesting after shift to the restrictive temperature

Viruses	Complementation group	Minus-strand RNA synthesis[a] (% of HR-RF)	Positive-strand RNA synthesis[b]		Reversibility
			Total % of HR	Molar ratio 42S:26S RNA	
ts-24[c]	A	110	87	0.77	+
ts-11	B	17	43	0.38	+
ts-6	F	28	6	0.32	+
ts-18	G	89	64	0.46	+
HR		100	100	0.28	+

[a] Minus-strand synthesis was measured as described by Sawicki et al. (submitted for publication). The cultures were incubated at 28°C for 3 hours followed by shift to 39°C and labeling with ³H uridine. ³H label in Sindbis HR double-strand RF was taken as 100%.

[b] Infected cultures were incubated at 28°C for 6 hours followed by shift to and labeling at 39°C as described (Keränen and Kääriäinen, 1979).

[c] This mutant fails to shut off minus strand RNA synthesis at 39°C (Sawicki et al., to be published).

that of 26S RNA is ^{7}mGpppAUUG (Pettersson *et al.*, 1980). This result strongly suggests that the transcription of 26S RNA takes place by internal initiation at about 4800 nucleotides from the 5' end of the 42S RNA negative strand (for review see Kääriänen and Söderlund, 1978). We have previously demonstrated that the same minus strand can be used as the template for either 42S or 26S RNA synthesis (Sawicki *et al.*, 1978). If the 26S RNA synthesis is shut off the minus-strand templates are recruited for the synthesis of 42S RNA. Once 26S RNA synthesis starts again, some of the template strands synthesizing 42S RNA are shifted for templates of 26S RNA.

What are the factors controlling the internal initiation of 26S RNA? Shift-up experiments, similar to those described for ts-6, were carried out with SFV and Sindbis virus RNA-negative mutants (Keränen and Kääriäinen, 1979). Sindbis-virus mutants from two complementation groups A and G showed a temperature-sensitive shut-off of 26S RNA synthesis (Table III). Again the defects were reversible and unaffected by inhibition of protein synthesis. Thus it seems that internal initiation of 26S RNA synthesis is controlled by two different gene products "A" and "G". Interestingly "A" is also involved in the regulation of minus-strand RNA synthesis. This means that this gene product has at least two different functions manifesting at different times during infection.

We have tried to identify the non-structural proteins involved in the late RNA synthesis (Ranki and Kääriäinen, 1979; Gomatos *et al.*, 1980). The proteins were labeled 2.5 to 3.5 hours post infection under conditions which allowed labeling of virus-specific proteins only (Kääriäinen *et al.*, 1978). The replication complex was labeled either by a short pulse of ^{3}H uridine *in vivo* (Gomatos *et al.*, 1980) or using post-mitochondrial pellet as a source for RNA polymerase *in vitro* (Ranki and Kääriäinen, 1979). The replication complex was purified either from isolated smooth membranes or directly from the post-mitochondrial pellet. In both cases the main non-structural protein was ns70 together with variable amounts of ns72 and ns86. Since the replication complex was able to synthesize double- but not single-stranded RNA we concluded that factors required for chain initiation and release had been preferentially lost during the purification (Gomatos *et al.*, 1980). The presence of ns70, as the major component, has been taken to suggest that this protein is probably the main polymerase component, respective to the gene product "F". The minor components ns86 and ns72 could well represent elements controlling the 26S RNA synthesis and thus could be equivalent to gene products "A" and "G" of Sindbis virus. The conclusion is supported by the accumulation of ns220 (consisting of ns70, ns86 and ns72) in cells infected with mutants displaying defects in the synthesis of 26S RNA (Bracha *et al.*, 1976; Kääriäinen *et al.*, 1978; Keränen and Kääriäinen, 1979). If we assume that mutation in either ns86 or ns72 can cause a cleavage defect in the non-structural polyprotein the accumulation of ns220 would be understandable.

SYNTHESIS AND PROCESSING OF STRUCTURAL PROTEINS

The synthesis of alphavirus structural proteins has been reviewed extensively recently (Kääriäinen and Renkonen, 1977; Kääriäinen and Söderlund, 1978; Simons *et al.*, 1978) and is discussed in detail in the paper by Garoff *et al.* (this volume). We will review some of our recent results relevant for the understanding of the processing and transport of SFV structural proteins.

The four structural proteins are translated from 26S RNA as a polyprotein in the following order: capsid-p62 (precursor to E3 and E2)-E1. The translation starts on free polysomes and nascent capsid protein binds to the polysomal ribosomes—preferably to the 60S subunit (Ulmanen *et al.*, 1979). After the cleavage of capsid protein the ribosome binds to the RER membrane and p62 and E1 are then sequestered into the lumen of RER, becoming glycosylated by high mannose-type oligosaccharide chains.

An SFV mutant ts-3 (Keränen and Kääriäinen, 1974) turned out to be useful in establishing the existence of a separate signal sequence for E1 protein. When ts-3 is grown at 39°C the cleavage between capsid protein and p62 is frequently inhibited, whereas cleavage between p62 and E1 takes place (Keränen and Kääriäinen, 1975; Lachmi *et al.*, 1975). As a result a fusion protein p86 and E1 accumulate. We have recently shown that E1 protein is glycosylated and segregated into the cysternal side of RER, whereas p86 remains at the cytoplasmic side as evidenced by its sensitivity to trypsin (Hashimoto *et al.*, to be published). This must mean that a signal sequence, distinct from that of p62, directs E1 through RER membrane. This signal sequence is not at the N-terminus of E1 which does not show an amino acid sequence typical for signal sequences (Kalkkinen *et al.*, 1980). A 6K polypeptide has been isolated from SFV-infected cells which has been proposed to be a signal sequence (Welch and Sefton, 1980). As reported in the chapter by Garoff *et al.* (this volume) there is a 60 amino acid residues long sequence between the C-terminus of E2 (Kalkkinen, 1980) and the N-terminus of E1 (Kalkkinen *et al.*, 1980), which must be the signal sequence for E1.

The pathway of envelope glycoproteins from their site of synthesis to the plasma membrane has recently been studied by the aid of RNA-positive temperature-sensitive mutants (Saraste *et al.*, 1979, 1980; Kääriäinen *et al.*, 1980). The mutants fall into two categories: (i) mutants showing positive surface immunofluorescence with anti-envelope serum; (ii) mutants with negative surface immunofluorescence, indicating that the viral glycoproteins have not been transported to the host-cell plasma membrane. Since the temperature-sensitive lesion of latter mutants turned out to be reversible, they have been used to study the transport and glycosylation of viral glycoproteins.

Existence of virus mutants with temperature-dependent transport of glycoproteins led us to postulate that virus envelope proteins have information for their own transport—a "transport signal" (Saraste *et al.*, 1979). In the case of Sindbis virus the transport-defective mutants fall into complementation group D (Saraste *et al.*, 1980), which is supposed to represent the E1 protein (for review see Kääriäinen and Söderlund, 1978). Thus the "transport signal" may reside in E1. Since the alphavirus glycoproteins are presumably transported as a dimer of p62-E1 (Bracha and Schlesinger, 1976; Jones *et al.*, 1976; K. Simons, personal communication) it may be sufficient to have the signal in only one of the proteins.

When cells infected with transport-defective mutants are maintained at 39°C, the envelope proteins remain in RER and have exclusively high mannose-type glycans (Kääriäinen *et al.*, 1980; Pesonen *et al.*, 1980). If the cultures are transferred to the permissive temperature, in the presence of cycloheximide, the virus glycoproteins appear on the cell surface (Saraste *et al.*, 1979, 1980). The transport is completed within 90 min during which they migrate most probably through the Golgi complex (Kääriäinen *et al.*, 1980) where part of the high mannose glycans are converted to complex ones (Pesonen *et al.*, 1980).

VIRUS ASSEMBLY

The nucleocapsid is assembled in the cytoplasm from 42S RNA and ribosome-associated capsid proteins. We have been unable to demonstrate free mono- or oligomeric capsid protein (Ulmanen *et al.*, 1976; Söderlund and Ulmanen, 1977). Excluding nucleocapsids, all capsid protein is ribosome-associated either in polysomes or monosomes. The transfer seems to take place directly from the polyribosomes or monosomes to the nucleocapsid pool (Ulmanen *et al.*, 1979). This unique assembly process is still poorly understood and is probably related to prevailing RNA-protein interactions of the nucleocapsid (for reviews see Söderlund *et al.*, 1975; Kääriäinen and Söderlund, 1978).

The virus maturation takes place at the plasma membrane where the nucleocapsid presumably recognizes the cytoplasmic C-terminal sequence of p62 (i.e. E2), resulting in the budding of virus through the plasma membrane (Garoff and Simons, 1974; Garoff *et al.*, this volume). The p62 is

TABLE IV

Function and genetic correlation of alphavirus gene products

Designation of protein	Molecular weight	Complementation group	Assigned functions	Additional properties
Structural				
Capsid (C)	30 000	C	Nucleocapsid	Binding to ribosome (60S subunit)
Envelope E1	49 000	D	Penetration, HA, fusion	Glycoprotein transport (?)
Envelope E2	52 000	E	Infectivity	Nucleocapsid binding (?)
Envelope E3	10 000	?	Non-essential	Not known
Non-structural				
ns70	60–70 000	F	RNA polymerization (?) (42S, 26S RNA)	Not known
ns86	86–90 000	A(?)	Regulation of 26S RNA synthesis. Shut off of minus-strand synthesis	Binding to ribosome
ns72	72 000	G(?)	Regulation of 26S RNA synthesis	Not known
ns60	60 000(?)	B(?)	Regulation of minus-strand synthesis	Short half-life, cleavage (?)

probably cleaved to E2 and E3 at this stage. In SFV E3 remains associated with the virion, whereas in Sindbis-virus-infected cells it is released into the medium (J. H. Strauss, personal communication).

CONCLUSIONS

Here we have made an attempt to ascribe functions to the alphavirus gene products at different times of virus infection (Table IV). The evidence must be taken as tentative in most of the cases and some of the correlations rely solely on genetic analysis.

Notwithstanding this, we can already say that many of the proteins have more than one biological function: e.g. E2 protein is required in the adsorption and later for the recognition of the nucleocapsid. In addition to its function in penetration, E1 protein is possibly needed in the transport of viral glycoproteins. Gene product "A" is needed in the shut-off of minus-strand RNA synthesis as well as in the regulation of 26S RNA synthesis. It may have an additional vital role early in infection since group A mutants have an RNA-negative phenotype, which is difficult to understand on the basis of their known defects.

Many evidently virus-specific functions such as shut-off of host and macromolecular syntheses, shut-off of non-structural protein synthesis and capping of viral RNAs have not yet been correlated with any of the gene functions or known proteins.

REFERENCES

Bell, J. R., Hunkapiller, M. W., Hood, L. E. and Strauss, J. H. (1978). *Proc. natn. Acad. Sci. U.S.A.* **75**, 2722–2726.
Bracha, M. and Schlesinger, M. J. (1976). *Virology* **74**, 441–449.
Bracha, M., Leone, A. and Schlesinger, M. J. (1976). *J. Virol.* **20**, 612–620.
Bruton, C. J. and Kennedy, S. I. T. (1975). *J. gen. Virol.* **28**, 111–128.
Brzeski, H. and Kennedy, S. I. T. (1977). *J. Virol.* **22**, 420–429.
Clarke, D. H. and Casals, J. (1958). *Am. J. trop. Med. Hyg.* **7**, 561–573.
Dalrymple, J. M., Schlesinger, S. and Russell, P. K. (1976). *Virology* **69**, 93–103.
Dubin, D. T., Stollar, V., Hsuchen, C.-C., Timko, K. and Guild, G. M. (1977). *Virology* **77**, 457–470.
Frey, T. K. and Strauss, J. H. (1978). *Virology* **86**, 494–506.
Garoff, H. and Simons, K. (1974). *Proc. natn. Acad. Sci. U.S.A.* **71**, 3988–3922.
Glanville, N. and Lachmi, B. (1977). *FEBS Lett.* **81**, 399–402.
Glanville, N., Ranki, M., Morser, J., Kääriäinen, L. and Smith, A. E. (1976). *Proc. natn. Acad. Sci. U.S.A.* **73**, 3059–3063.
Glanville, N., Lachmi, B., Smith, A. E. and Kääriäinen, L. (1978). *Biochim. biophys. Acta* **518**, 497–506.

Gomatos, P. J., Kääriäinen, L., Keränen, S., Ranki, M. and Sawicki, D. L. (1980). *J. gen. Virol.* **49**, 61–69.

Helenius, A., Fries, E., Garoff, H. and Simons, K. (1976). *Biochim. biophys. Acta* **436**, 319–334.

Helenius, A., Fries, E. and Kartenbeck, J. (1978a). *J. cell. Biol.* **75**, 866–880.

Helenius, A., Morein, B., Fries, E., Simons, K., Robinson, P., Schirrmacher, V., Terhorst, C. and Strominger, J. L. (1978b). *Proc. natn. Acad. Sci. U.S.A.* **75**, 3846–3850.

Helenius, A., Kartenbeck, J., Simons, K. and Fries, E. (1980a). *J. cell. Biol.* **84**, 404–420.

Helenius, A., Marsh, M. and White, J. (1980b). *Trends biochem. Sci.* **5**, 104–106.

Jones, K. J., Scupham, R. K., Pfeil, J. A., Wan, K., Sagik, B. P. and Bose, H. R. (1977). *J. Virol.* **21**, 778–787.

Kääriäinen, L. and Renkonen, O. (1977). *In* "Cell Surface Reviews" (G. Poste and G. L. Nicolson, eds) Vol. 4, pp. 741–801. Elsevier/North-Holland Biomedical Press, Amsterdam.

Kääriäinen, L. and Söderlund, H. (1978). *Curr. Topics Microbiol. Immunol.* **82**, 15–69.

Kääriäinen, L., Lachmi, B. and Glanville, N. (1976). *Ann. Microbiol. (Inst. Pasteur)* **127A**, 197–203.

Kääriäinen, L., Sawicki, D. and Gomatos, P. J. (1978). *J. gen. Virol.* **39**, 463–473.

Kääriäinen, L., Hashimoto, K., Saraste, J., Virtanen, I. and Penttinen, K. (1980). *J. cell. Biol.* **87**, 783–791.

Kalkkinen, N. (1980). *FEBS Lett.* **115**, 163–166.

Kalkkinen, N., Jörnvall, H., Söderlund, H. and Kääriäinen, L. (1980). *Eur. J. Biochem.* **108**, 31–37.

Keränen, S. and Kääriäinen, L. (1974). *Acta path. microbiol. scand. Sect.* B **82**, 810–820.

Keränen, S. and Kääriäinen, L. (1975). *J. Virol.* **16**, 388–396.

Keränen, S. and Kääriäinen, L. (1979). *J. Virol.* **32**, 19–29.

Lachmi, B. and Kääriäinen, L. (1976). *Proc. natn. Acad. Sci. U.S.A.* **73**, 1936–1940.

Lachmi, B., Glanville, N., Keränen, S. and Kääriäinen, L. (1975). *J. Virol.* **16**, 1615–1629.

Lehtovaara, P., Ulmanen, I., Kääriäinen, L., Keränen, S. and Philipson, L. (1980). *Eur. J. Biochem.* **112**, 461–468.

Marsh, M. and Helenius, A. (1980). *J. molec. Biol.* **142**, 439–454.

Pesonen, M., Saraste, J., Hashimoto, K. and Kääriäinen, L. (1981). *Virology.* **109**, 165–173.

Pettersson, R. F., Söderlund, H. and Kääriäinen, L. (1980). *Eur. J. Biochem.* **105**, 435–443.

Pfefferkorn, E. R. and Shapiro, D. (1974). *In* "Comprehensive Virology" (H. Fraenkel-Conrat and R. R. Wagner, eds) Vol. 2, pp. 171–230. Plenum Press, New York and London.

Ranki, M. and Kääriäinen, L. (1979). *Virology* **98**, 298–307.

Ranki, M., Ulmanen, I. and Kääriäinen, L. (1979). *FEBS Lett.* **108**, 299–302.

Saraste, J., von Bonsdorff, C.-H., Hashimoto, K., Kääriäinen, L. and Keränen, S. (1979). *Virology* **100**, 229–245.

Saraste, J., von Bonsdorff, C.-H., Hashimoto, K., Keränen, S. and Kääriäinen, L. (1980). *Cell Biol. int. Rep.* **4**, 279–286.

Sawicki, D. L. and Gomatos, P. J. (1976). *J. Virol.* **20**, 446–464.

Sawicki, D. L. and Sawicki, S. G. (1980). *J. Virol.* **34**, 108–118.
Sawicki, D. L., Kääriäinen, L., Lambeck, C. and Gomatos, P. J. (1978). *J. Virol.* **25**, 19–27.
Simmons, D. T. and Strauss, J. H. (1972a). *J. molec. Biol.* **71**, 599–613.
Simmons, D. T. and Strauss, J. H. (1972b). *J. molec. Biol.* **71**, 615–631.
Simmons, D. T. and Strauss, J. H. (1974). *J. molec. Biol.* **86**, 397–409.
Simons, K., Garoff, H., Helenius, A. and Ziemiecki, A. (1978). *In* "Frontiers in Physiochemical Biology" (B. Pullman, ed.) pp. 387–407. Academic Press, New York and London.
Söderlund, H. and Ulmanen, I. (1977). *J. Virol.* **24**, 907–909.
Söderlund, H., Kääriäinen, L. and von Bonsdorff, C.-H. (1975). *Med. Biol.* **53**, 412–417.
Strauss, J. H. and Strauss, E. G. (1977). *In* "The Molecular Biology of Animal Viruses" (D. P. Nayak, ed.) pp. 111–166. Marcel Dekker, New York.
Taylor, R. M. (1967). "Catalogue of Arthropod-borne Viruses of the World". Public Health Service Publication No. 1760, U.S. Dept. of Health, Education and Welfare, Washington.
Ulmanen, I., Söderlund, H. and Kääriäinen, L. (1976). *J. Virol.* **20**, 203–210.
Ulmanen, I., Söderlund, H. and Kääriäinen, L. (1979). *Virology* **99**, 265–276.
Väänänen, P. and Kääriäinen, L. (1979). *J. gen. Virol.* **43**, 593–601.
Väänänen, P. and Kääriäinen, L. (1980). *J. gen. Virol.* **46**, 467–475.
Welch, W. J. and Sefton, B. M. (1980). *J. Virol.* **33**, 230–237.
White, J. and Helenius, A. (1980). *Proc. natn. Acad. Sci. U.S.A.* **77**, 3273–3277.

Semliki Forest Virus—A Model System to Study Membrane Structure and Assembly

HENRIK GAROFF

European Molecular Biology Laboratory, Postfach 102209, 6900 Heidelberg,
Federal Republic of Germany

INTRODUCTION

Several enveloped animal viruses have during the last few years been used successfully to study the structure and synthesis of biological membranes (Kääriäinen and Renkonen, 1977; Lenard, 1978; Simons *et al.*, 1978; Patzer *et al.*, 1979). These include paramyxo viruses (e.g. Sendai virus), myxoviruses (e.g. influenza A virus), rhabdoviruses (e.g. vesicular stomatitis virus) and alphaviruses (e.g. Sindbis virus and Semliki Forest virus, SFV). All of these viruses consist of a nucleocapsid with the viral genome and a surrounding lipid bilayer with glycoprotein spikes. The viral membrane glycoproteins are synthesized in the rough endoplasmic reticulum (RER) of the host cell and then transported to the plasma membrane (PM) in the same way as normal plasma membrane proteins are. During infection, however, viral proteins are synthesized very efficiently whereas host protein synthesis is inhibited. This makes it possible to study in detail the synthesis and assembly of the viral membrane proteins within the host. The mRNAs of the viral glycoproteins are present in large amounts in the infected cell cytoplasm. These can be purified and translated *in vitro* in the presence of microsomal membranes giving information about how the membrane proteins are inserted into the RER membrane. The assembly of the viral membranes can also be studied using temperature-sensitive mutants of the enveloped viruses. Today there exist several mutants which show blocks in different stages of the assembly process.

The virus particle is formed at the plasma membrane by a budding process. In this process the viral nucleocapsid surrounds itself with a membrane that contains only viral spike glycoproteins. The released virus particles represent therefore a very homogeneous and simple membrane preparation the structure of which can be studied using sophisticated biochemical and physical techniques. Viral membrane glycoproteins can easily be isolated from a purified virus preparation and used for chemical characterization or antibody production. The genomic RNA of these viruses can after a single purification step be transcribed into cDNA, cloned and sequenced, giving information about the primary structure of the viral proteins. In the near future it will probably be possible to express the cloned viral membrane glycoprotein genome and its *in vitro* mutagenized forms in eukaryotic cells.

SEMLIKI FOREST VIRUS

The glycoprotein spikes of SFV are composed of three polypeptide chains, E1 (MW 50K), E2 (MW 50K) and E3 (MW 10K) (Figs 1 and 2) (Garoff *et al.*, 1974; Ziemiecki and Garoff, 1978). E1 and E2 are integral membrane proteins whereas E3 is a peripheral one (Utermann and Simons, 1974; Garoff and Söderlund, 1978). The fourth structural protein of SFV is the capsid protein, C (MW 30K) (Simons and Kääriäinen, 1970), which is complexed with the genomic 42S RNA in the viral nucleocapsid. Two hundred and forty copies of each protein are present in one virus particle (Laine *et al.*, 1973; Bonsdorff and Harrison, 1975).

A successful infection by SFV requires the release of the viral RNA into the cell cytoplasm. Recent work has shown that the SFV enters the cell by absorptive endocytosis (Helenius *et al.*, 1980) (Fig. 3). Virus particles are taken up into coated vesicles which transport the viruses to the lysosomes.

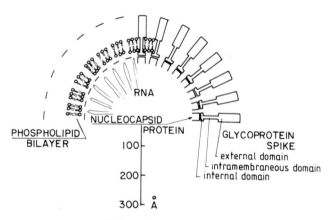

FIG. 1. Structure of SFV.

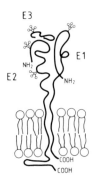

FIG. 2. Structure of the SFV spike glycoproteins. The external hydrophilic domain consists of E3 and most of E1 and E2 including their amino terminal ends. All protein-bound carbohydrate (⚓) is present in this domain. The membrane-binding domain consists of hydrophobic segments of E1 and E2. The carboxyterminal end of E2 makes up the internal domain of the spike glycoprotein.

The acidic pH within the lysosomes probably triggers fusion between the viral and the lysosomal membranes (White and Helenius, 1980), and as a result of this event the nucleocapsid enters the cytoplasm. Here the viral genome becomes uncoated so that translation of viral polymerase molecules can begin. These make more 42S RNA molecules and in addition smaller 26S RNA molecules. The latter molecules are homologous to the 3' end of the 42S RNA (Kennedy, 1976) and they function as messenger RNAs for the four structural proteins of SFV. The four proteins are read sequentially in the order C, E3, E2 and E1 from the mRNA using a single initiation site (Clegg, 1975; Glanville et al., 1976). The C protein associates with the 42S RNA into nucleocapsids in the cell cytoplasm whereas the membrane proteins become inserted into the RER membrane. Within the cell the E3 protein extends the amino terminus of E2 (Simons et al., 1973; Garoff et al., 1974) and the precursor protein is called p62. The membrane proteins are glycosylated in RER (Krag and Robbins, 1977; Sefton, 1977; Garoff et al., 1978); one core-sugar unit is added to E1 and three units to p62 (one to the E3 part and two to the E2 part) (Mattila et al., 1975; Personen and Renkonen, 1976). All carbohydrate groups are N-glycosidically linked to asparagine residues of the polypeptides. A p62-E1 spike complex is formed already in the RER and this is transported to the PM of the host cell where it is incorporated into the membrane of new virus particles in the budding process (Acheson and Tamm, 1967; Ziemiecki et al., 1980). The p62 protein is cleaved to E3 and E2 when it arrives at the PM.

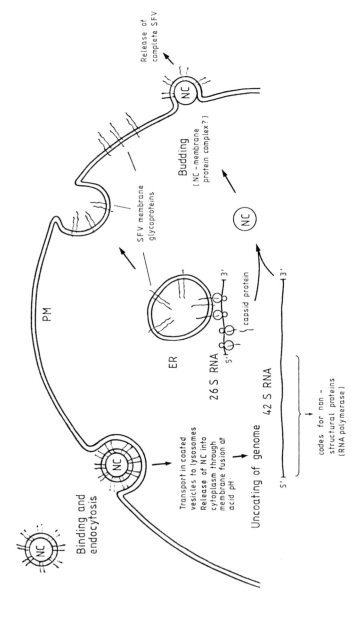

FIG. 3. Replication of SFV in baby hamster kidney cells.

THE C PROTEIN AND ASSEMBLY OF THE NUCLEOCAPSID

Studies *in vitro* have shown that the C protein is cleaved from the nascent polypeptide chain as soon as it is completed (Garoff *et al.*, 1978). It has been suggested that this cleavage is an autocatalytic event because the cleavage is brought about so effectively in all *in vitro* systems tested. An alternative explanation is that the C protein is cleaved by a very common host protease that is associated with the large subunit of the ribosome. In this context it is interesting to note that the C protein itself has been found to associate with the large ribosomal subunit before incorporation into nucleocapsids (Söderlund and Ulmanen, 1977). Pulse-chase experiments show that newly made C protein associates rapidly with 42S RNA into complete nucleocapsid structures. Only in the beginning of the infection cycle when the 42S RNA and the C proteins are still present in low concentrations within the cell, are nucleocapsid structures containing less than their full complement of C proteins found (Ulmanen, 1978). Nucleocapsid-like structures lacking 42S RNA or aggregates of C protein cannot be demonstrated at any stage of infection. This suggests that C protein–RNA interactions are of major importance for the formation of the SFV nucleocapsid whereas C protein–C protein interactions are less significant. The importance of protein–RNA bonds is also suggested by the sensitivity of the SFV nucleocapsid towards treatments with RNase and low concentrations of SDS (Söderlund *et al.*, 1979). The amino acid sequence of the C protein, which has recently been established, shows a striking clustering of basic amino acids and proline in the amino terminus of the polypeptide chain (Garoff *et al.*, 1980a). It is probable that strong protein–RNA interactions result from the formation of several salt bridges between the basic amino acid residues of these clusters and the phosphate groups of the RNA.

THE STRUCTURE OF THE VIRAL MEMBRANE

The topography of the spike glycoprotein in the lipid membrane has been studied using several different experimental approaches. These include protease treatment, surface labeling and protein–protein cross-linking. Protease treatment of virus particles removes most of the spike glycoproteins and all protein-bound carbohydrates, leaving only small hydrophobic peptide segments which remain in the membrane (Utermann and Simons, 1974). The peptides have been shown to be derived from the carboxyterminal region of E1 and E2 (Garoff and Söderlund, 1978). When microsomes which are isolated from SFV-infected cells are treated with a protease, only about 30 amino acids are cleaved from the carboxyterminus of E2, the rest of the spike glycoproteins being protected (Garoff and Söderlund, 1978). These results show that the SFV spike glycoproteins span the lipid bilayer and have a large external hydrophilic domain (digested in virus particles), a membrane-binding hydrophobic domain and a small internal hydrophilic domain (digested in microsomes) (see Fig. 1). The external domain contains the complete E3 protein and most of the E1 and E2 protein (see Fig. 2). The membrane-binding domain consists of hydrophobic peptides of E1 and E2 respectively. About 30 amino acids from the carboxyterminus of E2 constitute the internal domain.

Treatment of virus particles with short-ranged protein–protein cross-linking reagents like dimethylsuberimidate and dithiobispropionimidate have not only revealed the oligomeric structure of the spike glycoprotein but also demonstrated a close relationship between these and the underlying nucleocapsid (Garoff, 1974; Garoff and Simons, 1974; Ziemiecki and Garoff, 1978). More than half of the glycoproteins were linked to the nucleocapsid in such experiments. This is only possible by the formation of cross-links between internal domains of the glycoproteins and the nucleocapsid.

The internal portion of the spike glycoproteins has also been demonstrated by surface labeling with formyl-^{35}S-methionyl sulphone methylphosphate (^{35}S FMMP) (Garoff and Simons, 1974; Simons et al., 1980). Intact virus particles, and such particles where the membrane had been lysed by treatment with low concentration of Triton X-100, were labeled with ^{35}S FMMP. The glycoproteins of both preparations were isolated and the extent of their labeling was compared by analyzing their peptide maps. It was shown that the E2 glycoprotein from the lysed virus preparation contained one additional peptide with basic characteristics.

The amino acid sequence of E1 shows a stretch of 24 hydrophobic and neutral amino acids close to its carboxyterminus (see Fig. 4) (Garoff et al., 1980b). A stretch of 35 neutral and hydrophobic amino acids is found near the carboxyterminus of E2. These segments are unique in their length and hydrophobicity when compared to other uncharged peptides present in the structural proteins of SFV and they must represent the membrane-spanning peptides of E1 and E2. The peptides are similar to the membrane-spanning segments of glycophorin and the hemagglutinins of influenza A Victoria and FPV (see Fig. 4) (Tomita and Marchesi, 1975; Porter et al., 1979; Min Jou et al., 1980). The intramembraneous peptide of glycophorin contains 23 hydrophobic and neutral amino acids, the hemagglutinin of influenza A Victoria contains 27 such residues and that of FPV, 26. The comparison of the amino acid sequences of the membrane-spanning segments shows that in addition to Glu, Asp, Arg, Lys, Gln, Asn and His, Pro is also lacking from these peptides. The membrane segment of E1 is flanked by two arginine residues on its carboxyterminal side. These basic residues might fix the

```
                 INTRAMEMBRANEOUS SEGMENT                    CYTOPLASMIC/INTERNAL SEGMENT

                 GLYCOPHORIN
        S E I E I T  L I V F G V M A G V I G T I L L I S Y G I R R L I K K S P S D V K P L P S P D -
                                                                 T D V P L S S V E I E N P E T S D Q

                 HEMAGGLUTININ, INFLUENZA A/VICTORIA
        K D W I L W I S F A I S C F L L C V V L L G F I M W A C Q K G N I R C N I C I

                 HEMAGGLUTININ, INFLUENZA A/FPV
         K D V I L W F S F G A S C F L L L A I A V G L V F I C V K N G N M R C T I C I

                                                                 HEAVY CHAIN, HLA B7
                                                              M C R R K S S G G K G G S Y S Q A -
                                                                 A C S D S A Q G S D V S L T A
                 E2, SFV
  K P H G W P H Q I -
   V Q V V V G L Y P A A T V -
                  S A V V G M S L L A L I S I F A S C V M L V A A R S K C L T P Y A L T P G A A V -
                                                                 P W T L G I L C C A P R A H A

                 E1, SFV
               Q K I S G G L G A F A I G A I L V L V V V T C I G L R R
```

FIG. 4. Amino acid sequences of intramembraneous (underlined) and internal (cytoplasmic) parts of spanning membrane proteins.

membrane segment of E1 on the internal side of the lipid bilayer. An Arg-Ser-Lys sequence is present on the carboxyterminal side of the membrane segment of E2. One or several basic residues also flank the other spanning-membrane proteins on their carboxyterminal side. The 31 amino acids on the carboxyterminal side of the E2 membrane segment (Fig. 4) represent the internal domain of the SFV spike glycoprotein, which is of the same size as the internal domain of glycophorin and the histocompatibility antigens (Tomita and Marchesi, 1975; Robb et al., 1978).

ASSEMBLY OF THE VIRAL MEMBRANE

Translation of the 26S mRNA *in vitro* in the presence of microsomes derived from RER of dog pancreas has shown that both the p62 protein and the E1 protein are translocated across the lipid membrane cotranslationally (Garoff et al., 1978). The amino terminal end of the p62 protein, which is exposed after the capsid protein has been completed and cleaved off, binds the polysome to the microsomal membrane and initiates the translocation of the p62 protein. The amino acid sequence of the amino terminal end of p62 shows a stretch of 28 uncharged residues (no Glu, Asp, Arg or Lys) (Garoff et al., 1980b). It is, however, difficult to define the actual signal sequence within this region. The signal peptides that have been characterized so far for several secreted proteins and some membrane proteins consist of 15–30 amino acids (Austen, 1979; Emr et al., 1980). Characteristic for all of them is an un-charged region of at least 9 amino acids. Several signal peptides display a cluster of hydrophobic amino acids within this region. The signal peptide of p62 is not cleaved, as are most other signal peptides, but remains as a part of the E3 protein in the virus particle.

Translocation of the p62 protein proceeds until the membrane-spanning segment is made and locks the protein in the membrane. The basic amino acids that follow the synthesis of the membrane-spanning segment might cooperate in stopping the translocation process. After synthesis of the cytoplasmic portion of the p62 protein, translocation across the membrane must start again with the synthesis of the amino terminal end of E1. The 26S RNA nucleotide sequence shows that a small peptide (6K) containing 60 amino acids is translated after E2. In this peptide there are two potential regions which might act as E1 signal peptides. The former contains 15 residues and the latter 14 residues. Both regions have clusters of hydro-phobic amino acids. The 6K-E1 cleavage site (Ala–Tyr) is typical for many other signal peptides. These are mostly cleaved at the carboxyterminal side of an uncharged amino acid with a small side-chain.

The cotranslational translocation of the E1 protein ends with the synthesis of its membrane-spanning segment which remains in the lipid bilayer. The two last arginine residues probably bind to the cytoplasmic surface of the RER membrane. Figure 5 shows schematically the different events during the synthesis of the SFV membrane proteins.

The glycosylation of the membrane proteins most probably takes place by addition of core-sugar units to the nascent polypeptide chain (Sefton, 1977; Garoff et al., 1978). Unglycosylated forms of the membrane proteins have not been detected in vivo or in vitro (Garoff and Schwarz, 1978). The two sugar units of E2 are added to the asparagine residues in positions 200 and 262. The single-sugar unit of E1 is attached to asparagine in position 141. These positions are the only potential glycosylation sites found in the amino acid sequence of these proteins (Garoff et al., 1980b). E3 has two potential sites where the single-sugar unit could be attached. These are the asparagines at positions 13 and 60. During the transport of the spike glycoprotein from the RER to the PM the core-sugar units are trimmed to their final shape (Robbins et al., 1977).

The intracellular route by which the viral membrane proteins are transported from the RER to the PM is still unknown. When the membrane protein p62 arrives at the PM it is cleaved into E3 and E2. This cleavage

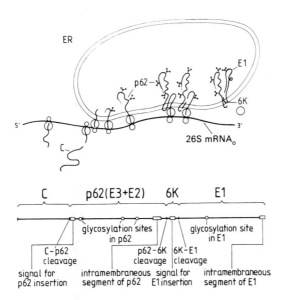

FIG. 5. Assembly of SFV membrane proteins in the RER membrane during synthesis. The C, p62 and E1 proteins are translated sequentially from the 26S mRNA (below). After p62 a small non-structural peptide (6K) is synthesized. Membrane-binding segments, signal peptides, proteolytic cleavage sites and glycosylation sites are indicated.

takes place on the carboxyterminal side of two arginine residues (Garoff *et al.*, 1980b). Another viral membrane protein, the hemagglutinin (HA) is also cleaved posttranslationally at the carboxyterminal side of two basic amino acid residues to yield the two hemagglutinin polypeptides HA_1 and HA_2 (Porter *et al.*, 1979; Min Jou *et al.*, 1980). Both of these cleavages resemble the late processing of several prohormones and proproteins like proinsulin (Kemmler *et al.*, 1973; Steiner, 1976), proglucagon (Tager and Steiner, 1973), proparathryreoideahormone (Habener *et al.*, 1977), progastrin (Gregory and Tracy, 1972) and proalbumin (Russel and Geller, 1975). All of these precursors are processed into the mature hormone or protein by cleavage at the carboxyterminal side of basic amino acid pairs.

The final assembly of the viral membrane takes place at the PM in the budding process. The viral nucleocapsid probably binds directly to the internal domains of the spike glycoproteins. In this way incorporation of only viral glycoproteins into the envelope is assured. Host proteins which will lack any affinity for the nucleocapsid template will be extruded from the viral glycoprotein patch. When a full complement of spikes (240 copies) has bound to the nucleocapsid the virus membrane pinches off from the host membrane and a new virus particle is formed.

REFERENCES

Acheson, N. H. and Tamm, I. (1967). *Virology* **32**, 128–143.
Austen, B. M. (1979). *FEBS Lett.* **103**, 308–313.
Bonsdorff, C.-H. and Harrison, S. C. (1975). *J. Virol.* **16**, 141–145.
Clegg, J. C. S. (1975). *Nature, Lond.* **254**, 454–455.
Emr, S. D., Hedgpeth, J., Clement, J.-M., Silhavy, T. J. and Hofnung, M. (1980). *Nature, Lond.* **285**, 82–85.
Garoff, H. (1974). *Virology* **62**, 385–392.
Garoff, H. and Schwarz, R. (1978). *Nature, Lond.* **274**, 487–490.
Garoff, H. and Simons, K. (1974). *Proc. natn. Acad. Sci. U.S.A.* **71**, 3988–3992.
Garoff, H. and Söderlund, H. (1978). *J. molec. Biol.* **124**, 535–549.
Garoff, H., Simons, K. and Renkonen, O. (1974). *Virology* **61**, 493–504.
Garoff, H., Simons, K. and Dobberstein, B. (1978). *J. molec. Biol.* **124**, 587–600.
Garoff, H., Frischauf, A.-M., Simons, K., Lehrach, H. and Delius, H. (1980a). *Proc. natn. Acad. Sci. U.S.A.* **77**, 6376–6380.
Garoff, H., Frischauf, A.-M., Simons, K., Lehrach, H. and Delius, H. (1980b). *Nature, Lond.* **288**, 236–241.
Glanville, N., Ranki, M., Morser, J., Kääriäinen, L. and Smith, A. E. (1976). *Proc. natn. Acad. Sci. U.S.A.* **73**, 3059–3063.
Gregory, R. A. and Tracy, H. J. (1972). *Lancet* **2**, 797.
Habener, J. F., Chang, H. T. and Potts, J. T. Jr. (1977). *Biochemistry* **16**, 3910–3917.
Helenius, A., Kartenbeck, J., Simons, K. and Fries, E. (1980). *J. Cell Biol.* **84**, 404–420.

Kääriäinen, L. and Renkonen, O. (1977). *In* "The Synthesis, Assembly and Turnover of Cell Surface Components" (G. Poste and G. L. Nicolson, eds) pp. 748–801. North Holland, Amsterdam.
Kemmler, W., Steiner, D. F. and Borg, J. (1973). *J. biol. Chem.* **248**, 4544–4551.
Kennedy, S. I. T. (1976). *J. molec. Biol.* **108**, 491–511.
Krag, S. S. and Robbins, P. W. (1977). *J. biol. Chem.* **252**, 2621–2629.
Laine, R., Söderlund, H. and Renkonen, O. (1973). *Intervirology* **1**, 110–118.
Lenard, J. (1978). *A. Rev. biophys. Bioeng.* **7**, 139.
Mattila, K., Luukkonen, A. and Renkonen, O. (1975). *Biochim. biophys. Acta* **419**, 435–444.
Min Jou, W., Verhoeyen, M., Devos, R., Saman, E., Fang, R., Huylebroeck, D., Fiers, W., Threlfall, G., Barber, C., Carey, N. and Emtage, S. (1980). *Cell* **19**, 683–696.
Patzer, E. J., Wagner, R. R. and Dubovi, E. J. (1979). *Critic Rev. Biochem.* **6**, 165–217.
Pesonen, M. and Renkonen, O. (1976). *Biochim. biophys. Acta* **455**, 510–525.
Porter, H. G., Barber, C., Carey, N. H., Hallewell, R. A., Threlfall, G. and Emtage, J. S. (1979). *Nature, Lond.* **282**, 471–477.
Robb, R. J., Terhorst, C. and Strominger, J. L. (1978). *J. biol. Chem.* **253**, 5319–5324.
Robbins, P. W., Hubbard, S. C., Turco, S. J. and Wirth, D. F. (1977). *Cell* **12**, 893–900.
Russel, J. H. and Geller, D. M. (1975). *J. biol. Chem.* **250**, 3409–3413.
Sefton, B. (1977). *Cell* **10**, 659–668.
Simons, K. and Kääriäinen, L. (1970). *Biochem. biophys. Res. Commun.* **5**, 981–988.
Simons, K., Keränen, S. and Kääriäinen, L. (1973). *FEBS Lett.* **29**, 87–91.
Simons, K., Garoff, H., Helenius, A. and Ziemiecky, A. (1978). *In* "Frontiers in Physiochemical Biology" (B. Pullman, ed.) pp. 387–407. Academic Press, New York and London.
Simons, K., Garoff, H. and Helenius, A. (1980). *In* "Togaviruses" (W. Schlesinger, ed.), pp. 317–341. Academic Press, New York and London.
Söderlund, H. and Ulmanen, I. (1977). *J. Virol.* **24**, 907–909.
Söderlund, H., von Bonsdorff, C.-H. and Ulmanen, I. (1979). *J. gen. Virol.* **45**, 15–26.
Steiner, D. F. (1976). *In* "Peptide Hormones" (J. A. Parsons, ed.), p. 49. Macmillan, London.
Tager, H. S. and Steiner, D. F. (1973). *Proc. natn. Acad. Sci. U.S.A.* **70**, 2321–2325.
Tomita, H. and Marchesi, V. T. (1975). *Proc. natn. Acad. Sci. U.S.A.* **72**, 2964–2968.
Ulmanen, I. (1978). *J. gen. Virol.* **41**, 353–365.
Utermann, G. and Simons, K. (1974). *J. molec. Biol.* **85**, 569–581.
White, J. and Helenius, A. (1980). *Proc. natn. Acad. Sci. U.S.A.* **77**, 3273–3277.
Ziemiecki, A. and Garoff, H. (1978). *J. molec. Biol.* **122**, 259–269.
Ziemiecki, A., Garoff, H. and Simons, K. (1980). *J. gen. Virol.* **50**, 111–123.

Heterogeneity of Semliki Forest Virus 18S Defective Interfering RNA Containing Heterogeneous 5′-terminal Nucleotide Sequences

RALF F. PETTERSSON and LEEVI KÄÄRIÄINEN

*Department of Virology, University of Helsinki, Haartmaninkatu 3, 00290
Helsinki 29, Finland*

INTRODUCTION

The defective interfering (DI) particles are deletion mutants containing truncated viral genomes packed into particles, which possess all the normal structural proteins. Such particles are generated in most viral systems when cultures are serially infected with undiluted inocula. They have the ability to interfere with the replication of "standard" virus and to become enriched during successive passaging (Huang and Baltimore, 1977). Only in a few cases has it been elucidated how these DI genomes are synthesized. For RNA viruses, at least two mechanisms appear to operate. In the case of poliovirus (Nomoto *et al.*, 1979; Lundqvist *et al.*, 1979), the alphaviruses (Bruton and Kennedy, 1976; Kennedy, 1976; Guild and Stollar, 1977; Stark and Kennedy, 1978; Dohner *et al.*, 1979) and some classes of vesicular stomatitis virus (VSV) (Perrault and Semler, 1979) DI RNAs, internal nucleotide sequences are deleted during replication. It is thought that the RNA polymerase dissociates from the template and reassociates with it at a new point closer to the 5′ end. Such dissociations and reassociations may occur at different points as shown for poliovirus (Lundqvist *et al.*, 1979). The second mechanism, so far shown to operate only in the generation of certain classes of VSV, and Sendai DI RNAs, also involves the dissociation of the RNA polymerase from the template. In these cases, however, the polymerase appears to reassociate with the newly synthesized strand and copy that strand backwards (Leppert *et al.*, 1977; Perrault and Semler, 1979). This event results in the synthesis of molecules with inverted complementary sequences at the 3′ and 5′ ends. They can therefore form circular hairpin structures with "stems".

In cells serially infected with undiluted inoculum of Sindbis or Semliki Forest virus, DI RNAs of different sizes accumulate. Simultaneously, the amount of the genomic 42S RNA and the subgenomic 26S RNA drastically decrease (Stark and Kennedy, 1978). Based on nucleic acid hybridization studies (Guild and Stollar, 1977) and oligonucleotide fingerprinting (Kennedy, 1976; Stark and Kennedy, 1978; Dohner et al., 1979) it has been concluded that the DI RNAs of alphaviruses have conserved 5' and 3' ends of the genome RNA with internal deletions. With the help of "standard" virus, these DI RNAs are able to replicate and to become encapsidated. Thus, they must contain recognition sequences—probably close to the 3' end—for the RNA polymerase. In addition, they apparently contain an "enucleation site" for the binding of the capsid protein. Since the 26S RNA—which is identical to the 3' third of the 42S RNA (Simmons and Strauss, 1972; Kennedy, 1976; Wengler and Wengler, 1976)—is not encapsidated, this sequence must be located in the 42S RNA-specific region (5' two-thirds) of the 42S RNA. To identify these regulatory sequences, we have initiated structural studies of the SFV DI RNAs. Here we summarize some recent results on the isolation of an 18S DI RNA from infected cells. We have found evidence that the RNA is heterogeneous and has a heterogeneous 5'-terminal nucleotide sequence not found in the 42S RNA.

ISOLATION AND CHARACTERIZATION OF 18S DI RNA

BHK21 cells in one liter roller bottles were infected with the prototype strain of Semliki Forest virus (Kääriäinen et al., 1969) at a multiplicity of 100 PFU/cell and incubated for 24 hours at 37°C. Passaging of the virus, using undiluted inoculum, was continued for 14 further passages. Infectivity and hemagglutination were recorded from the culture media after a constant 24 hour growth period. A 500-fold decrease in the yield of infectious virus had already been observed at the fifth passage, whereafter the yield of infectious virus fluctuated between 0.02 and 20 PFU/cell in the further passages. The HA/PFU ratio also varied, but was generally at least 10-fold higher than in the normal infection. The culture fluid from the sixth passage caused a 1000-fold reduction in the yield of standard SFV, indicating the presence of defective interfering particles.

The intracellular RNAs were isolated at different passages from cells labeled with ^{3}H uridine in the presence of actinomycin D. A drastic reduction of the synthesis of 42S and 26S RNAs, with a concomitant appearance of a new 18S single-stranded RNA species was observed in the fifth passage. Further passaging affected neither the size distribution nor the proportion of the 18S RNA relative to the 42S and 26S RNAs. Figure 1

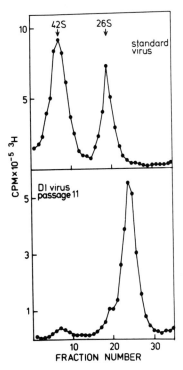

FIG. 1. Sucrose gradient sedimentation RNAs isolated from cells infected with either standard (upper panel) or the tenth undiluted passage (lower panel) of SFV. Intracellular RNAs were labeled with ³H uridine in the presence of actinomycin D. The cytoplasmic extract was prepared at 6 hours after infection and the RNA fractionated on a 15–30% sucrose gradient in a SW27 rotor.

shows the situation at the eleventh passage. Analogous to previous reports (Kennedy, 1976; Stark and Kennedy, 1978) we consider this 18S RNA to represent a defective interfering RNA.

The size of the RNA was also determined in a denaturing agarose gel containing 5 mM methyl mercury hydroxide (Bailey and Davidson, 1976). The DI RNA was uniform in size and comigrated with the 18S ribosomal RNA marker run in a separate gel (Fig. 2).

FINGERPRINT ANALYSIS OF THE RNA

It is thought that the extreme 5' and 3' ends of the 42S RNA are conserved in alphavirus DI RNAs, while internal sequences have been deleted. These conclusions have been drawn from both nucleic acid hybridization (Guild

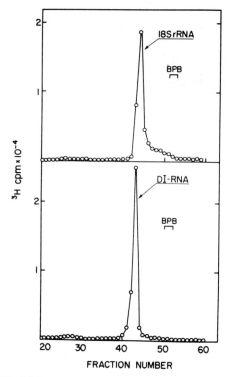

FIG. 2. Analysis of SFV DI RNA on a denaturing agarose gel. Purified ³H-uridine-labeled DI RNA was run in a cylindrical 1 % agarose gel containing 5 mM CH₃HgOH according to Bailey and Davidson (1976). ³H-uridine-labeled 18S ribosomal RNA was run in a separate gel. The gels were sliced in 2 mm segments and the radioactivity determined. BPB = bromphenol blue.

and Stollar, 1977) and oligonucleotide fingerprinting (Kennedy, 1976; Dohner *et al.*, 1979; Stark and Kennedy, 1978) studies. To study whether our 18S DI RNA had the same structure, we analyzed the ³²P-labeled RNase T1-oligonucleotides by two-dimensional polyacrylamide gel electrophoresis and compared the fingerprints with those obtained from 42S and 26S RNA. The complexity of the DI RNA (Fig. 3) was clearly lower than that of the 42S or 26S RNAs (Fig. 4). As has been shown previously (Kennedy, 1976; Wengler and Wengler, 1976) the 26S RNA spots representing long T1-oligonucleotides were found in the fingerprint of the 42S RNA. This is due to the fact that the 26S RNA is identical to the 3′ third of the 42S RNA. To determine the origin of the DI RNA oligonucleotides, they, as well as a large number of oligonucleotides from the 42S and 26S RNAs were eluted and redigested with RNase A. By comparing the digestion products

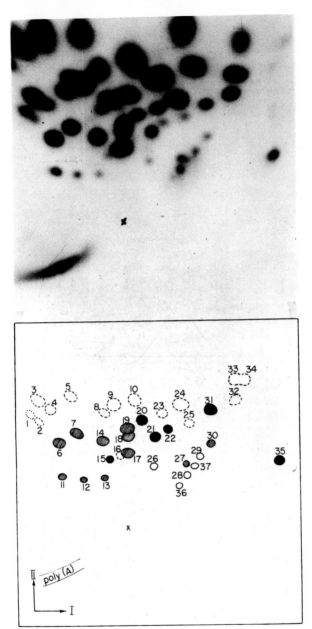

FIG. 3. Two-dimensional polyacrylamide gel analysis of T1-oligonucleotides from DI RNA. DI RNA, labeled with ^{32}P, was digested with RNase T1 and the oligonucleotides fractionated on a two-dimensional gel (upper panel) (De Wachter and Fiers, 1972; Pettersson et al., 1977). The oligonucleotides were eluted and further characterized by digestion with RNase A (Barrell, 1971; Pettersson et al., 1977). By comparing the digestion products with those obtained from oligonucleotides of the 42S and 26S RNAs (Fig. 4) the origin of the DI RNA spots could be determined (lower panel). Shaded circles = oligonucleotides derived from the 42S RNA-specific portion of the genome. Filled circles = oligonucleotides common to 42S and 26S RNAs (3′ third of the genome). Open circles = oligonucleotides not present in 42S or 26S RNA (DI RNA-specific spots). Broken circles = origin not determined. The spots have been arbitrarily numbered.

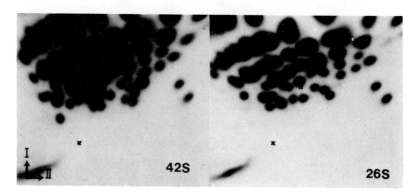

FIG. 4. Fingerprints of SFV 42S and 26S RNAs. ^{32}P-labeled 42S and 26S RNAs purified from SFV-infected cells were digested with RNase T1 and analyzed by two-dimensional gel electrophoresis.

analyzed by ionophoresis on DEAE paper, and the location of the spots, it was possible to assign several oligonucleotides (spots 6, 7, 11, 12, 13, 14, 17, 18, 19, 27 and 30 in Fig. 3, shaded circles) as being derived from the 42S RNA-specific region of the genome (5' two-thirds) and a few oligonucleotides (spots 15, 20, 21, 22, 31 and 35 in Fig. 3, filled circles) as being derived from the region common to both 42S and 26S RNAs. In addition, spot 26 and a cluster of spots (28, 29, 36 and 37 in Fig. 3, open circles) were absent from the 42S and 26S RNA fingerprints and therefore are DI RNA-specific (see below). Thus, the bulk of the DI RNA sequences was found to be derived from the 42S RNA-specific portion of the genome.

We have recently found that oligonucleotide 35 in the DI RNA is located about 20 nucleotides from the poly(A) tract (Söderlund et al., this volume). Since the DI RNA is only about 1800 nucleotides long, and the 26S RNA about 4500 nucleotides long, this clearly shows that the DI RNA contains nucleotide sequences derived from the extreme 3' end, as well as from some region of the 5'-terminal portion of the genome, with one or several large internal deletions comprising most of the 26S RNA region.

HETEROGENEITY OF DI RNA

The spots in the 42S and 26S RNA fingerprints representing long oligonucleotides had, with a few exceptions (double spots), roughly the same intensities in the autoradiogram (Fig. 4), indicating that they were present in equimolar ratios. This was also confirmed by direct quantitation of the radioactivity in the spots and determination of their molar yields. In contrast, the intensities of the spots in the DI RNA fingerprint varied markedly. Some long oligonucleotides (more than 10 bases) were present in large amounts (e.g. spots 6, 7, 14, 17, 18 and 19), whereas others were present in small amounts (e.g. spots 11, 12, 13, 15, 22 and 26). To calculate the molar yields of the oligonucleotides, they were cut out, the radioactivity determined and the length of the oligonucleotides estimated from the RNase A digestion products. Table I summarizes the results for some of the oligonucleotides. Spot 6 was taken as reference, since it was the most abundant. All spots listed in Table I were found to consist of only one species of oligonucleotide. These results confirmed the non-equimolarity of the oligonucleotides.

The results could be explained in several ways: (i) The DI RNA may have been contaminated with 42S RNA-specific sequences. This is highly unlikely, since the DI RNA was both size- and oligo(dT)cellulose-selected. In addition, very little 42S RNA was made under the conditions used (Fig. 1). Finally, only a selected set of 42S RNA-specific oligonucleotides were seen in the fingerprint. One would have expected to see all 42S RNA-specific spots if there had been random contamination with 42S RNA sequences. (ii) The DI RNA could have been either under- or overdigested by RNase T1, resulting in either partial digestion products or cleavage at bases other than G. This also seems excluded, since secondary digestions with RNase A revealed the presence of only one G residue per T1-oligonucleotide. Neither did changes in the enzyme:substrate ratio alter the

TABLE I

Molar yields of some RNase T1-oligonucleotides from
SFV DI RNA

Spot number	Length in nucleotides[a]	Origin[b]	Molar yield[c]
6	14	42S	1.0
11	20	42S	0.1
12	19	42S	0.1
14	13	42S	0.8
18	12	42S	0.3
19	11	42S	0.6
20	10	26S	0.2
21	12	26S	0.2
30	15	42S	0.1
31	9	26S	0.3
35	17	26S	0.2

[a] Determined from RNase A digestion products.
[b] 42S = 42S RNA-specific region (5' two-thirds of the genome), 26S = 3' third of the genome (common to 42S and 26S RNA).
[c] Relative to oligonucleotide 6, which is the most abundant.

results. (iii) Some of the oligonucleotides may be present in more than one copy per molecule. Such a situation would be seen if there are repeated sequences in the RNA. We have recently found that several of the supramolar oligonucleotides (e.g. spots 6, 7 and 14) are, indeed, included in a sequence that is repeated in the DI RNA (Söderlund et al., this volume). (iv) The DI RNA could be heterogeneous. There could be several species of DI RNAs all having roughly the same size. The supramolar oligonucleotides could be common to most or all of them, whereas the submolar ones could be unique to the different subspecies of RNAs. This interpretation would mean that the location of the internal deletion must vary from one DI RNA to another. Since the presence of repeats in the DI RNA does not satisfactorily explain the non-equimolar yield of the oligonucleotides (Table I), we interpret our results to mean that the DI RNA is in fact heterogeneous and consists of RNA molecules of about the same size, but with a partially different nucleotide sequence.

THE 5'-TERMINAL NUCLEOTIDE SEQUENCE OF DI RNA IS HETEROGENEOUS

We (Pettersson et al., 1980) and others (Wengler and Wengler, 1979) have recently shown that the 5' terminus of the 42S RNA is ^{7}mGpppAUG. If the DI RNA did indeed contain a conserved extreme 5' end of the 42S RNA as has been proposed (Guild and Stollar, 1977; Stark and Kennedy, 1978) then it should also have the same 5'-terminal structure. To study this, we isolated and characterized the 5'-terminal RNase T1-oligonucleotide of the DI RNA. First, we determined whether the DI RNA was capped using methods described previously (Pettersson et al., 1980). We found that all RNA molecules had a cap of the same structure as the 42S and 26S RNAs (^{7}mGpppAp) (Pettersson, 1981). Using this cap as a marker, we then searched for the cap-containing T1-oligonucleotide (T1-cap) in the fingerprint. This was done by cutting out and eluting the various oligonucleotides from the two-dimensional gel, followed by digestion with a mixture of RNase T1, T2 and A and ionophoresis on DEAE paper of the digestion products. No T1-caps migrating to the position of the ^{7}mGpppAUG (42S) or ^{7}mGpppAUUG (26S) (Fig. 5, upper right corner) were found in the DI RNA fingerprint. Instead, the cap was recovered from a series of spots, which migrated slower in both dimensions than the 42S and 26S RNA T1-caps (Fig. 5, indicated by arrows). These spots were absent from the 42S and 26S RNA fingerprints (Fig. 4). This suggested that the 5'-terminal nucleotide sequence was heterogeneous and different from that of the 42S RNA. To determine the sequence of the T1-caps, they were eluted from the gel and

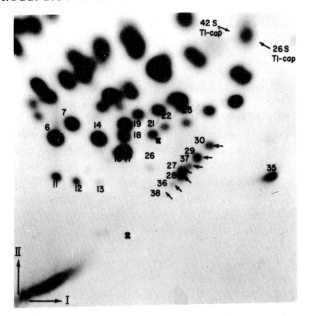

FIG. 5. Fingerprints of DI RNA showing the position of the cap-containing RNase T1-oligonucleotides (T1-caps) (numbered spots indicated by arrows). The position of the T1-caps from the 42S and 26S RNAs are as indicated.

digested with either RNase A or U2. The sequences deduced from these digestions are summarized in Table II. The T1-caps all had the same general sequence: cap-AU(AU)$_n$CAUG and differed from each other only in the number of AU-residues ($n = 4$–8). The exact location of the C residue could not be determined, since the enzyme digestion results were compatible with the C being located between any of the AU residues. We have tentatively

TABLE II

Sequence of the cap-containing T1-oligonucleotides (T1-caps) of SFV DI RNA

Spot number	Sequence[a]	Length[b]	Relative molar yield
29	cap-AU(AU)$_4$CAUG	14	1.00
37	cap-AU(AU)$_5$CAUG	16	0.36
28	cap-AU(AU)$_6$CAUG	18	0.79
36	cap-AU(AU)$_7$CAUG	20	0.17
38	cap-AU(AU)$_8$CAUG	22	0.02

[a] Deduced from the RNase A and U2 digestion products; cap = ^{7}mGppp.
[b] Length in nucleotides excluding the ^{7}mG in the cap.

placed it between the two most 3'-located AU residues. Table II also shows the relative molar yield of the T1-caps. Oligonucleotides 29 and 28 ($n = 4$ and 6) were the most abundant and comprised about 75% of the total cap recovered. Oligonucleotide 38 was present in very small amounts.

At present we do not know how such aberrant 5'-terminal sequences could arise. One possible way by which a variable number of AU residues could be synthesized is by the polymerase initiating on the template RNA at different positions of an uninterrupted AU stretch. This would result in a variable number of AU residues proximal to the cap. Since both the RNA population and the 5'-terminal sequence are heterogeneous, it is possible that each sub-species of RNA possesses one of the different 5'-terminal T1-caps. It will be interesting to see whether the results reported here also apply to alphavirus DI RNAs isolated in other laboratories or whether our 18S RNA represents a unique case.

CONCLUSIONS

Here we summarize studies on the structure of an 18S, about 1800 nucleotides long, defective interfering RNA isolated from Semliki Forest virus-infected BHK21 cells. In conformity with previous reports (Kennedy, 1976; Stark and Kennedy, 1978), the DI RNA contained sequences derived from the extreme 3' end of the 42S RNA and from some region(s) located in the 5' two-thirds of the 42S RNA. Most of the 26S RNA and a substantial portion of the 42S RNA sequences have therefore been deleted. In contrast to previous reports, we have found heterogeneity in our DI RNA preparation. Although the DI RNA is rather uniform in size, it appears to contain a mixture of RNA species with partially different nucleotide sequences. These RNAs may contain sequences common to most or all of the RNA species, whereas some sequences may be unique to each RNA subspecies. This would mean that the location of the internal deletion(s) varies from one DI RNA to another.

The 5'-terminal cap-containing RNase T1-oligonucleotide (T1-cap) was also found to be heterogeneous, having the general sequence cap-$AU(AU)_n CAUG$ ($n = 4–8$). The two most abundant T1-caps ($n = 4$ and 6) comprised about 75% of the total yield of T1-caps. To our knowledge such a heterogeneous 5'-terminal sequence has not been reported previously for alphavirus DI RNAs. It remains to be seen whether these findings are more general among alphavirus DI RNAs, or whether our 18S RNA represents a special situation. Since the T1-caps are different from those found at the 5' end of the genomic 42S RNA (cap-AUG), it appears that the extreme 5'-terminal nucleotide sequence of the 42S RNA is not conserved in the DI RNA.

REFERENCES

Bailey, J. M. and Davidson, N. (1976). *Analyt. Biochem.* **70**, 75–85.

Barrell, B. G. (1971). *In* "Procedures in Nucleic Acids Research", 2 (G. L. Cantoni and D. R. Davies, eds) pp. 751–828. Harper and Row, New York.

Bruton, C. J. and Kennedy, S. I. T. (1976). *J. gen. Virol.* **31**, 383–395.

De Wachter, R. and Fiers, W. (1972). *Analyt. Biochem.* **49**, 184–197.

Dohner, D., Monroe, S., Weiss, B. and Schlesinger, S. (1979). *J. Virol.* **29**, 794–798.

Guild, G. M. and Stollar, V. (1977). *Virology* **77**, 175–188.

Huang, A. and Baltimore, D. (1977). *In* "Comprehensive Virology", 10 (H. Fraenkel-Conrat and R. R. Wagner, eds) pp. 73–116. Plenum Press, New York.

Kääriäinen, L., Simons, K. and von Bonsdorff, C.-H. (1969). *Ann. Med. exp. Biol. Fenn.* **47**, 235–248.

Kennedy, S. I. T. (1976). *J. molec. Biol.* **108**, 491–511.

Leppert, M., Kort, L. and Kolakovsky, D. (1977). *Cell* **12**, 539–552.

Lundqvist, R. E., Sullivan, M. and Maizel, J. V. Jr. (1979). *Cell* **18**, 759–769.

Nomoto, A., Jacobson, A., Lee, Y. F., Dunn, J. and Wimmer, E. (1979). *J. molec. Biol.* **128**, 179–196.

Perrault, J. and Semler, B. L. (1979). *Proc. natn. Acad. Sci. U.S.A.* **76**, 6191–6195.

Petterrson, R. F. (1981). *Proc. natn. Acad. Sci. U.S.A.* **78**, 115–119.

Pettersson, R. F., Hewlett, M. J., Baltimore, D. and Coffin, J. M. (1977). *Cell* **11**, 51–63.

Pettersson, R. F., Söderlund, H. and Kääriäinen, L. (1980). *Eur. J. Biochem.* **105**, 435–443.

Simmons, D. T. and Strauss, J. H. (1972). *J. molec. Biol.* **71**, 615–631.

Stark, C. and Kennedy, S. I. T. (1978). *Virology* **89**, 285–299.

Wengler, G. and Wengler, G. (1976). *Virology* **73**, 190–199.

Wengler, G. and Wengler, G. (1979). *Nature, Lond.* **282**, 754–756.

Structural Analysis of Semliki Forest Virus Defective Interfering RNA Using DNA Copies

HANS SÖDERLUND, SIRKKA KERÄNEN,
PÄIVI LEHTOVAARA, ILKKA PALVA and
RALF F. PETTERSSON

*Recombinant DNA Laboratory, Department of Bacteriology and Immunology,
University of Helsinki, Haartmaninkatu 3, 00290 Helsinki 29, Finland*

INTRODUCTION

The genome of Semliki Forest virus is a single-stranded RNA molecule of about 13 000 nucleotides in length. This molecule serves as the mRNA for the virus-specific enzymes and regulatory proteins required for viral RNA synthesis. In normal productive infection another major RNA molecule is also synthesized; this 26S RNA is a copy of the 3' third of the genome and codes for the viral structural proteins (Kääriäinen and Söderlund, 1978). Additionally, under certain conditions defective interfering RNAs are formed in the infected cells (Pettersson and Kääriäinen, this volume). These RNAs are considered to be molecules created by deletion of up to 85% of the genome. In contrast to the 26S RNA, the DI RNA molecules, however, have retained the capacity to act as a template for the RNA polymerase and to bind capsid protein in nucleocapsid assembly. These properties make the structure of the DI RNAs quite interesting.

Consequently we recently initiated a study with the aim of sequencing cloned cDNA copies of DI RNA. We have partially sequenced one such cloned DI cDNA molecule. This molecule is about 1650 nucleotides long and contains a triplicated linear repeat. Its 3' end is identical to that of the standard viral genome. Restriction endonuclease analysis of cloned and uncloned cDNA indicates, however, that several non-identical DI RNA structures have been generated in the infected cells, and that the DI RNA population thus is heterogeneous.

CLONING OF A DI RNA cDNA COPY

Defective interfering (DI) RNA was extracted from BHK-21 cells which had been infected with a preparation of Semliki Forest virus containing DI particles. The 18S DI RNA, corresponding to that described in the previous chapter by Pettersson and Kääriäinen, was then purified by sucrose gradient centrifugation. Using this RNA as template, a cDNA copy was synthesized using reverse transcriptase and oligo dT_{14-18} as primer. An aliquot of the product was analyzed by electrophoresis in an alkaline agarose gel. Two major bands, about 1950 and 1700 nucleotides long were seen, in addition to small-sized material. The bulk of the material was treated with alkali to destroy the RNA template and a second strand synthesized with reverse transcriptase. The duplex DNA, which again contained the two species of about 1700 and 1950 base pairs, was then treated with nuclease S1. Thereafter large-sized DNA, sedimenting at 10–15S was selected on a sucrose gradient. The removal of the single-stranded regions by S1 nuclease somewhat decreased the size of the double-stranded cDNA material (Fig. 1).

The DNA prepared in this manner was then inserted into the Pst I site of pBR322 using the dG–dC tailing method. The hybrid plasmid was used to transform *E. coli* $\chi1776$ (kindly obtained from Roy Curtiss III). Transformants containing SFV-specific DNA was selected by colony hybridization (Grünstein and Hogness, 1975), and one clone, pKTH301 containing a 1650 bp insert, was chosen for further characterization.

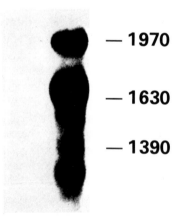

— 1970

— 1630

— 1390

FIG. 1. Analysis of double-stranded cDNA synthesized with reverse transcriptase from Semliki Forest virus 18S defective interfering RNA. After digestion with nuclease S1, 10–15S material was selected on a sucrose gradient. The sample was electrophoresed on a neutral 1.4% agarose gel. The molecular weight scale bars (in base-pairs) are from an Eco RI/Hind III digest of phage λ DNA run in parallel.

PARTIAL SEQUENCING OF THE INSERT

After the preliminary characterization the recombinant plasmid was transformed into *E. coli* HB 101 for propagation. A restriction endonuclease map was constructed by comparing the size of single- and double-digestion products, and by a second digestion of isolated fragments. The orientation of the insert was determined by Southern blot hybridization (Southern, 1975) of the Pst I fragments against ^{32}P-labeled 26S RNA. This probe gives the 3′ end (in RNA-polarity) of the insert (Fig. 2).

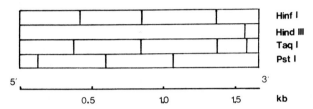

FIG. 2. The restriction endonuclease map of the DI RNA-specific insert in pKTH301. The 5′ end of the RNA is to the left.

Sequencing, by the Maxam and Gilbert procedure (1980), from the Hind III site shows that the 3′ end of the DI RNA cloned in pKTH301 is identical and coterminal with the genome RNA (Fig. 3). The A residue before the C_{13} tail is equivalent to the first A in the poly(A) tract of the 3′ end of the genomic RNA (Ou *et al.*, 1980).

However, the region of direct identity between the 3′ ends of cloned DI RNA and the 42S/26S RNA, is maximally the (roughly) 100 nucleotides

A G C T T A A T T C G A C G A A T A A T T G G A T T T T T A

T T T T A T T T T G C A A T T G G T T T T T A A T A T T T C

C C A C_{13}

FIG. 3. The nucleotide sequence of the 3′ end of the DI RNA-specific insert in pKTH301.

preceding the poly(A) tract. At this point, an Ava I site is present in cloned 26S RNA (Garoff *et al.*, 1980), where it is lacking in the DI RNA clone pKTH301 (Fig. 4).

The restriction map (Fig. 2) shows a striking regularity between some sites. For instance, a Taq I site is directly to the left of all three Hinf I sites. Sequencing from the Taq I sites gave the explanation for this: the Taq I sites at position 380, 850 and 1370 are flanked by identical (or nearly

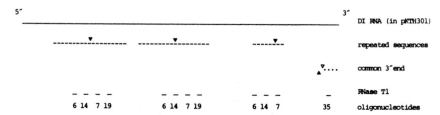

FIG. 4. Some of the structural features of the DI RNA-specific insert in pKTH301, schematically presented. The (———) lines represent the triple repeat around the common "fix point", a Taq I site (▼). The (··) line represents the 3' part of the molecule which is identical with the 3' end of the viral genome and 26S RNAs (Fig. 3). The (▽) is the Hind III site common to genome and DI RNA while the (▲) stands for the Ava I site present only in genome and 26S RNA. The numbered dashes (-) show the localization of DNA sequences which correspond to certain T1-oligonucleotides of DI RNA (Table I).

identical) sequences. Thus, the central part of the insert in pKTH301 is composed of a triplicated linear repeat which is apparently several hundred base-pairs long (Fig. 4).

ARE THE DI RNA AND THE CLONED cDNA COLINEAR?

The peculiar structure of the cloned cDNA raised the question of whether the structure actually represents that of the DI RNA. It is known that at least inverted duplicates may be created in the process of molecular cloning (Derynck *et al.*, 1980). We have, however, reasons to believe that the plasmid pKTH301 contains a faithful copy of a DI RNA molecule.

One set of evidence is obtained by an S1 nuclease protection experiment. A fragment of pKTH301, containing the entire insert together with some 870 bp of the flanking pBR322 sequences was isolated. This was then hybridized to ^{32}P-labeled DI RNA in conditions in which only DNA–RNA hybrids are stable (Casey and Davidson, 1977) (70% formamide, 0.4M NaCl, 0.1M MOPS pH7 and 0.01M EDTA at 56°C). The hybridization mixture was cooled on ice, diluted 10-fold in 0.3M NaCl, 0.03M Na-acetate, 3 mM ZnCl$_2$ and extensively digested with nuclease S1. The DNA–RNA hybrid was then electrophoresed in an agarose gel. Autoradiography revealed one main band with an apparent size of 1720 bp, compared to ds DNA markers. Several bands of smaller material could also be detected. The result suggests that a 1650-bp-long uninterrupted hybrid can be created between DI RNA and pKTH301 DNA.

Another set of evidence is obtained from comparison of the sequence with the T1-oligonucleotide composition of DI RNA preparations. The composition of these spots was determined by analyzing their RNase A digestion products. At the very 3' end there is a 17-nucleotides-long potential T1-oligonucleotide in the DNA sequence. This sequence matches perfectly the composition of T1 spot 35. However, the spot appears in a molar ratio of about 0.2 compared to spot 6 (Pettersson and Kääriäinen, this volume) which is the most abundant one. Spot 35 is close to the region which Ou *et al.* (1980) postulate to be essential for RNA replication or encapsidation. Thus it may be present in every DI RNA molecule even if the DI population is heterogeneous (see below). On the other hand, three of the spots present in hypermolar amounts compared to spot 35 can be found in the triplicated region (Table I, Fig. 4). These data therefore suggest that the triplicate is present in the DI RNA itself.

TABLE I

Comparison of the nucleotide sequences in DI RNA and in pKTH301

Sequence found in pKTH301	Composition of RNase T1-oligonucleotide[a]	Spot number[b]	Relative molar yield[c]
GATTAACCACCCAGG	AAC, AU, $(AC)_2$, C_3, U, G	6	5
GTCACACCAAATG	AAAU, $(AC)_2$, C_2, U, G	7	3
GCCCATTCATCAAG	$(AU)_2$, C_5, U, AAG	14	4
GCTACCAAATTG	AAAU, AC, C_3, U_2, G	19	3
GATTTTTATTTTATTTTG	$(AU)_4$, U_{10}, G	35	1

[a] ^{32}P DI RNA was digested with RNase T1 and the oligonucleotides fractionated by two-dimensional electrophoresis. Large oligonucleotides were eluted and redigested with RNase A. The products of this digestion were then separated and quantitated (Pettersson, 1980).

[b] "Spot number" refers to the numbers in the T1 fingerprints presented in the previous chapter (Pettersson and Kääriäinen).

[c] The molar ratios were determined from the ^{32}P activity in each spot. Unequal labeling and varying yields introduce some uncertainty in the values.

HETEROGENEITY OF THE cDNA

T1 fingerprint analysis has suggested that the DI RNA of SFV is heterogeneous (Pettersson and Kääriäinen, this volume; Pettersson, 1980). Additional information on the heterogeneity should be obtained from restriction enzyme patterns of uncloned double-stranded cDNA prepared from DI RNA. If the RNA population (and therefore the ds cDNA) is indeed heterogeneous, a complex restriction enzyme digestion pattern should be obtained. A prerequisite for this approach is that the cDNA preparations contain only minor amounts of contaminating reverse transcripts of host-cell RNA.

To exclude this source of error we isolated the two major ^{32}P-labeled single-stranded cDNA species from an alkaline agarose gel and hybridized them against an excess of viral 42S RNA. After this the amount of cDNA resistant to nuclease S1 was determined. According to this assay, more than 90% of the cDNA is virus-specific.

TABLE II

Heterogeneity of DI RNA as estimated by restriction endonuclease digestion of the 1650 bp cDNA.

Fragment size (base pairs)	Molar ratio[a]	Present in pKTH301[b]
Msp I		
1100	0.9	—
870	2.5	—
740	2.8	—
610	2.0	—
520	6.3	+
480	4.0	+
430	4.2	+
350	1.1	—
290	2.2	—
220	3.4	+
Pst I		
1650	1.8	—
1300	0.9	—
1000	1.0	—
780	3.8	—
620	5.3	—
580	4.2	+
520	3.1	—
470	7.9	+ +
290	2.6	—
170	2.0	—
120	3.6	+
Bgl I		
1650	9.3	+
920	6.7	—
740	6.4	—

[a] This figure is obtained by dividing the ^{32}P activity in each fragment by its size in base-pairs.

[b] (+) Indicates fragments found in the DI RNA-specific insert in pKTH301 (cf. Fig. 2). (+ +) Indicates a band containing two fragments from pKTH301.

The labeled double-stranded cDNA of the two classes 1700 bp and 1900 bp was then isolated on a neutral agarose gel. The DNA was eluted from the gel and digested with different restriction enzymes. Unlabeled pBR322 or phage λ DNA was mixed with the samples as an internal control for complete digestion. The sizes and relative amounts of restriction fragments were estimated after electrophoresis and autoradiography.

The sum of the fragment sizes was greater than the 1700 bp–1900 bp expected for a complete digest of a homogeneous molecule. Neither were the fragments present in molar ratios (Table II). This result is in accordance with the idea that the DI population is heterogeneous.

CONCLUSIONS

Our data suggests that the DI RNA of Semliki Forest virus is actually a mixture of several RNAs of essentially the same size. This conclusion is based on the analysis of uncloned double-stranded cDNA, so the heterogeneity may be an artefact created by the reverse transcription. Both the single-stranded and the double-stranded cDNA contain, however, the major 1700- and 1900-base molecules. Reverse transcription artefacts presumably arise at the stage of second-strand synthesis (W. Fiers, presented at this symposium). Furthermore, the T1 fingerprints of the DI RNA also indicate heterogeneity (Pettersson and Kääriäinen, this volume). The digestion of the cDNA with Bgl I divides the population in two; only about one third of the molecules contain a Bgl I site, two-thirds lack this site (Table II). The other restriction enzymes used, however, show a more complex pattern which cannot directly be recognized as the one created from a mixture of only two or three DI species. In fact there may be a large number of different RNA species within the DI RNA population.

Does the cloned cDNA then represent any significant part of the heterogeneous DI RNA? We assume that it does because all of the large RNase T1-oligonucleotides which can be identified from the sequence of the cDNA, have also been found in the T1 fingerprint. They are also present in high molar ratios (Table I). This indicates that at least significant parts of the cDNA base sequence are present in many DI RNA molecules.

The triple linear repeat of the cloned DI RNA is one sequence segment which is probably present in most of the DI RNAs, since T1-oligonucleotides from this repeat are obviously over-represented in the fingerprint. The significance of this repeated nucleotide sequence remains, however, obscure. It could equally well be essential for the replicative cycle of the DI particles involved in the mechanism giving rise to the large deletions. That the extreme 3′ end is identical with that of the SFV genome is in line with the finding that this part is conserved in several alphaviruses (Ou *et al.*, 1980), and strengthens the conclusion that this segment is involved in either RNA-replication or encapsidation.

ACKNOWLEDGEMENTS

This investigation was funded by the Finnish National Fund for Research and Development. The cloning experiments were performed under P2/EK2 containments until the clones were defined; then under P2/EK1 according to the instructions of the Finnish National Board of Health Committee for Recombinant DNA research.

REFERENCES

Casey, J. and Davidson, N. (1977). *Nucl. Acids Res.* **4**, 1539–1552.
Derycnk, R., Content, J., DeCleracq, E., Volckaert, G., Tavernier, J., Devos, R. and Fiers, W. (1980). *Nature, Lond.* **285**, 542–547.
Garoff, H., Frischauf, A.-M., Simons, K., Lehrach, H. and Delius, H. (1980). *Nature, Lond.* **288**, 236–241.
Grünstein, M. and Hogness, D. S. (1975). *Proc. natn. Acad. Sci. U.S.A.* **72**, 3961–3965.
Kääriäinen, L. and Söderlund, H. (1978). *Curr. Top. Microbiol. Immunol.* **82**, 15–69.
Maxam, A. M. and Gilbert, W. (1980). *Meth. Enzymol.* **65**, 499–560.
Ou, J.-H., Strauss, E. G. and Strauss, J. H. (1980). *Virology* **109**, 281–289.
Pettersson, R. F. (1980). *Proc. natn. Acad. Sci. U.S.A.* **78**, 115–119.
Southern, E. M. (1975). *J. molec. Biol.* **98**, 503–517.

IV

Structure and Expression of Eukaryotic Genes

Structure, Organization and Developmental Expression of the Chorion Multigene Families

F. C. KAFATOS

Department of Cellular and Developmental Biology, Harvard University, Cambridge, Massachusetts 02138, U.S.A.

INTRODUCTION

My laboratory has been studying eukaryotic gene evolution and developmental expression using the silk-moth chorion (eggshell) as a model system. The chorion is produced, towards the end of oogenesis, by a monolayer of follicular epithelial cells surrounding the single oocyte of each follicle. We have studied the proteins that constitute the chorion, their biosynthesis, and the structure, organization and evolution of the corresponding structural genes. I shall summarize here our main findings.

BIOCHEMICAL COMPLEXITY OF THE SILK-MOTH CHORION AND ITS GENETIC BASIS

Because of its large size and solubility properties, the silk-moth chorion can easily be purified, both from individual eggs and on a mass scale. Typically, it is approximately 2 mm in diameter and has a dry weight of approximately 0.5 mg. The chorion is insoluble unless treated with both a denaturant and a reducing agent. Thus, the rest of the egg can be washed thoroughly away before the pure chorion is collected and dissolved for biochemical analysis (Paul et al., 1972; Kafatos et al., 1977).

The chorion is almost completely proteinaceous (c. 96%). When the chorion proteins are resolved on high-resolution two-dimensional gels, an amazing complexity becomes evident. As many as 186 protein spots have been recognized from a single chorion of *Antheraea polyphemus* (Regier, Mazur and Kafatos, 1980). These are largely clustered within certain molecular weight ranges, and have been classified accordingly into classes A (*c*. 9000), B (*c*. 13 000), C (*c*. 18 000) and D. An additional class, E, has been defined recently on the basis of unusual solubility properties (Regier *et al.*, 1980; Mazur *et al.*, 1980).

Chorion components, like other secretory proteins, are synthesized in the form of precursors bearing N-terminal signal peptides (Thireos and Kafatos, 1980). These peptides appear to be processed away during or very shortly after translation, since the precursors are not observed in *in vivo* experiments (they have only been detected by cell-free translation of chorion mRNAs). In addition to removal of the signal peptide, many chorion proteins (including most of the Bs but probably none of the As) are subjected to further, and somewhat slower, post-translational modification: this modification alters the charge but not the molecular weight of these proteins, and apparently corresponds to cyclization of an N-terminal glutamine (Regier and Kafatos, 1981).

Despite the existence of these modifications, the large number of proteins recognizable in mature chorions is genetically based. Indirect evidence in favor of this interpretation came from biosynthetic studies, including pulse-chase and developmental specificity experiments (see below). Definitive proof came from sequencing studies on the A and B families (Regier *et al.*, 1978a; Regier *et al.*, 1978b; Rodakis, 1978; Rodakis *et al.*, 1981). Eleven A proteins and five B proteins of *A. polyphemus* have been sequenced for 18 or more residues each, and proven to differ by amino acid replacements or insertions/deletions. Therefore, these proteins, and presumably other major chorion proteins as well, must be encoded by distinct DNA sequences. Equally important, however, is the observation that the sequences of each size class are highly related, i.e., must be evolutionarily homologous. This leads us to the concept that the chorion proteins are encoded by families of reduplicated genes (multigene families). The A and B proteins are sufficiently different from each other to be considered products of different multigene families. The amino acid compositions of most C and E proteins are quite different from the compositions of A and B proteins, suggesting the existence of additional multigene families (Regier *et al.*, 1980).

A significant proportion of chorion proteins show polymorphism between individuals in a wild population of silk moths, in terms of charge, size or quantity (Regier *et al.*, 1978a; Rodakis, 1978). Similarly, inbred strains of the cultivated silk moth, *Bombyx mori*, have clearly distinct chorion protein patterns, although their chorion morphology is normal (Goldsmith and Basehoar, 1978). A genetic basis for chorion protein polymorphism has also been suggested by sequencing studies. The single most abundant protein of the *A. polyphemus* chorion, B6--f1, has been sequenced three times, both from individual moths and from pooled samples; each of the determined sequences differed from the other two by a small number of amino acid replacements (Jones *et al.*, 1979).

DEVELOPMENTAL CONTROL OF SPECIFIC CHORION PROTEIN SYNTHESIS

In terms of protein synthesis, the multiple chorion genes are not all expressed in parallel during development. When a series of follicles at progressively more advanced developmental stages are labeled with an amino acid label, the newly synthesized chorion proteins of each show a distinctive and characteristic pattern (Kafatos *et al.*, 1977; Paul and Kafatos, 1975). Individual chorion proteins are synthesized in overlapping succession, during characteristic developmental periods, ranging from a few hours to more than half of the total 51-hour period of choriogenesis. In broad terms, four major developmental groups of proteins have been recognized: early middle, late and very late (Sim *et al.*, 1979). For example, most C proteins are early, although a few are very late. B and A proteins account for the bulk of the middle and late proteins, although early and very late components also exist within these size classes. Throughout most of choriogenesis, chorion protein synthesis accounts for approximately 95% of total protein synthesis (Paul *et al.*, 1972; Kafatos *et al.*, 1977; Paul and Kafatos, 1975).

Presumably, the developmental specificity of the chorion proteins is related to their function. For example (Regier *et al.*, 1980), the very late proteins are regionally localized, and appear to form the protruding crowns of respiratory openings (aeropyles). Proteins of the E class constitute an architectural "filler" which helps mold the intricate shape of the aeropyle crowns (Mazur *et al.*, 1980).

In summary, the silk-moth chorion is encoded by a large number of developmentally regulated genes, belonging to several multigene families. Work at the nucleic acid level, to be reviewed in the following sections, has illuminated the detailed structure of these genes and the manner in which they are organized within the chromosome—leading to specific ideas about how they evolve and how they are regulated during development.

GENERATION OF A LIBRARY OF CLONED cDNA PROBES, AND THEIR EVOLUTIONARY AND DEVELOPMENTAL CHARACTERIZATION

Because of the high degree of biological specialization of the secretory follicular cells, chorion mRNAs can be purified as a mixture without difficulty (Gelinas and Kafatos, 1973; Gelinas and Kafatos, 1977). By contrast, resolution of these mRNAs into individual components is impossible by physical means because of the multigene nature of the system: the mRNAs are simply too numerous and too similar in size and composition. Recombinant DNA procedures are ideally suited, however, to purification of individual components from this mixture (Sim et al., 1977).

Starting with total chorion mRNA as a template, a population of double-stranded DNA copies (ds-cDNA) was synthesized, using the enzymes reverse transcriptase, DNA polymerase and S1 nuclease (the method was initially developed using the rabbit β-globin mRNA as model; Efstratiadis et al., 1976; Maniatis et al., 1976). These DNAs were inserted in a plasmid vector and cloned in bacteria. By the very nature of this procedure, each transformed bacterial colony contained hybrid plasmids derived from a single molecule, i.e., represented in homogeneity by a single chorion mRNA species. These chorion cDNA clones were analyzed in detail by sequencing and hybridization methods.

Complete sequencing of several clones permitted us to recognize specific features of each multigene family. Both A and B proteins have relatively conservative middle regions flanked by more variable segments (Jones et al., 1979; Tsitilou et al., 1980). The extreme carboxy terminal region is also conservative, however, consisting of tandem Cys·Gly repeats (Cys residues are also clustered near the amino terminus in both families). Computer prediction of secondary structure indicates that the proteins of both families are largely in the β conformation (Hamodrakas et al., 1981). From EM studies (Kafatos et al., 1977; Smith et al., 1971) it is known that the chorion proteins are arranged in fibers; presumably they are cross-linked via their Cys-rich ends.

The complete sequences obtained from the cDNA clones confirmed that A and B are distinct multigene families (Jones et al., 1979; Tsitilou et al., 1980). Certain interfamily similarities are observed, however, in both conservative and variable regions. It is premature to speculate whether these similarities reflect distant evolutionary homology or convergent evolution. Particularly intriguing is the similarity of the amino terminal variable region of the As with the carboxy terminal variable region of the Bs. If this similarity is indicative of homology, it would imply rearrangement of sequence segments of an ancestral gene in the early stages of formation of the two gene families (Jones et al., 1979).

Within each family (Jones et al., 1979; Sim et al., 1979; Jones, 1980), mRNA sequences range from highly homologous (c. 1% mismatching) to highly divergent (c. 50% mismatching; no detectable cross-hybridization under normal conditions). Sequence diversification occurs both by base substitution and by small deletions/insertions, related to tandem and non-tandem direct repeats. A convenient "dot-hybridization" procedure (Kafatos *et al.*, 1979) has been developed for estimating sequence homologies. DNAs from many different clones are spotted on the same nitrocellulose filter, in dots of equal diameter and DNA content, and are then hybridized with a radioactive probe representing one of the sequences. The degree of hybridization, as estimated by autoradiography, is related to the degree of sequence homology (Sim *et al.*, 1979). Our sequence data (Jones *et al.*, 1979; Jones, 1980) have validated the information obtained from dot-hybridization. In short we now have extensive information on the relatedness of a large number of chorion structural genes.

The same technique has been used to determine the developmental specificity of chorion sequences (Sim *et al.*, 1979). Stage-specific mRNA preparations were end-labeled *in vitro* and used as probes in dot-hybridization experiments. From the intensities of the autoradiograms, we could determine when during development each of the genes is expressed most abundantly in mRNAs, and could thus classify them as developmentally early, middle, late or very late.

CHROMOSOMAL ORGANIZATION OF CHORION GENES

Genetic analysis of the chorion protein polymorphisms characteristic of inbred strains of *B. mori* (Goldsmith and Basehoer, 1978; Goldsmith and Clermont-Rattner, 1979a, b) revealed that the chorion structural genes are clustered in a small region of a single chromosome ($n = 28$). The detailed organization of the genes was revealed by analysis of cloned chromosomal DNA segments.

A chromosomal DNA "library" of *A. polyphemus* was constructed (Maniatis *et al.*, 1978), using the Charon 4 derivative of phage λ as vector. A large number of clones containing chorion sequences were obtained. Two clones, each hybridizing with two different, well-characterized cDNA clones were selected and studied in detail (Maniatis *et al.*, 1978; Jones and Kafatos, 1980).

One of these clones, APc110, contains two copies of the B gene, 401, and three copies of the A gene, 18. The 401 and 18 genes are coordinately expressed, and are developmentally late. The other chromosomal clone, APc173, contains two copies of the B gene, 10, and two copies of the A gene, 292. The 10 and 292 genes are coordinately expressed, and are developmentally middle. We see, then, that chorion genes are organized according to the developmental period of their expression, rather than being segregated by evolutionary family.

The A and B genes in each clone alternate and are transcribed from opposite strands. In each case, the genes are very closely linked in divergent pairs (A ↔ B), with their 5' termini separated by only 264 or 325 bp. Each short 5' flanking region, interposed between the A and B genes of a pair, contains two "Hogness boxes" $\left(\text{TATA} {}^{\text{T}}_{\text{A}} \text{AA} \right)$, each located 21 to 23 bp upstream from the corresponding 5' gene terminus. Presumably, therefore, each short 5' flanking region contains promoters for both genes in the pair.

In each clone, the fundamental unit appears to be the gene pair. Two or three copies of each pair are directly linked in the clones discussed thus far. Southern analysis suggests that the haploid genome contains 15 ± 5 copies of the 401/18 gene pair (Jones and Kafatos, 1980). Analysis of additional genomic clones suggests that all of these copies may be linked in a single array (Jones, 1980). The sequence divergence of the gene copies analyzed to date is in the range of 1–6%, both for the genes themselves and for the 5' flanking region. Divergence is due to base substitutions and small insertions/deletions, invariably associated with direct repeats.

Each gene contains a single intron, invariably found in the same location, between codons -4 and -5 of the signal peptide sequence. The introns of the gene copies are very similar, but the introns of different genes (even those belonging to the same family: 401 and 10, or 18 and 292) do not have a recognizable sequence homology. The DNAs immediately flanking the 3' ends of gene copies also show considerable similarity. Sequence comparisons of relatively more divergent gene copies (such as the gene 10 or gene 292 copies) indicate that the divergence is:

$$\text{exons} < 5' \text{ flanking regions} < \text{introns} \approx 3' \text{ flanking regions}$$

The long (1.8–5.5 kb) DNA sequences between copies of gene pairs (i.e., the complete 3' flanking regions) are also recognizably homologous. As a result, the chorion locus consists of tandemly arranged segments approximately 4–8 kb long. Each segment contains a pair of coordinately expressed genes (from two different families), including their introns, a short 5' flanking region, and long 3' flanking regions. These tandem segments are copies of each other, i.e., must have originated by reduplication, but have undergone some divergence.

The most striking differences between adjacent segments are large insertions/deletions (0.065–2.9 kb) which are found within the 3' flanking regions. We will consider the significance of these insertions/deletions in the next section.

We do not know what exists between the tandem arrays of copies of chorion gene-bearing segments. From the genetically determined size of the chorion locus (Goldsmith and Clermont-Rattner, 1979b) and the large number of chorion genes, as well as from "walking" along the chromosome by isolation of overlapping clones (Eickbush *et al.*, 1981), we infer that the tandem arrays of copies are not very distant from each other (we cannot exclude the possibility that they are contiguous). We do suspect, however, that coordinately expressed genes tend to be found close together in the chromosome: in *B. mori* a deletion has been shown to eliminate many different chorion structural genes which are predominantly expressed in parallel during development (Nadel *et al.*, 1980; Iatrou *et al.*, 1980).

SOME SPECULATIONS ON CHORION GENE EVOLUTION AND DEVELOPMENTAL EXPRESSION

From the information we now have on chorion gene organization, we may put forth the following plausible speculations.

Chorion gene arrays undergo expansion and contraction by unequal crossing-over. This process is relatively infrequent, however, permitting accumulation of sequence differences between adjacent copies of the array (Smith, 1973, 1976). The differences may be useful for rapid evolution of the chorion genes, leading to rapid evolution of the chorion structure (Kafatos *et al.*, 1977). The large insertions/deletions in the 3' flanking regions may be involved in modulating downwards the rate of unequal crossing-over. When sufficient differences have accumulated, gene copies can be recognized as distinct genes. Such a mechanism would explain both the large multiplicity of chorion proteins, and their extensive polymorphism.

Although individual genes appear to possess their own promoter, the clustering of coordinately expressed genes, both within a pair and between arrays, suggests the existence of a gross level of gene regulation. For example, long regions of DNA (chromomere-size?) may undergo a coherent change in chromatin structure, becoming available for transcription at the same developmental period. The finer, quantitative control of expression of each gene would reside in the individual promoter.

If such a gross level of regulation operates, one may visualize specific genes changing their temporal characteristics by translocation into a different "chromomere". Such a process would appear to be rare, since homologous sequences are expressed at the same temporal period in both *A. polyphemus* and *A. pernyi* (Moschonas, 1980).

It remains to be seen whether these speculations are valid. Clearly, however, the revolution brought about by recombinant DNA and DNA sequencing procedures has made it possible to formulate and answer important questions on the evolution and developmental regulation of multigene families.

ACKNOWLEDGEMENTS

I am grateful to all members of my laboratory for their participation in this work. Individual contributions are identified by the respective research publications which are listed below. I wish to acknowledge in particular the contributions of Drs C. W. Jones and J. C. Regier, which are extensively reviewed here. Our work has been supported by grants from NSF and NIH.

REFERENCES

Efstratiadis, A., Kafatos, F. C., Maxam, A. M. and Maniatis, T. (1976). *Cell*, **7**, 279–288.

Eickbush, T., Jones, C. W. and Kafatos, F. C. (1981). *In* "Developmental Biology Using Purified Genes" (D. Brown and C. F. Fox, eds) ICN-UCLA Symposia 23. Academic Press, New York and London.

Gelinas, R. E. and Kafatos, F. C. (1973). *Proc. natn. Acad. Sci. U.S.A.* **70**, 3764–3768.

Gelinas, R. E. and Kafatos, F. C. (1977). *Devel. Biol.* **55**, 179–190.

Goldsmith, M. R. and Basehoar, G. (1978). *Genetics* **90**, 291–310.

Goldsmith, M. R. and Clermont-Rattner, E. (1979a). *Genetics* **91**, S40–41.

Goldsmith, M. R. and Clermont-Rattner, E. (1979b). *Genetics* **92**, 1173–1185.

Hamodrakas, S. J., Jones, C. W. and Kafatos, F. C. Submitted for publication.

Iatrou, K., Tsitilou, S. G., Goldsmith, M. R. and Kafatos, F. C. (1980). *Cell* **20**, 659–669.

Jones, C. W. (1980). Ph.D. Thesis, Harvard University.

Jones, C. W. and Kafatos, F. C. (1980). *Nature, Lond.* **284**, 635–638.

Jones, C. W., Rosenthal, N., Rodakis, G. C. and Kafatos, F. C. (1979). *Cell* **18**, 1317–1332.

Kafatos, F. C., Regier, J. C., Mazur, G. D., Nadel, M. R., Blau, H. M., Petri, W. H., Wyman, A. R., Gelinas, R. E., Moore, P. B., Paul, M., Efstratiadis, A., Vournakis, J. N., Goldsmith, M. R., Hunsley, J. R., Baker, B., Nardi, J. and Koehler, M. (1977). *In* "Results and Problems in Cell Differentiation" (W. Beermann, ed.) **8**, 45–145. Springer-Verlag, Berlin.

Kafatos, F. C., Jones, C. W. and Efstradiadis, A. (1979). *Nucl. Acids Res.* **7**, 1541–1552.

Maniatis, T., Sim, G.-K., Efstratiadis A. and Kafatos, F. C. (1976). *Cell* **8**, 163–182.

Maniatis, T., Hardison, R. C., Lacy, E., Lauer, J., O'Connell, C., Quon, D., Sim, G.-K. and Efstratiadis, A. (1978). *Cell* **15**, 687–701.

Mazur, G. D., Regier, J. C. and Kafatos, F. C. (1980). *Devel. Biol.* **76**, 305–321.

Moschonas, N. (1980). Ph.D. Thesis, University of Athens.

Nadel, M. R., Thireos, G. and Kafatos, F. C. (1980). *Cell* **20**, 649–658.

Paul, M. and Kafatos, F. C. (1975). *Devel. Biol.* **42**, 141–159.

Paul, M., Goldsmith, M. R., Hunsley, J. R. and Kafatos, F. C. (1972). *J. Cell Biol.* **55**, 653–680.

Regier, J. C. and Kafatos, F. C. (forthcoming). *J. biol. Chem.*

Regier, J. C., Kafatos, F. C., Goodfliesh, R. and Hood, L. (1978a). *Proc. natn. Acad. Sci. U.S.A.* **75**, 390–394.

Regier, J. C., Kafatos, F. C., Kramer, K. J., Heinrikson, R. L. and Keim, P. S. (1978b). *J. biol. Chem.* **253**, 1305–1314.

Regier, J. C., Mazur, G. D. and Kafatos, F. C. (1980). *Devel. Biol.* **76**, 286–304.

Rodakis, G. C. (1978). Ph.D. Thesis, University of Athens.

Rodakis, G. C., Regier, J. C. and Kafatos, F. C. (1981).

Sim, G.-K., Efstratiadis, A., Jones, C. W., Kafatos, F. C., Koehler, M., Kronenberg, H. M., Maniatis, T., Regier, J. C., Roberts, B. F. and Rosenthal, N. (1977). *Cold Spring Harb. Symp. quant. Biol.* **42**, 933–945.

Sim, G.-K., Kafatos, F. C., Jones, C. W., Koehler, M. D., Efstratiadis, A. and Maniatis, T. (1979). *Cell* **18**, 1303–1316.

Smith, D. S., Telfer, W. H. and Neville, A. C. (1971). *Tissue Cell* **3**, 321–334.

Smith, G. P. (1973). *Cold Spring Harb. Symp. quant. Biol.* **38**, 507–513.

Smith, G. P. (1976). *Science, N.Y.* **191**, 528–535.

Thireos, G. and Kafatos, F. C. (1980). *Devl. Biol.* **78**, 36–46.

Tsitilou, S. G., Regier, J. C. and Kafatos, F. C. (1980). *Nucl. Acids Res.* **8**, 1987–1997.

Gene Amplification and Methotrexate Resistance in Cultured Cells

ROBERT T. SCHIMKE

Department of Biological Sciences, Stanford University, Stanford, California 94305, U.S.A.

INTRODUCTION

In this report we shall summarize the current status of studies on the mechanism(s) whereby cultured animal cells become resistant to the 4-amino analog of folic acid, methotrexate, as a result of stepwise selection, a resistance due to elevated levels of dihydrofolate reductase and associated amplification of DNA sequences coding for dihydrofolate reductase.

SELECTION OF CELLS WITH ELEVATED DIHYDROFOLATE REDUCTASE LEVELS

Methotrexate (MTX) selectively inhibits dihydrofolate reductase (DHFR), an enzyme necessary for the generation of key precursors of nucleic acids. Cells can become resistant to MTX by three mechanisms: (1) an alteration in the enzyme such that MTX does not inhibit; (2) an alteration in inward MTX transport; or (3) an increased capacity for cells to synthesize DHFR. The latter type of resistance is obtained characteristically upon stepwise selection of cells in progressively increasing concentrations of MTX in the medium. The resistant cells contain increased levels of DHFR proportional to the level of resistance (Alt *et al.*, 1976). The elevated DHFR levels result from a corresponding increase in the DHFR gene copy number, i.e. gene amplification (Alt *et al.*, 1978). This is observed in a number of cell lines of both hamster and mouse origin, and in highly aneuploid cells as well as those having a relatively stable karyotype (Schimke *et al.*, 1979). We have proposed that DHFR gene amplification is an infrequent event and that various DNA sequences can undergo amplification randomly, and that amplification of specific sequences is detected only when cells are placed under appropriate selection conditions, i.e. in the presence of MTX. The

requirement for stepwise selection results from the fact that the amplifications occur in small steps, and hence cells with progressively more DHFR
genes (and MTX resistance) are obtained only by the stepwise selection
process (Schimke et al., 1978).

The amplified DHFR genes, increased DHFR levels, and MTX resistance
can occur in either a stable or an unstable state. When cells with amplified
genes are grown in the absence of MTX, some cell lines lose resistance
rapidly, such that within 20 cell-doublings the cell population has lost
50% of its resistance (and gene copy number). In other cell lines the
genes persist when cells are grown in the absence of MTX (Alt et al., 1978).
When cells are first selected for MTX resistance, the amplified genes
are characteristically unstable, whereas when cells are grown for long periods
under selection pressure, the population of cells that emerges has more cells
with stably amplified DHFR genes.

LOCALIZATION OF AMPLIFIED DHFR GENES IN STABLY AND UNSTABLY AMPLIFIED CELLS

In stably amplified cell lines the DHFR genes are localized to a single region
of a chromosome (Fig. 1). In a Chinese hamster ovary cell line (Nunberg et
al., 1978) and a mouse L5178Y cell line (Dolnick et al., 1979) the amplified
genes are present on an expanded region called a "homogeneously staining
region" (HSR) to which the amplified genes can be localized by in situ
hybridization with DNA complementary to DHFR mRNA. Where it has
been possible to determine, only one of the two homologous chromosomes
contains the HSR (Beidler and Spengler, 1976; Dolnick et al., 1979).
Preliminary results done in collaboration with Dr Frank Ruddle of Yale
University indicate that the resident (non-amplified) DHFR gene resides on
the long arm of chromosome 2 in Chinese hamster cells, the position of the
HSR in the cell line we have studied (Nunberg et al., 1978). Thus we
conclude that the chromosomally amplified genes are present at the site of
the resident gene.

In unstably amplified cell lines the DHFR genes are present as extrachromosomal elements, called "double-minute chromosomes" (DMs) (Fig.
2). Double-minute chromosomes are self-replicating chromosomal elements
that do not contain centromers (Levan and Levan, 1978; Barker and Hsu,
1979) and have been reported in tumor cell lines of human origin (Balaban-
Malenbaum and Gilbert, 1976; Barker and Hsu, 1979).

Double-minute chromosomes (containing DHFR genes) can be lost from
cells by two different mechanisms. Since DMs do not contain centromers,
they have the potential for unequal segregation into daughter cells at the

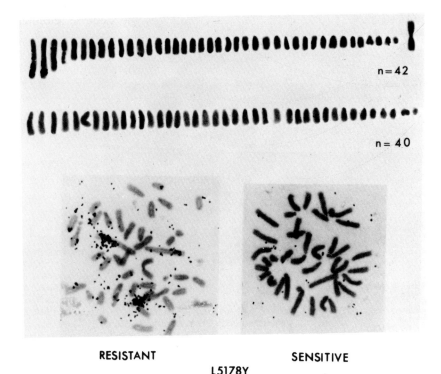

RESISTANT   **SENSITIVE**

L5178Y

FIG. 1. Karyotype and localization of amplified DHFR genes in a stably methotrexate-resistant L5178Y cell line by *in situ* hybridization. This particular cell line has undergone a second alteration in karyotype in which the chromosome with the amplified genes has been reduplicated, most likely as a result of mitotic non-dysjunction (Schimke *et al.*, 1980). Standard metaphase spreads of sensitive and methotrexate-resistant L5178Y murine lymphoblastoid cells were prepared (Dolnick *et al.*, 1979). Slides were treated sequentially with RNase A (previously heated to 80°C for 10 min, 100 µg/ml in 2XSSC, pH 7.0), acetic anhydride (Hayashi *et al.*, 1978), formamide (50 % v/v deionized formamide, 2XSSC pH 7.0 at 70°C for 10 min), and immediately dehydrated in an ethanol series (70 %, 80 %, 95 %, 100 % ethanol), and air-dried. Hybridization with approximately 10^5 dpm of ^{125}I-labeled DHFR cDNA prepared by nick-translation was performed under a sealed coverslip at 42°C overnight in 50 % deionized formamide 2XSSC, pH 7.0, 1X Denhardt's, 25 mM Na_2HPO_4, 1.5 mM $Na_4P_2O_7$, 10 % dextran sulfate, 10 µg/ml of single-stranded, sheared salmon sperm DNA, and 1 mM unlabeled 5'-Iodo-dCTP. Following hybridization, slides were washed extensively, dipped in Kodak NTB-2 (diluted 1:1 with 2 % glycerol), exposed for three days and then developed in Dektol. Slides were then stained with Giemsa and photographed.

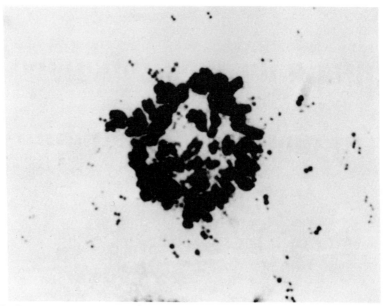

FIG. 2. Metaphase chromosome preparation of an unstably resistant murine S-180 cell. Chromosome spreads from the unstable R_2 clone of methotrexate-resistant murine S-180 cells (Schimke *et al.*, 1979) were prepared without prior colcemid arrest or hypotonic treatment. Fixed cells were stained with Giemsa and photographed.

time of mitosis. We have shown in mouse S-180 cells that individual cells with a lower DHFR gene-copy number grow more rapidly over a number of cell-doublings in the absence of MTX. Thus those cells with progressively fewer DHFR genes will become dominant in the population (Schimke *et al.*, 1979). In addition, extrachromosomal elements can undergo micronucleation (the nuclear membrane can reassemble around a packet of DMs) and be lost from cells as a single-step process (Levan *et al.*, 1976). We have observed in the progeny of single cells undergoing gene-loss cells which have lost virtually all of the amplified genes, as well as cells that show marked heterogeneity with respect to the gene-copy number (Schimke *et al.*, 1980).

HOW COMMON IS GENE AMPLIFICATION?

Gene amplification is considered to be a major mechanism for the increase in size and complexity of the genome during evolution. In addition, selective gene amplification occurs in certain developmentally controlled cases (e.g. ribosomal genes in amphibian oocytes) and as compensation in instances where certain genes have been deleted (ribosomal genes in *Drosophila*). The gene amplification involving MTX resistance is a relatively infrequent event, but such a process is not limited to MTX and DHFR genes. Stark and his colleagues (Wahl *et al.*, 1979) have shown that resistance to a highly specific inhibitor of aspartyltranscarbamylase (PALA), generated by stepwise selection also involves gene amplification. In addition there are other examples of stepwise selection for drug resistance that is either stable or unstable and results from elevated levels of specific enzymes (Schimke *et al.*, 1978). Certain reports of the properties of insecticide resistances also have these same characteristics. Most recently our laboratory has been studying several patients with tumors resistant to MTX, and we have preliminary evidence that this resistance results from elevated DHFR levels and that there is selective amplification of DHFR genes (Dower, Horns and Schimke, preliminary results). Thus we conclude that gene amplification may be relatively common, and that it occurs in clinical circumstances as well as in the experimental cell-culture systems we have been studying.

POSSIBLE MECHANISMS FOR AMPLIFICATION OF DHFR GENES

Our studies raise a large number of questions. Does such gene amplification occur in normal cells? Are there some DNA sequences that can be amplified more readily than others? Do the biochemical events leading to amplifications have any relationship to alterations in chromosome structure observed in various malignant cells? We are currently examining the molecular structure of the DHFR gene using recombinant DNA techniques (Nunberg *et al.*, 1980), and characterizing the DNA sequences present in DMs and their potential relationship to the amplified DNA sequences in HSRs.

We are also currently considering three different possible means for generating amplified DHFR genes: (1) DNA uptake from killed cells. If DNA segments are engulfed by a few remaining cells, and if such DNA can replicate, it would constitute DMs. If they subsequently integrated into a specific region of a chromosome, they could generate HSRs. (2) Unequal mitotic sister chromatid exchange. This process would generate cells with progressively expanding chromosomally localized genes, and if excised to produce self-replicating elements, would constitute DMs. (3) Saltatory (disproportionate) replication. If there are origins of DNA replication where multiple initiations of replication occur within a given S-phase ("hot spots") relative to adjacent replication forks, amplified DNA sequences would be generated, and they could be removed from the chromosome to constitute DMs, or

SALTATORY REPLICATION MODEL

LINEAR

LOOP

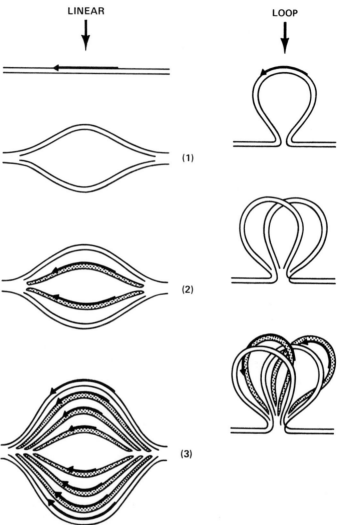

(1)

(2)

(3)

FIG. 3. Schematic representation of gene amplification by saltatory replication. The left represents three rounds of replication initiating at a single origin of replication, and presents a chromosome (DNA) segment in a linear configuration. The right represents two rounds of replication in which the DNA segment is depicted as a loop structure (Stubblefield, 1973; Marsden and Laemmli, 1979). The consequence of an additional round of replication (two rounds) is the existence of two DNA strands that are unattached within the chromosome. If such DNA structures were capable of replication, they could constitute DMs and their generation would not involve excision of the resident gene. Two additional components are required to generate chromosomally localized tandemly duplicated gene sequences. The first involves 3' to 5' end ligation, or its equivalent, of the free DNA strands. The second would involve a recombination event in which the ligated strands are recombined into the chromosome (Schimke *et al.*, 1980). The arrows indicate the direction of transcription from the sense strand of a hypothetical gene.

could recombine into the chromosome to generate an HSR (Fig. 3) (see Schimke *et al.*, 1980 for a discussion of these possible mechanisms and why mechanism No. 3 most generally fits a single means for various types of gene amplification).

All three of these mechanisms may occur in cultured cells to generate various cell lines with differing properties of gene amplification. The mechanisms are not mutually exclusive. The hypotheses are amenable to experimental testing, and an understanding of the generation of HSRs and DMs may provide information concerning abnormal replication processes in malignancy, as well as about how organisms (and cells) become resistant to various drugs by a gene amplification mechanism.

ACKNOWLEDGEMENTS

I would like to acknowledge the many collaborators who have assisted in these studies, including Dr J. Bertino of Yale University, Dr L. Chasin of Columbia University, Dr S. Cohen of Stanford University, as well as members of my laboratory group: F. Alt, R. Kellems, R. Kaufman, J. Nunberg, P. Brown, M. McGrogan, D. Slate, G. Crouse, W. Dower, R. Horns and C. Simonsen. This work is supported by research grants from the American Cancer Society, the National Institute of General Medical Sciences, and the National Cancer Institute.

REFERENCES

Alt, F. W., Kellems, R. E. and Schimke, R. T. (1976). *J. biol. Chem.* **251**, 3063–3074.

Alt, F. W., Kellems, R. E., Bertino, J. R. and Schimke, R. T. (1978). *J. biol. Chem.* **253**, 1357–1370.

Balaban-Malenbaum, G. and Gilbert, F. (1977). *Science, N.Y.* **198**, 739–742.

Barker, P. E. and Hsu, T. C. (1979). *J. natn. Cancer Inst.* **62**, 257–261.

Beidler, J. L. and Spengler, B. A. (1976). *Science, N.Y.* **191**, 185–188.

Dolnick, B. J., Berenson, R. J., Bertino, J. R., Kaufman, R. J., Nunberg, J. H. and Schimke, R. T. (1979). *J. cell. Biol.* **83**, 394–402.

Kaufman, R. J., Brown, P. C. and Schimke, R. T. (1979). *Proc. natn. Acad. Sci. U.S.A.* **76**, 5669–5673.

Levan, A. and Levan, G. (1978). *Hereditas* **88**, 81–92.

Levan, G., Mandahl, N., Bregula, V., Klein, G. and Levan, A. (1976). *Hereditas* **83**, 83–90.

Nunberg, J. M., Kaufman, R. J., Schimke, R. T., Urlaub, G. and Chasin, L. A. (1978). *Proc. natn. Acad. Sci. U.S.A.* **75**, 5553–5556.

Nunberg, J. H., Kaufman, R. J., Chang, A. C. Y., Cohen, S. N. and Schimke, R. T. (1980). *Cell* **19**, 355–364.

Schimke, R. T., Kaufman, R. J., Alt, F. W. and Kellems, R. E. (1978). *Science, N.Y.* **202**, 1051–1055.

Schimke, R. T., Kaufman, R. J., Nunberg, J. H. and Dana, S. L. (1979). *Cold Spring Harb. Symp. quant. Biol.* **43**, 1297–1303.

Schimke, R. T., Brown, P. C., Kaufman, R. J., McGrogan, M. and Slate, D. L. (1980). *Cold Spring Harb. Symp. quant. Biol.* In press.

Wahl, G. M., Padgett, R. A. and Stark, G. R. (1979). *J. biol. Chem.* **254**, 8679–8689.

Three Classes of Small RNA-protein Complexes Precipitated by *Lupus* Antibodies

J. A. STEITZ,[1] M. R. LERNER,[1] J. A. BOYLE,[1]
N. C. ANDREWS,[1] J. P. HENDRICK[1] and I. G. MILLER[2]

*Departments of [1]Molecular Biophysics and Biochemistry and [2]Pediatrics,
Yale University, New Haven, Connecticut 06510, U.S.A.*

INTRODUCTION

Systemic *Lupus erythematosus* is a rheumatic disease of unknown etiology in which patients develop autoantibodies against a variety of cellular antigens. (See Provost (1979) for a review.) Previously, we established the molecular identity of the *Lupus* antigens called RNP and Sm as a class of small nuclear ribonucleoproteins (snRNPs) (Lerner and Steitz, 1979). Anti-RNP antibodies precipitate the most abundant snRNPs, those containing U1 RNA—a previously analyzed low molecular weight nuclear RNA (Reddy *et al.*, 1974) (see Fig. 1, lane 2). Anti-Sm precipitates the U1-containing snRNPs and four others containing the snRNAs U2 (Shibata *et al.*, 1975), U4, U5 and U6 (Epstein *et al.*, 1980) (see Fig. 1, lane 1). Both antibodies precipitate the same constellation of seven abundant nuclear polypeptides (mol. wt 12 000–32 000); their presence is essential for antigenicity since the purified RNAs do not react with either antibody. The extreme conservation of snRNPs across species, their abundance (about 10^6/cell) and their localization in cells and nuclear extracts led us (Lerner *et al.*, 1980) to hypothesize that snRNPs play a role in the biogenesis of eukaryotic messenger RNAs. More specifically, the complementarity between sequences at the 5' end of U1 RNA and those at splice junctions in hnRNA suggested that the U1-containing snRNP could serve to align the two ends of an intron for precise splicing.

Here, we update our studies on the structure and function of snRNPs. We also describe two other non-overlapping classes of small RNA-protein complexes that are specifically precipitated by other *Lupus* antisera. Although these new ribonucleoproteins are slightly less abundant than the snRNPs precipitated by anti-RNP and anti-Sm, it seems likely that they too play central roles in the metabolism of mammalian cells.

RECENT PROGRESS ON snRNPs

A most curious finding concerning the protein composition of snRNPs is the fact that both anti-RNP (which reacts with only U1-containing particles) and anti-Sm (which recognizes five different RNA-protein complexes) precipitate the same collection of seven small polypeptides (labeled A–G, Lerner and Steitz, 1979). Does this mean that U1, U2, U4, U5 and U6 each bind an identical set of nuclear proteins? We do not yet have the complete answer but we can say more about both the uniqueness of the U1-containing snRNP and the identity of the Sm-antigenic determinant. First, experiments in which mouse Ehrlich ascites cell nuclear extracts are precleared of the U1-snRNP by precipitating with saturating amounts of anti-RNP and then reacted with anti-Sm show loss of polypeptides A and C (data not shown). Thus, these two proteins appear to be specific for the U1-snRNP, and the possible Sm-antigenic determinant is narrowed to polypeptides B, D, E, F or G. Second, a mouse–mouse hybridoma with anti-Sm specificity (as assayed by the spectrum of snRNAs it precipitates) has recently been constructed by E. Lerner and M. Lerner (unpublished). Its existence demonstrates conclusively the presence of a single antigenic determinant on all of the snRNPs and should allow us soon to identify exactly which of the five common proteins is the Sm-antigen.

To pursue further the hypothesis that the U1-containing snRNP participates directly in hnRNA processing, we have asked whether anti-RNP and/or anti-Sm antibodies can inhibit the splicing reaction *in vitro*. Recently, Yang *et al.* (1980) developed a system in which nuclei isolated from adenovirus-infected HeLa cells continue faithful synthesis and processing of viral transcripts *in vitro*. When our anti-RNP or anti-Sm IgG preparations are added, Drs V. Yang and J. Flint (personal communication) observe that the *in vitro* splicing, but not the synthesis, of RNAs encoded by several portions of the adeno early region is dramatically inhibited! Conversely, IgGs prepared from normal sera or from patients possessing anti-Ro or anti-La activity (see below) have no effect. Since the snRNP containing U1 RNA is the only particle reactive with both anti-RNP and anti-Sm, we conclude that the U1-snRNP does play a crucial role in the splicing mechanism. Elucidation of its exact mode of action awaits development of more refined *in vitro* splicing systems. Unfortunately, these results do not further enlighten us as to whether the snRNPs containing U2, U4, U5 and U6 are active in aligning minor sequence variants of splice junctions or play a totally different role in mRNA biogenesis.

Some progress has also been made towards obtaining the U1-snRNP in purified form, a prerequisite for further functional and structural studies. Whereas most protein antigens can be straightforwardly prepared using antibody affinity columns, in this case one confronts the problem that elution conditions must be able to break the antigen–antibody bonds without destroying the specific interactions between the U1 RNA and its seven bound polypeptides. Fortunately, the U1-snRNP is a surprisingly stable entity and can be recovered (in a still antigenic form) from anti-RNP columns by elution with 4M urea/2M LiCl! None the less, two of the proteins (A and C) are reduced in yield and further work is necessary to devise isolation procedures for the intact particle. Our observation that the antigenic U1-snRNP survives conditions that disassemble ribosomes, however, suggests that these complexes are exceedingly stable particles which will be easily manipulable in future physical as well as biological studies.

scRNPs PRECIPITATED BY ANTI-Ro

In the course of screening additional *Lupus* sera, we discovered that anti-Ro precipitates a distinctly different set of small ribonucleoproteins (see Fig. 1, lanes 3 and 4). Again proteins are required for antigenicity, and indirect immunofluorescence studies reveal that the Ro complexes are located exclusively in the cytoplasm of mammalian cells. The Ro RNAs are designated Y1–Y3 to distinguish them from the nuclear U series. The function of the Ro-precipitated small cytoplasmic ribonucleoproteins (scRNPs) is currently obscure. However, the recent observation of P. Courtoy (personal communication) that they are specifically associated with peroxisomes (small membrane-bound organelles containing various oxidases) suggests certain directions that may be fruitfully pursued.

ANTI-La RECOGNIZES BOTH HOST AND VIRAL RNPs

A third non-overlapping family of small RNAs is precipitated by yet another *Lupus* antibody, anti-La (Fig. 1, lane 5). This heterogeneous spectrum contains RNAs ranging in size from about 80 to 140 nucleotides; the pattern is different, but equally complex in HeLa cells (not shown) compared to mouse Ehrlich ascites cells (Fig. 1). Unlike U1, U2, U4, U5 and U6 (which possess 5′ caps), the La RNAs have 5′ triphosphates and in most cases lack internal modified nucleotides. Complexed protein(s) are essential for antigenicity. Unfortunately we have not yet succeeded in either identifying the associated polypeptide(s) or defining the subcellular location of the La-RNPs, primarily because the antisera available to us are apparently not monospecific. None the less, our current best guess (based on both immunofluorescence and cellular fractionation) is that the La-RNPs are mostly nuclear.

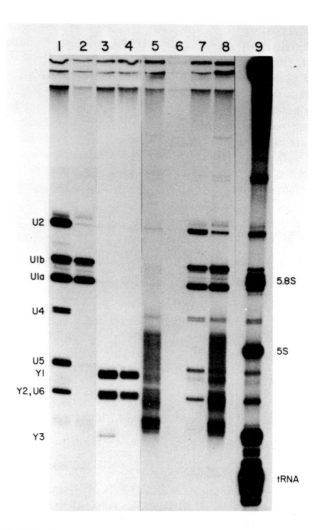

FIG. 1. Small RNAs from Ehrlich ascites cells precipitated by SLE antibodies. Small RNAs were isolated from immune precipitates or extracts of ^{32}P-labeled Ehrlich ascites cells and separated on a 400 mm wide, 0.5 mm thick gel made of 10% polyacrylamide, 0.38% bisacrylamide, 7 M urea, 1 mM EDTA and 50 mM Tris-borate, pH 8.3, as described previously (Lerner and Steitz, 1979; Lerner et al., 1980). Lane 9 shows small RNAs from a total Ehrlich ascites cell extract. Lanes 1–8 show small RNAs precipitated by IgG isolated from sera containing: (1) anti-Sm; (2) anti-RNP; (3) anti-Ro[b]; (4) anti-Ro[a]; (5) anti-La; (6) normal serum; (7) serum containing anti-Sm and -Ro[a] and (8) serum containing anti-Sm, -RNP, -Ro[a] and -La.

An intriguing property of anti-La is its ability to precipitate not only the cellular RNPs described above but also certain virus-specific RNAs in the form of small ribonucleoprotein complexes. Figure 2 shows that one such RNA is the VA RNA (Ohe and Weissman, 1971) encoded by the adenovirus genome. This approximately 150-nucleotide-long RNA of unknown function can be specifically and quantitatively precipitated from adeno-infected HeLa cell extracts by all three available sera containing anti-La.

Likewise, cultured primate cell lines transformed by Epstein-Barr virus (EBV) produce large amounts (about 10^7 molecules/cell) of two previously undescribed small RNAs reactive in RNP form with anti-La (Fig. 3). These RNAs are not found in a continuous B cell line lacking EBV genomes (BJAB) or in a continuous T-lymphocyte line transformed by *Herpes ateles* virus. The RNAs are both about 200 residues long and lack polyA and internal modified nucleotides, but have distinctly different fingerprints. Southern hybridization analyses reveal that at least one of these small RNAs is encoded by the Eco RI-J fragment of EBV strain B95-8 DNA; this fragment appears to be the most actively transcribed portion of the EBV genome in Burkitt lymphoma-derived cells (Rymo, 1979).

A clue to the potential functioning of the La family of RNPs (including those containing VA RNA or the two small EBV RNAs) comes from the realization that at least one of the uninfected cell La RNAs exhibits sequence overlap with the Alu family of highly repetitive interspersed sequences in mammalian genomes (Jelinek et al., 1980). Specifically, a murine 4.5S RNA first found hydrogen-bonded to hnRNA and mRNA by Jelinek and Leinwand (1978) and later sequenced by Harada and Kato (1980) can be identified by fingerprint analysis as one of the La RNAs in mouse Ehrlich ascites cells. Many Alu sequences can be transcribed *in vitro* by RNA polymerase III (Duncan et al., 1979; C. Duncan, personal communication); the presence of 5′ triphosphate moieties on La RNAs is suggestive that they are likewise synthesized *in vivo* by Pol III, as is VA RNA (Söderlund et al., 1976). The Alu family of sequences has been hypothesized by Jelinek et al. (1980) to be involved in the replication of mammalian genomes. Using anti-La antibodies it should be possible to test the involvement of La-RNPs in various aspects of the replication process.

Whatever their function, it is significant that the La family of small RNPs includes viral-specific RNAs. Construction of a hybridoma with La specificity will hopefully soon allow identification of polypeptide(s) (presumably host in origin) that confer antigenicity. Also worth investigating is the possibility of a direct link between viral infection and the onset of certain types of *Lupus*, which could dramatically increase our understanding of the pathogenesis of this human disease.

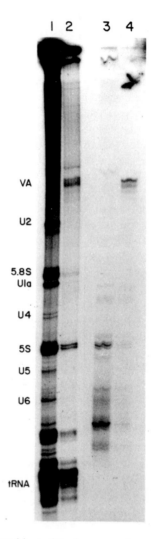

FIG. 2. Small RNAs precipitated by anti-La from normal or adenovirus-infected HeLa cells. Small RNAs isolated from immune precipitates and extracts of ^{32}P-labeled uninfected or adenovirus 2-infected HeLa cells were fractionated as described in Fig. 1. Lanes 1 and 2 show total small RNAs from extracts of HeLa and adenovirus-infected HeLa cells, respectively. Lanes 3 and 4 show small RNAs precipitated by anti-La from extracts of HeLa and adenovirus-infected HeLa cells, respectively. The triple-band pattern for VA$_I$ RNA is consistent with its known length heterogeneity (Akusjärvi et al., 1980).

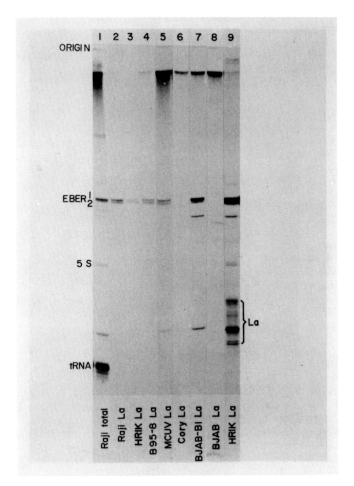

FIG. 3. Detection of EBV-encoded RNAs in anti-La precipitates from EBV genome positive cell lines. Cells were labeled in culture with $^{32}PO_4$ for 24 hours. RNAs isolated from total (lane 1) or anti-La precipitated (lanes 2–9) extracts were prepared and fractionated as described in Fig. 1. Lane 1 shows total RNAs in Raji cells (Burkitt's lymphoma-derived, EBV positive human line) including the two EBV-encoded RNAs (EBER 1 and EBFR 2), 5.8S RNA and tRNAs. Comparative intensities of EBER v. 5.8S bands allow an estimate of approximately 10^7 EBER RNA molecules/cell. Lane 2 shows Raji cell RNAs which are specifically precipitated by anti-La. Lanes 3–5 show the anti-La patterns obtained from the following cell lines: HR1K (another Burkitt's lymphoma-derived, EBV positive human line); B95-8 (marmoset B-lymphocytes transformed *in vitro* by EBV of infectious mononucleosis origin); and MCUV (another marmoset B-lymphocyte line). Lane 6 shows a parallel precipitation from Cory cells, an EBV genome negative marmoset T-lymphocyte cell line transformed by *Herpes ateles*. Lanes 7 and 8 show anti-La precipitates from the *in vitro* EBV-infected cell line BJAB-B1, and from its uninfected parent cell line BJAB (human B-lymphocytes which are EBV genome negative). Only the EBV genome positive cells show EBER 1 and EBER 2. Finally, lane 9 is a longer exposure of lane 3, which also shows the uninfected human cell anti-La RNA spectrum (note that these RNAs are less plentiful than EBER 1 and EBER 2).

264   J. A. STEITZ *et al.*

ACKNOWLEDGEMENTS

This work was supported by grants from the NIH and NSF; the *Lupus* sera used were the kind gifts of Drs J Hardin and M. Reichlin.

REFERENCES

Akusjärvi, G., Mathews, M. B., Andersson, P., Vennström, B. and Pettersson, U. (1980). *Proc. natn. Acad. Sci. U.S.A.* **77**, 2424–2428.
Duncan, C., Biro, P. A., Choudary, P. V., Elder, J. T., Wang, R. C., Forget, B. G., DeRiel, J. K. and Weissman, S. M. (1979). *Proc. natn. Acad. Sci. U.S.A.* **76**, 5095–5099.
Epstein, P., Reddy, R., Henning, D. and Busch, H. (1980). *J. biol. Chem.* **255**, 8901–8906.
Harada, F. and Kato, N. (1980). *Nucl. Acids Res.* **8**, 1273–1285.
Jelinek, W. R. and Leinwand, L. (1978). *Cell* **15**, 205–214.
Jelinek, W. R., Toomey, T. P., Leinwand, L., Duncan, C. H., Choudary, P. V., Biro, P. A., Weissman, S. M., Rubin, C. M., Houch, C. M., Deininger, P. L. and Schmid, C. W. (1980). *Proc. natn. Acad. Sci. U.S.A.* **77**, 1398–1402.
Lerner, M. R. and Steitz, J. A. (1979). *Proc. natn. Acad. Sci. U.S.A.* **76**, 5495–5499.
Lerner, M. R., Boyle, J. A., Mount, S. M., Wolin, S. L. and Steitz, J. A. (1980). *Nature, Lond.* **283**, 220–222.
Ohe, K. and Weissman, S. M. (1971). *J. biol. Chem.* **246**, 6991–7009.
Provost, T. T. (1979). *J. invest. Derm.* **72**, 110–113.
Reddy, R., Ro-Choi, T. S., Henning, D. and Busch, H. (1974). *J. biol. Chem.* **249**, 6486–6494.
Rymo, L. (1979). *J. Virol.* **32**, 8–19.
Shibata, H., Ro-Choi, T. S., Reddy, R., Choi, Y. C., Henning, D. and Busch, H. (1975). *J. biol. Chem.* **250**, 3909–3920.
Söderlund, H., Pettersson, U., Vennström, B., Philipson, L. and Mathews, M. B. (1976). *Cell* **7**, 585–593.
Weinmann, R., Brendler, T. G., Raskas, H. J. and Roeder, R. G. (1976). *Cell* **7**, 557–566.
Yang, V. W., Binger, M.-H. and Flint, S. J. (1980). *J. biol. Chem.* **255**, 2097–2108.

A Search for the Promoter Sequences Involved in Initiation of Transcription of Eukaryotic Genes Transcribed by RNA Polymerase B

B. WASYLYK, A. BUCHWALDER, P. SASSONE-CORSI,
C. KÉDINGER, J. CORDEN and P. CHAMBON

Laboratoire de Génétique Moléculaire des Eucaryotes du CNRS, Unité 184 de Biologie Moléculaire et de Génie Génétique de l'INSERM, Institut de Chimie Biologique, Faculté de Médecine, Strasbourg, France

INTRODUCTION

The ability to switch the expression of their genes on and off, for instance in response to extracellular signals, is a basic property of all living cells. It is well established that in prokaryotes this switching is mostly controlled at the level of RNA transcription. In complex eukaryotic organisms, although the expression of many different genes is turned on and off during development from the egg, and this switching continues in the differentiated cells, the importance of control of gene expression at the transcriptional versus posttranscriptional level is still a matter of controversy. There is, however, unequivocal evidence that, in higher eukaryotes, the expression of at least some genes is controlled at the level of RNA transcription. For instance, the transcription of the ovalbumin and conalbumin (ovotransferrin) genes is turned on by the steroid hormones, oestradiol or progesterone, in the tubular gland cells of the magnum portion of the chick oviduct, whereas these genes, like the genes coding for the other egg-white proteins, are not transcribed in the absence of these steroid hormones (McKnight and Palmiter, 1979; Perrin et al., 1979).

During the past twenty years some of the basic mechanisms involved in regulation of transcription in prokaryotes and their viruses have been elucidated in molecular terms. It has been learnt that transcription is regulated by modulation of the efficiency with which RNA polymerase can recognize and interact with specific DNA signal sequences, promoters and terminators, which specify starting or stopping sites and are involved in the promotion and termination of RNA transcription (see Rosenberg and Court, 1979 and references therein). The genetic approach has been invaluable in these studies. For instance, it is primarily on genetic evidence that Jacob et al. (1964) first defined the promoter as an initiating element indispensable for the expression of bacterial structural genes. Further progress was made possible by the availability of in vitro cell-free systems, reconstructed from purified components, in which the selective in vivo transcription events could be accurately duplicated. In addition to purified RNA polymerase and well-defined templates, such in vitro studies obviously require a detailed knowledge of the in vivo transcription unit to determine whether correct initiation and termination of transcription are occurring. The use of such in vitro systems, the possibility of purifying specific wild-type and mutant prokaryotic genes and their RNA products and the availability of DNA and RNA sequencing methods, have subsequently allowed one to analyze their structure and function in great detail and to show that prokaryotic promoters are regions of DNA 5' to the structural genes (for references see Rosenberg and Court, 1979; Losick and Chamberlin, 1976). The mRNA start points (the position on a DNA sequence which codes for the first nucleotide of an RNA) of many prokaryotic transcription units have been precisely located by genetic analysis and in vitro transcription, and the DNA sequences of these regions have been determined. Pribnow (1975a, b) and Schaller et al. (1975) first noted a sequence homology, related to 5'-TATAATG-3', located about 10 bp upstream from mRNA start points.* A second region of homology, the "recognition region", has also been noted in some promoters in an area centered about 35 bp upstream from the mRNA start point (see Rosenberg and Court, 1979 for references). DNase protection and chemical modification experiments have been used to show that RNA polymerase binds to these regions (see Rosenberg and Court, 1979, for references). Furthermore, several promoter mutants have been sequenced and their locations within the homologous sequences has established that these regions fit the original definition of a promoter.

In contrast to prokaryotic cells, the molecular mechanisms which underlie the regulation of transcription in eukaryotic cells are still largely unknown, notably because the classical genetic approach is usually not possible. That these mechanisms may not be identical to those in prokaryotes was first suggested ten years ago by the discovery of the multiplicity of eukaryotic RNA polymerases by our group and that of Rutter (reviewed in Chambon,

---

* By convention we give in this paper only the anti-sense (non-coding) DNA (5' → 3') strand and therefore transcription proceeds from left to right. DNA sequences in the direction of transcription (downstream) are numbered with positive integers while sequences 5' to the start point (upstream) are given negative values.

1975). It was subsequently firmly established that cells of both higher and lower eukaryotes contain three structurally and functionally distinct classes of RNA polymerase which are localized in different subcellular fractions. Class A or I catalyzes the synthesis of ribosomal RNA, class B or II that of mRNA, and class C or III that of tRNA and 5S RNA (reviewed in Chambon, 1975; Roeder, 1976). Although highly purified preparations of these enzymes, particularly RNA polymerase B, were available shortly after their discovery, progress has been extremely slow in analyzing their role in the control of transcription. The lack of meaningful *in vitro* cell-free transcription systems mostly accounts for this failure. Indeed, due to the complexity of the eukaryotic genome, there was no means to study the *in vitro* transcription of a given gene by incubating the total cellular DNA with purified RNA polymerase. In addition, even when well-defined viral DNA templates, like the SV40 and adenovirus 2 (ad2) genomes, were available, the primary transcription products were unknown, precluding any valid analysis of the factors involved in the control of transcription. Furthermore, intact viral DNAs proved to be very poor *in vitro* templates for the purified RNA polymerase B, which was known to transcribe SV40 and ad2 genomes *in vivo* (see Chambon, 1975 for review). Several technical breakthroughs were clearly required.

The discovery of restriction enzymes and reverse transcriptase, followed by the advent of molecular cloning and of methods for separating, visualizing and rapidly sequencing DNA and RNA molecules, have made it possible to study, at the nucleotide level, the anatomy of eukaryotic cellular and viral genes and of their primary RNA transcripts (Chambon, 1978; Breathnach and Chambon, 1981). It has now been shown in several instances (see for example Ziff and Evans, 1978; Wasylyk *et al.*, 1980b and references therein for the ad2 major late and ovalbumin transcription units, respectively) that the 5' ends of the RNA primary transcripts and of the mature mRNAs coincide and therefore that the start point of transcription corresponds to the base-coding for the 5'-terminal nucleotide of the mRNAs. By analogy to the bacterial case, we would expect eukaryotic promoters to be located in the region adjacent to the 5' end of the transcription unit. Indeed, the comparison of several cellular and viral genes has revealed the existence of an AT-rich region of homology centered about 25 bp upstream from the mRNA start points (Gannon *et al.*, 1979; Cochet *et al.*, 1979; Benoist *et al.*, 1980). This sequence, which is known as the "TATA" box, was first noticed by Goldberg and Hogness and bears some sequence resemblance to the Pribnow box of prokaryotic promoters (Gannon *et al.*, 1979; Cochet *et al.*, 1979; see also Fig. 1). Additional sequence homologies have been noticed at positions −70 to −80 upstream from the mRNA start sites of several

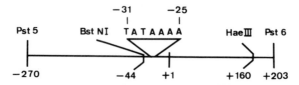

FIG. 1. Restriction enzyme map of the conalbumin Pst 5-Pst 6 fragment. Pst 5 and Pst 6 refer to Pst I restriction sites previously defined (Cochet *et al.*, 1979). DNA sequences in the direction of transcription (downstream) are numbered with positive integers, whilst sequences 5′ to the start point (upstream) are given negative values. +1 represents the position of the base-coding for the first nucleotide of conalbumin mRNA. TATAAAA is the sequence of the non-coding DNA strand between nucleotides −31 and −25, and is the conalbumin sequence equivalent to the "TATA" box consensus sequence found in many eukaryotic genes transcribed by RNA polymerase B (Gannon *et al.*, 1979).

cellular and viral protein-coding genes (Benoist *et al.*, 1980). However, these homologous sequences have not been found upstream from the start site of genes transcribed by RNA polymerases A (Sollner-Webb and Reader, 1979) and C (Sakonju *et al.*, 1980; Kressmann *et al.*, 1979), indicating that the specific transcription of different classes of genes by the distinct classes of eukaryotic RNA polymerases could be due to the specific recognition of sequences characteristic of a class of genes.

Although the recognition of sequence homologies is important in suggesting the location of control regions, it is obvious that the actual role of homologous sequences cannot be established without a functional assay, for instance a cell-free system capable of accurate *in vitro* transcription. As discussed above, such eukaryotic systems have been lacking until recently. The technical breakthrough was the establishment in 1978 by Wu (1978) and by Birkenmeier *et al.* (1978) of accurate cell-free transcription systems for viral and cloned cellular genes transcribed *in vivo* by RNA polymerase C. In these systems the necessary factor(s) which is (are) lacking in the purified RNA polymerase C are supplied by a KB cell cytoplasmic fraction (Wu, 1978) or by a *Xenopus oocyte* nuclear extract (Birkenmeier *et al.*, 1978). Unexpectedly, it was found by Brown and his collaborators (Sakonju *et al.*, 1980; Birkenmeier *et al.*, 1978) and by the group of Birnstiel (Kressmann *et al.*, 1979) that the essential information for 5S RNA and tRNA transcription by RNA polymerase C is contained in an intragenic control region, in a position strikingly different from that of promoter regions in prokaryotes. These observations obviously raised the question whether all eukaryotic promoters are similarly located or whether this location is particular to genes transcribed by RNA polymerase C. Subsequently, Weil *et al.* (1979) found that a system similar to that of Wu can also be used as a source of factor(s) to promote accurate initiation of transcription by purified RNA polymerase B at the major late ad2 promoter. Briefly, they used a "truncated template" assay which contains, in addition to a cytoplasmic HeLa cell extract (S100) and RNA polymerase B, a restriction-enzyme-cut DNA fragment containing, at a well-mapped position, the promoter region of the major late ad2 transcription unit. The *in vitro* synthesized RNAs are labeled with P^{32}-α-nucleoside triphosphates and then separated by gel electrophoresis. Discrete-size RNA products are produced by "run-off" termination whenever specific initiation occurs. From the length of the "run-off" transcripts the position of the region coding for the 5' end of the *in vitro*-synthesized RNAs can be deduced and compared with the position of the 5' end of the *in vivo* transcription unit. Since, at about the same time, we had completed an extensive study of the anatomy of the cloned chicken ovalbumin and conalbumin genes and of their transcription units (see Gannon *et al.*, 1979; Cochet *et al.*, 1979; Benoist *et al.*, 1980 for references), we used this technique to demonstrate specific *in vitro* initiation of transcription of these cellular genes at the sites corresponding to the 5'-terminal nucleotide of the *in vivo* primary transcripts (Wasylyk, 1980b). Furthermore, we showed by specific fragmentation of the conalbumin gene DNA that a short segment, situated between positions -8 to -44 upstream from the initiation site and containing the "TATA" region of homology, was required for specific *in vitro* transcription. These results suggested to us that, in contrast to the case of RNA polymerase C, the promoter sequences from RNA polymerase B are located upstream from the mRNA start sites, as in prokaryotes.

Additional support for the "promoter" role of these sequences was provided by comparison of the transcriptional efficiencies of conalbumin and ovalbumin genes with ad2 early (E1A) and major late genes. It is indeed noteworthy that the conalbumin and ad2 major late genes, which share an extensive 12 bp homology in their "TATA" box regions (Cochet *et al.*, 1979) are transcribed *in vitro* with the same efficiency. These genes were more strongly transcribed than those of adenovirus E1A and ovalbumin, which differ in the sequence of their "TATA" box (Benoist *et al.*, 1980). These observations suggest the existence of promoter sequences with different strengths (Wasylyk, 1980). In this respect it is interesting to recall that base changes in the Pribnow box (including $A \rightarrow T$ and $T \rightarrow A$) have marked effects on the efficiency of prokaryotic promoters (Rosenberg and Court, 1979).

In vitro genetic experiments were then designed to define the minimum sequence necessary to promote specific *in vitro* transcription by RNA polymerase B (Corden *et al.*, 1980). To this end, deletion mutants of either conalbumin or ad2 major late genes (because of their 5'-end upstream sequence homologies (see below and Cochet *et al.*, 1979), we used these two genes interchangeably) were constructed *in vitro*, propagated as plasmid DNA and used as templates for *in vitro* transcription reactions. These studies have conclusively demonstrated that sequences located between -44 (i.e. 44 nucleotides upstream from the start point for the mRNA ($+1$), see Fig. 1) and -10 for the conalbumin gene, and between -32 and -12 for the ad2 major late transcription unit (Corden *et al.*, 1980), are essential to promote specific *in vitro* transcription. These two regions contain in common a sequence of 12 base-pairs, 5'-CTATAAAAGGGG-3 (the C is at position -32 (see Cochet *et al.*, 1979), encompassing the "TATA" "Goldberg-Hogness" box. These results, however, must be interpreted with some caution. In deletion mutants the DNA upstream from the deletion end point is different from the original DNA, making it impossible to rule out that the observed effects are due to some inhibitory effect of the replacing sequences, rather than to an alteration of the promoter. In order to obtain unequivocal evidence that the "TATA" box is implicated in the promotion of transcription, we have used *in vitro* site-directed mutagenesis to make a single base-pair substitution in the "TATA" box. In this chapter we report the effect on specific *in vitro* transcription of converting to a G the second T of the conalbumin "TATA" box.

THE CONSTRUCTION OF A MUTANT CONTAINING A SINGLE BASE-PAIR SUBSTITUTION IN THE CONALBUMIN "TATA" BOX

It has been shown (Hutchinson *et al.*, 1978; Gillam *et al.*, 1979; Gillam and Smith, 1978a, b; Razin *et al.*, 1978) that a specific point mutation can be obtained with a synthetic oligonucleotide differing in one base from a wild-type DNA sequence. This oligonucleotide is first annealed to the complementary wild-type strand contained in a circular single-stranded (ss) DNA vector, extended with DNA polymerase I and finally ligated with DNA ligase to complete the other strand. Transfection with these molecules gives rise to clones containing DNA with the desired sequence change. We used a fd103 phage recombinant containing the conalbumin Pst 5-Pst 6 fragment (Fig. 1) and a synthetic undecanucleotide to construct a mutant containing a single base-substitution in the conalbumin "TATA" box (Fig. 2).

The non-coding strand of the conalbumin Pst 5-Pst 6 fragment (Fig. 1) was cloned in fd103 ssDNA at the position of the unique Pst I site of the RF DNA. The undecanucleotide, complementary to the "TATA" box region except for a C (boxed nucleotide in the upper left circle in Fig. 2) was annealed to the ssDNA, and used to synthesize the complementary strand

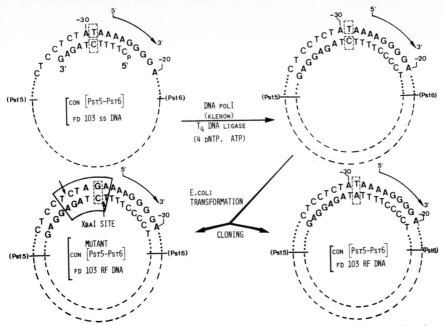

FIG. 2. Preparation of a mutant containing a single base-pair substitution in the conalbumin "TATA" box. The undecadeoxyribonucleotide d-CTTTTCTAGAG was prepared and purified as described elsewhere (Wasylyk et al., 1980). The isolation of single-stranded bacteriophage fd103 DNA containing the non-coding strand of the Pst 5-Pst 6 fragment is reported in Wasylyk et al. (1980a).

One hundred pmol of undecanucleotide primer was 5'-phosphorylated in a reaction (50 µl) containing 50 mM Tris-HCl pH 9.5, 10 mM MgCl$_2$, 5 mM dithiothreitol, 5% glycerol, 500 µCi ^{32}P-γ-ATP (NEN) and 5 units T4 polynucleotide kinase (Boehringer). After 1 hour at 37°C, additional T4 kinase (5 units) and ATP (to 100 µM) were added and incubation continued for 30 min at 37°C. Labeled primer was isolated by chromatography on a 10 ml Sephadex G25 column in 50 mM triethylammonium bicarbonate pH 7.9. The primer pool was lyophilized, resuspended in water and relyophilized a second time. 5 µl of a solution containing 40 pmol labeled primer, 0.6 pmol of ssDNA fd103 recombinant (containing the non-coding strand of the Pst 5-Pst 6 conalbumin fragment), 40 mM Tris-HCl pH 7.5, 20 mM MgCl$_2$, 0.1 M NaCl and 2 mM β-mercaptoethanol was incubated for 3 min at 80°C followed by 60 min at 0°C. 7.5 µl of a solution containing 22 mM Tris-HCl pH 7.5, 11 mM MgCl$_2$, 1 mM β-mercaptoethanol, 0.83 mM each dATP, dGTP, dCTP, TTP, 0.4 mM ATP, 2.5 units E. coli DNA polymerase I (Klenow fragment, Boehringer Mannheim) and 0.5 units T4 DNA ligase (Biolabs) was added and incubation continued for 30 min at 0°C and 5 hours at 23°C. 1 µl samples were analyzed either directly on 1% agarose gels (to check for conversion to double-stranded circles) and, after digestion with Pst I or Msp I, on 5% acrylamide urea gels (to check by autoradiography for correct priming and efficient ligation). 1 µl was diluted 10-fold to 30 mM sodium acetate pH 4.3, 4.5 mM ZnCl$_2$, 0.28 M NaCl and digested with 1 unit S1 nuclease (Miles) for 15 min at 23°C. Then, 2.5 µl 2 M Tris-HCl pH 7.9 was added and after extraction with phenol followed by ether and extensive dialysis against 100 mM Tris-HCl pH 7.9, the DNA was transfected into E. coli K514 as described above. Approximately 10 plaques/ng of starting ssDNA were obtained. A phage pool from approximately 200 plaques was used to infect E. coli KB35 and, after plating on chloramphenicol-agar plates, clear lysate DNA from individual colonies was screened for the presence of an Xba I site in the Pst 5-Pst 6 conalbumin fragment. In the figure conalbumin and fd103 ssDNA are represented by dotted and dashed lines respectively. The individually boxed nucleotides are mismatched (upper circles) and the boxed base-pairs (lower circles) are matched. The boxed hexanucleotide sequence (lower left) is the Xba I recognition site. DNA pol. I (Klenow) = DNA polymerase I (Klenow fragment); DNTP = deoxyribonucleoside triphosphate; con = conalbumin; Pst 5 and Pst 6, see Fig. 1 and text; ss = single-stranded; RF = replicative form.

(Hutchinson *et al.*, 1978; Gillam *et al.*, 1979; Gillam and Smith, 1979a, b; see legend to Fig. 2). After cloning, RF DNAs from 50 individual colonies were screened for the presence of an Xba I site, which is conveniently formed by the mutation (see Fig. 2). The DNA of two colonies contained an Xba I site. The mutant Pst 5-Pst 6 fragment was isolated from the fd103 RF DNA and inserted into the Pst I site of pBR322. Recombinants corresponding to the two original clones were selected and were found, on subsequent characterization, to give essentially identical results.

CHARACTERIZATION OF THE CONALBUMIN "TATA" BOX MUTANT

Figures 3 and 4 demonstrate that the conalbumin mutant differs from the wild-type DNA by the change of the second T of the "TATA" box to a G. In Fig. 3 we have compared the restriction enzyme digestion patterns of mutant and wild-type DNAs. It is clear from Fig. 3A, lanes 7–10, that only the mutant, and not wild-type pBR322 recombinant, contains an Xba I site. From lanes 2–5 it is seen that only the Pst 5-Pst 6 fragment (470 bp band) from the mutant DNA contains an Xba I site (giving the bands at 235 bp). To show that the mutant DNA is not contaminated with wild-type fragment we digested 0.5 µg mutant Pst 5-Pst 6 fragment with increasing amounts of Xba I (Fig. 3B, lanes 4–7). Comparison of lane 4 (corresponding to the largest amount of Xba I tested) with lanes 1–3, which contained 0.005, 0.01 and 0.05 µg of undigested fragment, shows that more than 99% of the mutant Pst 5-Pst 6 fragment contains the expected Xba I site. The T → G transversion was the only modification observed in this region, as shown by the sequence analysis in Fig. 4 and other unpublished sequencing results.

IN VITRO TRANSCRIPTION OF THE CONALBUMIN "TATA" BOX MUTANT

We have previously demonstrated by S1 nuclease mapping and by the "run-off" transcription method that specific initiation occurs *in vitro* on conalbumin gene DNA (Wasylyk *et al.*, 1980b). In the "run-off" assay, discrete-size RNAs are produced on restriction enzyme DNA fragments by "run-off" termination whenever specific initiation occurs. From the length of the "run-off" transcripts the position of the 5′ end of the RNA can be deduced. As expected (see Fig. 1), a major "run-off" transcript of about 200 nucleotides (see band labeled with an arrowhead in Fig. 5A, lane 3 and also Wasylyk *et al.*, 1980b; Corden *et al.*, 1980) is obtained with the Pst 5-Pst 6

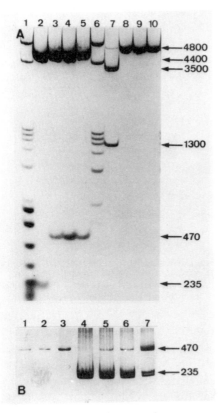

FIG. 3. Restriction enzyme mapping of mutant and wild-type recombinants. The mutant conalbumin Pst 5-Pst 6 fragment was cloned in the Pst I site of pBR322. (A) The purified recombinant DNA was digested with restriction enzymes and analyzed on a 1.5% agarose gel. Lanes 1 and 6: DNA size-markers; lanes 2, 4, 7 and 9: mutant recombinant digested with Pst I and Xba I, Pst I, Xba I and Bam HI, and Bam HI respectively; lanes 3, 5, 8 and 10: wild-type recombinant digested with Pst I and Xba I, Pst I, Xba I and Bam HI, and Bam HI respectively. (B) The mutant conalbumin Pst 5-Pst 6 fragment was isolated by sucrose density gradient centrifugation, digested with different quantities of Xba I and analyzed on a 5% poly-acrylamide gel. Lanes 1, 2 and 3: 0.005 μg, 0.01 μg and 0.05 μg of mutant Pst 5-Pst 6 fragment, respectively, before digestion; lanes 4, 5, 6 and 7: 0.5 μg of mutant fragment digested with 20, 12.5, 7.5 and 2.5 units of Xba I respectively for 1 hour at 37°C.

fragment as a template. Similarly, for the Hae III-digested Pst 5-Pst 6 fragment a specific "run-off" fragment of about 160 nucleotides is obtained (Fig. 5A, lane 4, and 5B, lanes 1 and 3). Specific "run-off" transcripts of identical mobility, but weaker intensity, were obtained (Fig. 5A) for the mutant Pst 5-Pst 6 fragment which is intact (lane 1) or digested (lane 2) with Hae III. Electrophoresis of the "run-off" transcript obtained with the

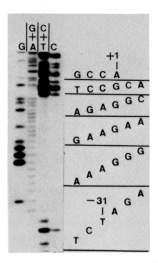

FIG. 4. Sequence of mutant DNA. The Bst NI (−44) to Hae III (160) fragment (Fig. 1), 5′-end labeled at the Bst NI site, was sequenced by the method of Maxam and Gilbert (1980) and electrophoresed on a 20% acrylamide urea gel. −31 and +1 refer to the number of nucleotides from the mRNA start point (see Fig. 1).

mutant Pst 5-Pst 6 fragment cut with Hae III next to the wild-type transcript on a 5% polyacrylamide-urea sequencing gel (Fig. 5B, lanes 1 and 2, and 3 and 4) demonstrates that both RNAs have the same mobility. Thus it appears that the transcription start site is the same for wild-type and mutant conalbumin genes. By scanning various exposures of the autoradiograms we found that specific transcription on the mutant fragment was about 5% of the wild type. Similar results were obtained using a variety of DNA concentrations (10–50 µg/ml in the assay) and DNA preparations (from both the fd103 and pBR322 recombinants).

DISCUSSION

We have previously established that a 20 bp region containing the "TATA" box of both the conalbumin and ad2 major late transcription units is required for specific initiation of *in vitro* transcription (see Introduction and Wasylyk *et al.*, 1980b; Corden *et al.*, 1980). We show here that a single base-pair transversion (T to G) at position −29 in the "TATA" box drastically decreases the efficiency of specific transcription of conalbumin DNA. This down-mutation, together with our previous results (Wasylyk *et al.*, 1980b; Corden *et al.*, 1980), demonstrates that the "TATA" box is an indispensable

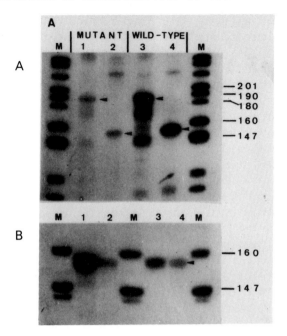

FIG. 5. *In vitro* transcription of wild-type and mutant conalbumin gene DNA. The wild-type and mutant conalbumin DNA fragments were transcribed *in vitro* as described previously using purified calf-thymus RNA polymerase B and HeLa cell S100 extract (Wasylyk *et al.*, 1980b), except that the nucleotide concentrations were 100 mM ATP, 100 mM GTP, 100 mM UTP and 10 μM CTP, and 0.13 mg/ml creatine kinase and 0.1 mM creatine phosphate were included in the reaction. (A) The ^{32}P-labeled RNA was displayed on a 5 % polyacrylamide urea gel. Lanes 1 and 3: 0.5 μg of mutant and wild-type Pst 5-Pst 6 fragment, respectively; lanes 2 and 4: 0.5 μg of mutant and wild-type Pst 5-Pst 6 fragments digested with Hae III, respectively (see Fig. 1). (B) The RNA synthesized on Pst 5-Pst 6 fragment digested with Hae III was electrophoresed on a 5 % polyacrylamide urea 40 cm-thin gel (0.3 mm). Lane 1: RNA synthesized on the wild-type fragment; lanes 2 and 4: RNA synthesized on the mutant fragment; lane 3: one fifth of the amount of RNA run in lane 1. M = ^{32}P-labeled DNA size markers. The arrowheads point to the bands discussed in the text.

element for initiation of specific *in vitro* transcription. Therefore, it appears that the "TATA" box fulfils at least one of the criteria used to define prokaryotic promoters. However, we cannot definitely conclude that the "TATA" box is part of a eukaryotic promoter region since prokaryotic promoters were defined as regions of the DNA which are indispensable for initiation of transcription and to which RNA polymerase binds (see Bogenhagen *et al.*, 1980; Rosenberg and Court, 1979; Siebenlist *et al.*, 1980 and references therein).

The "TATA" box shares both sequence and functional homologies to the

Pribnow box (Sakonju *et al.*, 1980; Corden *et al.*, 1980) of prokaryotic promoters (see Introduction). The effects of a variety of mutations in the Pribnow box have been recently reviewed (Rosenborg and Court, 1979; Siebenlist *et al.*, 1980). Strikingly, all A-T to G-C base-substitutions are promoter-down mutations, but G-C to A-T base-substitution is a promoter-up mutation, suggesting that the Pribnow box mutations are exerting some effect on local DNA melting. However, since A/T to T/A base-changes also influence transcription (for references, see Rosenberg and Court, 1979; Siebenlist *et al.*, 1980), this is probably not the major effect. These observations raise the question of whether the mutation we have constructed in the "TATA" box prevents a similar DNA-opening event in eukaryotic transcription or decreases the affinity of RNA polymerase B or factor(s) in the S100 extract, for this DNA sequence. In preliminary mixing experiments, using crude S100 extracts, we have been unable to demonstrate any competition between wild-type and mutant conalbumin DNAs for factor(s) present in the *in vitro* transcription system. However, since we cannot rigorously show that the DNA is in excess over essential factor(s) in the system, we cannot draw any firm conclusion about the mechanism by which the T to G transversion affects the efficiency of specific *in vitro* transcription of the conalbumin gene. Studies of the interaction of purified RNA polymerase B and factor(s) present in the S100 extract with the "TATA" box region will be necessary to answer this question.

Point mutations have unequivocally demonstrated the *in vivo* promoter role of the Pribnow box. Although it is tempting to speculate that the "TATA" box will also exhibit a promoter role *in vivo*, no direct evidence is available at the present time to support this assumption. Two reports dealing with this problem have recently been published. First, Grosschedl and Birnstiel (1980) found that for the sea-urchin histone H2A gene injected into *Xenopus oocytes*, deletion of the "TATA" sequence did not abolish transcription. Instead a number of new start sites of lower efficiency were generated. Second, Benoist and Chambon (1980) showed that a recombinant plasmid, constructed by inserting SV40 early genes into pBR322, expressed the early genes when introduced into eukaryotic cells. A recombinant with a 60 bp deletion, which removes the "TATA" box, still expressed the SV40 early genes. Along the same lines it should be recalled that there are two notable exceptions to the universal presence of the "TATA" box, the papovavirus late genes and the ad2 early region 2 genes (Baker *et al.*, 1979). These exceptions are accompanied by the occurrence of multiple start sites, suggesting that there may be more than one class of promoter for RNA polymerase B. However, the above *in vivo* experiments with deletion mutants do not exclude that the "TATA" box plays an essential role *in vivo*. For the experiments of Benoist and Chambon it is possible that, in the mutant lacking the "TATA" box, minor promoters, initiating upstream from the major cap sites, were responsible for the transcription of the early genes. In the experiments of Grosschedl and Birnstiel one cannot exclude that some of the sequences replacing the deleted DNA are mimicking the original "TATA" box function. It is clear that definite evidence for the role of the "TATA" box in promoting transcription will await *in vitro* studies with the purified components required for specific *in vitro* transcription and evaluation of the effect of point mutations in the "TATA" box region on *in vivo* transcription after introduction of the modified genes into *Xenopus oocytes* or living cells.

ACKNOWLEDGEMENTS

We would like to thank R. Herrmann, H. Schaller and C. Kédinger for gifts of materials, K. O'Hare for advise on cloning, C. Branlant and M. Wintzerith for help with the two-dimensional mobility shift analysis, F. Perrin for the electron microscopy and J. M. Bornert and C. Wasylyk for excellent technical assistance. This work was supported by grants from the CNRS (ATPs 4160, 3907 and 3558), the INSERM (ATP 72.79.104), la Fondation pour la Recherche Medicale Française and la Fondation Simone et Cino del Duca.

REFERENCES

Baker, C. C., Herisse, J., Courtois, G., Galibert, F. and Ziff, E. (1979). *Cell* **18**, 569–580.
Benoist, C. and Chambon, P. (1980). *Proc. natn. Acad. Sci. U.S.A.* In press.
Benoist, C., O'Hare, K., Breathnach, R. and Chambon, P. (1980). *Nucl. Acids Res.* **8**, 127–142.
Birkenmeier, E. H., Brown, D. and Jordan, E. (1978). *Cell* **15**, 1077–1086.
Bogenhagen, D. F., Sakonju, S. and Brown, D. D. (1980). *Cell* **19**, 27–35.
Breathnach, R. and Chambon, P. (1981). *A. Rev. Biochem.* **50**, in press.
Chambon, P. (1975). *A. Rev. Biochem.* **44**, 613–638.
Chambon, P. (1978). *Cold Spring Harb. Symp. quant. Biol.* **XLII**, 1209–1234.
Cochet, M., Gannon, F., Hen, R., Maroteaux, L., Perrin, F. and Chambon, P. (1979). *Nature, Lond.* **282**, 567–574.
Corden, J., Wasylyk, B., Buchwalder, A., Sassone-Corsi, P., Kédinger, C. and Chambon, P. (1980). *Science, N.Y.* **209**, 1406–1414.
Gannon, F., O'Hare, K., Perrin, F., LePennec, J. P., Benoist, C., Cochet, M., Breathnach, R., Royal, A., Garapin, A., Cami, B. and Chambon, P. (1979). *Nature, Lond.* **278**, 428–434.
Gillam, S. and Smith, M. (1979a). *Gene* **8**, 81–97.
Gillam, S. and Smith, M. (1979b). *Gene* **8**, 99–106.

Gillam, S., Jahnke, P., Astell, C., Phillips, S., Hutchinson, C. and Smith, M. (1979). *Nucl. Acids Res.* **6**, 2973–2985.

Grosschedl, R. and Birnstiel, M. L. (1980). *Proc. natn. Acad. Sci. U.S.A.* **77**, 1432–1436.

Hutchinson, C. A., Phillips, S., Edgell, M. H., Gillam, S., Jahnke, P. and Smith, M. (1978). *J. biol. Chem.* **253**, 6551–6560.

Jacob, F., Ullman, A. and Monod, J. (1969). *C.N. Acad. Sci., Paris* **258**, 3125–3131.

Kressmann, A., Hofstetter, H., Di Capua, E., Grosschedl, R. and Birnstiel, M. (1979). *Nucl. Acids Res.* **7**, 1749–1763.

Losick, R. and Chamberlin, M. (eds) (1976). "RNA Polymerase". Cold Spring Harbor Laboratory, New York.

Maxam, A. M. and Gilbert, W. (1980). *Meth. Enzymol.* **65**, 499–560.

McKnight, G. S. and Palmiter, R. D. (1979). *J. biol. Chem.* **254**, 9050–9058.

Perrin, F., Cochet, M., Gerlinger, P., Cami, B., LePennec, J. P. and Chambon, P. (1979). *Nucl. Acids Res.* **6**, 2731–2748.

Pribnow, D. (1975a). *Proc. natn. Acad. Sci. U.S.A.* **72**, 784–788.

Pribnow, D. (1975b). *J. molec. Biol.* **99**, 419–443.

Razin, A., Hirose, T., Itakura, K. and Riggs, A. D. (1978). *Proc. natn. Acad. Sci. U.S.A.* **75**, 4268–4270.

Roeder, R. G. (1976). *In* "RNA Polymerase" (R. Losick and M. Chamberlin, eds), pp. 285–329. Cold Spring Harbor Laboratory, New York.

Rosenberg, M. and Court, D. (1979). *A. Rev. Genet.* **13**, 319–353.

Sakonju, S., Bogenhagen, D. F. and Brown, D. D. (1980). *Cell* **19**, 13–25.

Schaller, H., Gray, C. and Herrmann, K. *Proc. natn. Acad. Sci. U.S.A.* **72**, 737–741.

Siebenlist, U., Simpson, R. B. and Gilbert, W. (1980). *Cell* **20**, 269–281.

Sollner-Webb, B. and Reeder, R. H. (1979). *Cell* **18**, 485–499.

Wasylyk, B., Derbyshire, R., Guy, A., Molko, D., Roget, A., Téoule, R. and Chambon, P. (1980a). *Proc. natn. Acad. Sci. U.S.A.* **77**, 7024–7028.

Wasylyk, B., Kédinger, C., Corden, J., Brison, O. and Chambon, P. (1980b). *Nature, Lond.* **285**, 367–373.

Weil, P. A., Luse, D. S., Segall, J. and Roeder, R. G. (1979). *Cell* **18**, 469–484.

Wu, G. J. (1978). *Proc. natn. Acad. Sci. U.S.A.* **75**, 2175–2179.

Ziff, E. B. and Evans, R. M. (1978). *Cell* **15**, 1463–1475.

Specific 5'-flanking Sequences Promote Initiation of Cell-free Transcription of the Ovalbumin Gene

MING-JER TSAI, SOPHIA Y. TSAI and BERT W. O'MALLEY

Department of Cell Biology, Baylor College of Medicine,
Houston, Texas 77030, U.S.A.

INTRODUCTION

The mechanisms which permit the transcription of eukaryotic genes are poorly understood. Development of a DNA-dependent, cell-free, *in vitro* transcription system would facilitate the delineation of the steps and possible control mechanism involved in the transcription process. In 1978, Wu demonstrated that accurate synthesis of a 5.5S RNA could be carried out using a soluble extract from KB cells containing polymerase III and adenovirus 2 (ad2) DNA (Wu, 1978). Subsequently, specific transcription of *Xenopus* 5S RNA and yeast tRNA was reported in an oocyte cell-free system (Birkenmeier *et al.*, 1978; Schmitt *et al.*, 1978; Ng *et al.*, 1979) and a KB cell-reconstituted system (Weil *et al.*, 1979b). With the construction of 5' and 3' deletion mutants and an *in vitro* assay system, Sakonju *et al.* (1980) and Bogenhagen *et al.* (1980) established that an internal fragment between 41–87 nucleotides of the *Xenopus* 5S gene contains sufficient information to direct the initiation of specific transcription of the 5S gene by RNA polymerase III. Recently, Engelke *et al.* (1980) isolated a protein factor which specifically interacts with the internal control region (45–96) of the 5S gene and facilitates accurate transcription of that gene.

Contrary to the rapid progress using the polymerase III transcription system, little is known about the initiation of synthesis of mRNA by RNA polymerase II. Recently, Weil *et al.* (1979a) demonstrated the selective initiation of transcription at a major late promoter of ad2 DNA using crude extracts from KB cells supplemented with purified RNA polymerase II. Similarly, Manley *et al.* (1980) showed the specific initiation of transcription at late promoters of ad2 DNA using a whole-cell extract from HeLa cells. Aided by these observations, we undertook to develop an *in vitro* transcrip-

279

tion system for the ovalbumin gene which would permit a definition of specific features in the DNA sequences which are essential for accurate initiation of transcription of the ovalbumin gene. Since we have previously defined the transcription unit of the ovalbumin gene (Roop *et al.*, 1980; Tsai *et al.*, 1980) and the sequences of the entire ovalbumin natural gene and the flanking DNA bordering the 5′ end of the cap site (Woo *et al.*, 1981), our search for the regulatory element required for initiation of transcription of the ovalbumin gene was facilitated. In this manuscript we report the establishment of an *in vitro* transcription system, using cloned ovalbumin DNA as template and HeLa-cell crude extract as a source for RNA polymerase II and other factors, which provides proper initiation of the ovalbumin DNA. Using this transcription system and deletion mutants generated by exonuclease III and S1 nuclease trimming techniques (Sakonju *et al.*, 1980), a region upstream from the cap site which included the Hogness box was found to be essential for correct initiation of the ovalbumin DNA.

TRANSCRIPTION OF OVALBUMIN DNA *IN VITRO*

A cloned ovalbumin gene fragment, OV1.7, which contains 5′ flanking sequences, the first structural sequence region and a portion of the first intervening sequences (Fig. 1) was used as template. The HeLa-cell crude extract prepared according to Manley *et al.* (1980) was used as the source of RNA polymerase and initiation factors. Since the 5′ end of ovalbumin mRNA is well defined (Roop *et al.*, 1980; Tsai *et al.*, 1980), one should observe an RNA product of 393 nucleotides in length if transcription starts at the cap site, identical to the *in vivo* start site. As shown in Fig. 2, a major radioactive labeled band with a length of 393 ± 5 was observed in lane 1. This is consistent with the notion that transcription of the ovalbumin DNA

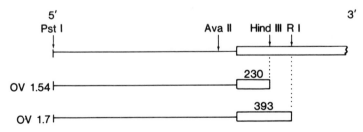

FIG. 1. Map of the 5′ end of the ovalbumin gene and its flanking sequences displaying the restriction endonuclease cleavage sites.

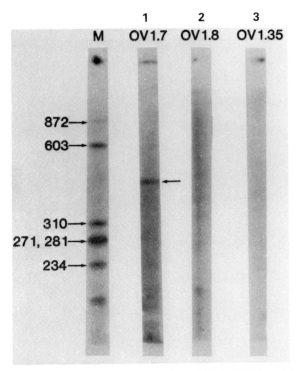

FIG. 2. Analysis of RNA synthesized *in vitro* from OV1.7, OV1.8 and OV1.35 by gel electrophoresis. RNAs synthesized from standard reaction mixture were denatured in formamide and loaded on 7M urea, 4% polyacrylamide gels. A standard 100 μl reaction mixture contained 12 mM *Hepes* pH 7.9, 3–5 mM $MgCl_2$, 60 mM KCl, 1.5 mM DTT, 10% glycerol, 0.2 mM EDTA, 500 μM ATP, CTP and UTP, 50 μM GTP containing 50 μCi $\alpha^{32}P$ GTP, 2.5 μg DNA and 60 μl of crude extract from HeLa cells. The HeLa crude extract was prepared according to the method of Manley *et al.* (1980) except that it was dialyzed against a buffer containing 5 mM $MgCl_2$ prior to storage at −70°C. The reaction mixtures were incubated at 30°C for 45 min and RNA isolated as described by Tsai *et al.* (1978). The autoradiograms show the products synthesized by various templates. Arrows indicate the position of the specific runoff transcript. End-labeled Hae III-digested fragments of φX174 were used as size markers to determine the length of the specific *in vivo* transcripts.

fragment begins at the cap site. When an internal fragment from the ovalbumin gene, OV1.8 (Eco RI-Eco RI fragment (Dugaiczyk *et al.*, 1978)) was used as template, no specific product could be detected (Fig. 2, lane 2). Similarly, another internal fragment of ovalbumin gene, OV1.35 (Hinf 1-Hinf 1) (Dugaiczyk *et al.*, 1979), did not support specific transcription of the ovalbumin gene (Fig. 2, lane 3). These results suggested that OV1.7 might contain some sequences which are important for initiation of transcription of the ovalbumin gene.

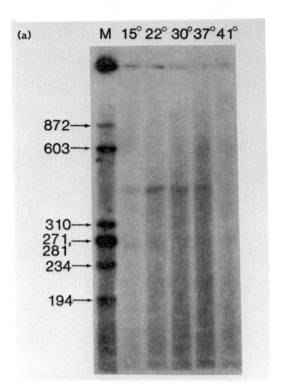

(a)

M   15° 22° 30° 37° 41°

872 →
603 →

310 →
271, →
281
234 →

194 →

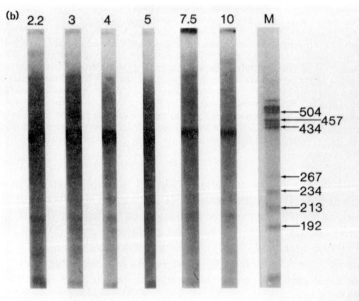

(b)   2.2   3   4   5   7.5   10   M

504
457
434

267
234
213
192

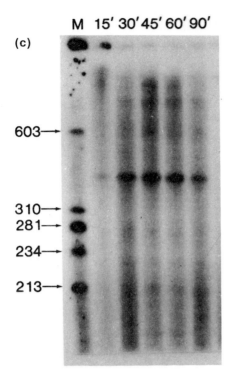

FIG. 3. Conditions for optimal RNA synthesis from OV1.7. RNA synthesis was carried out as described in Fig. 2 except that (a) various temperatures (°C), (b) concentrations of $MgCl_2$ (μg/ml), and (c) incubation times (min) were used.

The optimum conditions for specific *in vitro* transcription were determined. Synthesis of the 395 nucleotides RNA product was optimum between 22–30°C (Fig. 3a) at a concentration of 3–5 mM Mg^{2+} (Fig. 3b) and at a DNA concentration of 25 μg/ml for 45 min (Fig. 3c). These conditions were then used for all the following experiments.

We next examined which class of RNA polymerase was responsible for the specific transcription of ovalbumin DNA templates in this *in vitro* system. RNA was synthesized in the presence or absence of low concentrations of α-amanitin (2 μg/ml) using both OV1.7 and OV1.54 as templates. As shown in Fig. 4, the presence of α-amanitin completely abolished the synthesis of the specific RNA products indicating that these RNAs were products of RNA polymerase II.

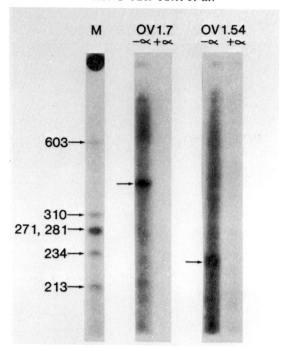

FIG. 4. Electrophoretic analysis of RNAs synthesized from truncated ovalbumin DNA in the presence and absence of α-amanitin. 2.5 μg of OV1.7 or OV1.54 were used as template. Reaction conditions were as described in the legend of Fig. 2. RNA samples were resuspended in 99% formamide, denatured by heating at 100°C for 5 min and then loaded on a 4% gel. Gel electrophoresis was carried out at 20 mA/4 hour as described by Maniatis *et al.* (1976). Autoradiograms were obtained by exposure to Kodak XR-p film with an intensifier screen at −20°C. The specific run-off transcripts as discussed in the text are indicated by arrows. α-amanitin sensitivity was measured in the presence of 2 μg/ml of α-amanitin.

SELECTIVE INITIATION OF OVALBUMIN DNA TEMPLATES THAT ARE TRUNCATED AT THE 3′ END

To ascertain that the RNA product transcribed from the OV1.7 DNA fragment was indeed initiated at the cap site, different restriction fragments of ovalbumin DNA, having identical 5′ ends but varying lengths of 3′ sequences (Fig. 1) were used as templates. If transcription initiates at the cap site and extends rightward towards the 3′ end of the DNA fragment, the truncated templates OV1.7 (Pst I-Eco RI) and OV1.54 (Pst I-Hind III) should yield run-off RNA products of 393 and 230 nucleotides in length, respectively. As shown in Fig. 4, a major RNA species with a size of 393 nucleotides can be seen in the autoradiogram when OV1.7 (Pst I-Eco RI) fragment was used as template. The shorter fragment OV1.54 (Pst I-Hind III) yielded a major product of 230 nucleotides in length (Fig. 4) and no apparent product of 393 nucleotides could be detected. The size of the RNA products synthesized by the truncated templates closely agreed with the predicted values, 395 ± 5 versus 393 for OV1.7 and 230 ± 3 versus 230 for OV1.54 (the average value of at least six sets of experiments). These results strongly suggest that transcription initiates at a specific site *in vitro*, elongates in a rightward direction, and terminates at the 3′ end of the truncated DNA templates.

To further substantiate this conclusion, we carried out S1 mapping studies using the RNA products of the reaction and radiolabeled ovalbumin DNA fragments. S1 nuclease digestion of hybrids between the 393 nucleotide RNA product of OV1.7 and OV0.36 (Ava II-Hind III, Fig. 1) should yield a single RNA band at nucleotide position 230 if RNA synthesis was initiated at the cap site of the ovalbumin gene. The result of this experiment is shown in Fig. 5. As expected, the ^{32}P RNA was trimmed from 393

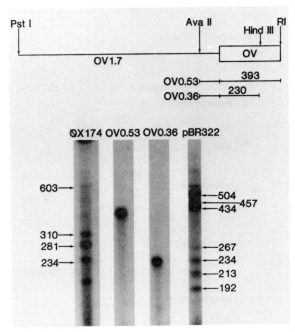

FIG. 5. S1 mapping of 393 nucleotides transcription product of OV1.7. ^{32}P RNA was synthesized from OV1.7 according to the procedure described in Fig. 2. After electrophoresis in 4% acrylamide gel, the 393 nucleotide band was extracted according to the procedure described by Maxam and Gilbert. Isolated 393 nucleotides/^{32}P/RNA was then hybridized to OV0.36 on OV0.53 DNAs as described by Roop *et al.* (1980) except that NaCl was at 0.5 M. After S1 nuclease digestion, the resistant ^{32}P RNAs were then run in 4% acrylamide gel and autoradiographed.

nucleotides to 230 nucleotides in length following hybridization to the OV0.36 DNA fragment and S1 nuclease digestion. Therefore, initiation of synthesis of the major RNA product in this *in vitro* transcription system starts at the *in vivo* cap site.

CONSTRUCTION OF 5′ DELETION MUTANTS

To test whether sequences flanking the 5′ end of the ovalbumin gene are important for initiation of the specific RNA *in vitro*, cloned pOV1.7 plasmid DNA was first digested with Bam HI and Ava II to yield a 900-nucleotide DNA fragment. This fragment was then subjected to trimming by exonuclease III to generate DNA with various lengths (Sakonju *et al.*, 1980) as illustrated in Fig. 6. Single-stranded DNA tails were then digested with S1

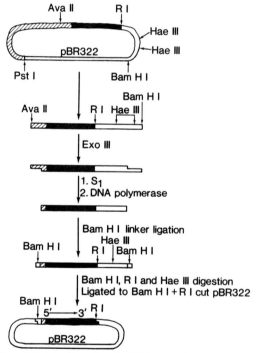

FIG. 6. Construction of 5′ deleted ovalbumin DNA according to the procedure of Sakonju *et al.* (1980). The blunt-ended DNAs were ligated to Bam H1 linkers and recloned as Bam H1/Eco RI inserts in pBR322. Transformation and identification of the recombinant plasmids were carried out according to Stein *et al.* (1978). Exact end points of the deletion were determined by DNA sequencing using the procedure of Maxam and Gilbert (1977). pOV0.2 DNA was cloned by inserting the Bam HI linker-linked 5′-end ovalbumin DNA fragment (boundary between Ava II and Taq I sites) in the Bam HI site of pBR322 DNA.

nuclease and DNA polymerase was used to fill in any residual single-strand regions. The blunt-ended DNAs were ligated to Bam HI linkers, digested with Bam HI, Eco RI and Hae III and then recloned at Bam HI and Eco RI sites in pBR322. The clones were then digested with Bam HI, Eco RI and Hind III and subjected to gel electrophoresis. The size of the Bam HI-Eco RI fragments depended on the length of the sequences deleted during exonuclease III treatment. The Bam HI-Eco RI fragments were labeled at the 3' end and then sequenced to locate the exact location site (Maxam and Gilbert, 1977). Some of the mutants obtained are diagrammed in Fig. 7. Three of them contained deletions up to a position 5' to TATATAT sequences. In one, all of the TATATAT nucleotides except the last T were deleted. The rest of them contained deletions into the internal gene sequences.

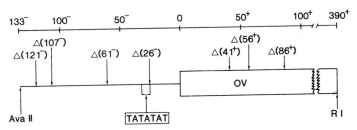

FIG. 7. Ovalbumin clones containing 5' deleted sequences. The 5' deleted ovalbumin DNA ($\Delta 5'$) is a series of clones in which the direction of deletion approaches from the 5' end towards the 3' end. The precise end point of deletion is determined by DNA sequencing. For instance, 121^- indicates that the ovalbumin flanking DNA is deleted to the left of nucleotide 121^- which itself is 121 nucleotides upstream from the cap site of the ovalbumin gene.

TRANSCRIPTION OF THE 5' AND 3' DELETION MUTANTS

DNAs from the above 5' deletion mutants were digested with Ava I and Eco RI. The Ava I-Eco RI fragments contained the ovalbumin DNA sequences and 1.05 kb of plasmid sequences upstream from the ovalbumin DNA. These fragments were purified by agarose gel electrophoresis and then used as templates for *in vitro* transcription. As shown in Fig. 8, the specific run-off product of the ovalbumin DNA (393 nucleotides in length) was synthesized by deletion mutants $OV\Delta 5'(121^-)$, $OV\Delta 5'(107^-)$ and $OV\Delta 5'(61^-)$. Thus, deletion of any DNA sequences upstream from position 61^- did not affect the initiation of transcription of the ovalbumin DNA *in vitro*. By contrast, transcription of the specific product was completely abolished when $OV\Delta 5'(26^-)$, $OV\Delta 5'(41^+)$, $OV\Delta 5'(56^+)$ and $OV\Delta 5'(86^+)$ mutants were

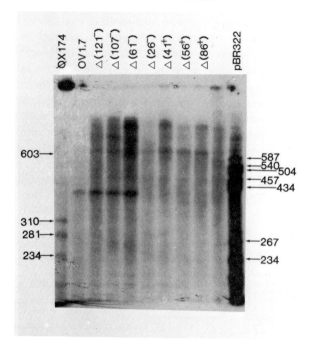

FIG. 8. Electrophoretic analysis of transcripts from cloned 5' deleted ovalbumin DNA. DNAs were isolated from various deletion mutants described in Fig. 7 by Ava I and Eco RI restriction digestion. ^{32}P RNA synthesis was carried out as described in Fig. 2. The autoradiograms show *in vitro* transcribed ^{32}P RNA from various clones. The size markers are ^{32}P end-labeled Hae III fragments of ϕX174 and Hae III fragments of pBR322.

used as templates. It should be noted that some higher molecular weight products were seen in the autoradiogram when the deletion mutant DNA was used as template. These appeared to result from transcription of the plasmid DNA but no attempt was made to characterize them. The deletion of sequences between 61$^-$ and 26$^-$ completely eliminated the initiation of ovalbumin DNA and suggests that this region of DNA is essential for initiation of ovalbumin DNA *in vitro*. Interestingly, OV 5'(26$^-$) contains only the last nucleotide of the Hogness box (TATATAT in the case of ovalbumin) which exists in most eukaryotic genes and has been postulated to function as a possible recognition site by RNA polymerase II.

In order to define further the important region which determines the initiation of RNA synthesis, we cloned a DNA fragment generated by Ava II and Taq I restriction digestion (pOV0.2). This DNA fragment contains 133 nucleotides of 5'-flanking DNA but only 43 nucleotides of ovalbumin gene sequence. If internal gene sequences 3' to position 43 are required for the initiation of RNA synthesis, we would not have expected to detect an RNA product initiated at the cap site when this DNA was used as template. As shown in Fig. 9, when a DNA fragment generated by Sal I and Eco RI from pOV0.2 was used as template, an RNA product with the expected size (420 nucleotides) was detected. Therefore, it appears as if there were sufficient informational sequences in this DNA fragment to direct faithful RNA initiation at the cap site. Thus, we conclude that sequences 3' to position 43 are not essential for initiation of transcription of the ovalbumin gene.

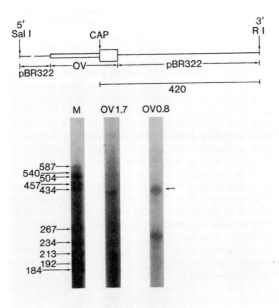

FIG. 9. Analysis of RNA synthesized *in vitro* from OV0.8 and OV1.7. pOV0.2, chimeric plasmid containing 5' ovalbumin gene, Bam HI linker-linked ovalbumin DNA fragment (Ava II(at position 133⁻)-Taq I(at position 43⁺) was constructed at the Bam HI site of pBR322. OV0.8 was obtained from pOV0.2 by Sal I and Eco RI digestion and electrophoresis in agarose gel. Synthesis and analysis of ^{32}P RNA transcription products of OV0.8 and OV1.7 were carried out as described in Fig. 2.

IS THE HOGNESS BOX A SUFFICIENT SIGNAL FOR SPECIFIC INITIATION OF TRANSCRIPTION OF THE OVALBUMIN GENE?

In addition to the presumptive promoter, the Hogness box (TATATAT) at 32⁻ position of the ovalbumin DNA, there are several other similar sequences present in the OV1.7 DNA fragment (Table I). If these sequences can be recognized by RNA polymerase as promoters, we would expect to find RNA products of OV1.7 having the sizes listed in Table I.

As shown in Fig. 9 and also previous figures, the only major RNA product observed was 393 nucleotides in length. We conclude that specific transcription of the ovalbumin gene is directed solely by the Hogness box situated at −32 position. One might argue that RNA polymerase or factors in our *in vitro* system may concentrate on the strong promoter (at position −32) and leave out other weak ones unattended. Thus we obtained a DNA fragment restricted by Hae III (OV0.7) which was devoid of most of the Hogness boxes including the one at position −32. With this DNA fragment we might expect that without competition from a stronger promoter, the RNA polymerase could concentrate at this promoter region if it is possible to be used. As shown in Fig. 10, no major RNA band with a predicted size of 310 nucleotides could be synthesized from this DNA template (OV0.7). Therefore, most of the Hogness boxes other than the one located at position −32 cannot be used as meaningful promoters for *in vitro* RNA synthesis. Taken together, our entire results suggest that the Hogness box is essential but not sufficient for specific initiation of RNA synthesis of the ovalbumin gene.

TABLE I

Existence of false Hogness boxes in the OV1.7 DNA fragment

Position	Sequences			Expected RNA product
	Hogness box		Possible initiation site	
−1090	[1]TATATAT	. . . . .	[30]AAACAAA	∼1451
−825	TATATTA	. . . . .	ACTCCTC	∼1186
−624	TATAAAG	. . . . .	GATTTAA	∼985
−607	CATAAAA	. . . . .	TATAGGA	∼968
−32[a]	TATATAT	. . . . .	TGTACAT	393
+78	TATATTA	. . . . .	AAAAAAT	∼283
+174	TATAAAA	. . . . .	TCTGCAC	∼187

[a] Only the 393 nucleotides band was observed experimentally.

DISCUSSION

Our present study has demonstrated that specific transcription of ovalbumin DNA can be initiated *in vitro* by using cloned ovalbumin DNA and a crude total-cell extract from HeLa cells. Initiation of *in vitro* transcription appears to be accurate as judged by S1 mapping and by the size of RNA products synthesized from various ovalbumin DNA templates that were truncated at the 3' end. Thus, *in vitro* initiation of transcription occurs at a unique site, the *in vivo* cap site, and extends rightwards to the 3' end of the ovalbumin DNA fragment.

The efficiency of the *in vitro* initiation of transcription for the ovalbumin gene is comparable to that of the ad2 late promoter as described by Manley *et al.* (1980) and Weil *et al.* (1979a). Approximately 2×10^{-3} pmol of specific ovalbumin RNA transcripts were calculated to be synthesized in a 50 μl reaction mixture in a period of 45 min. For the ad2 late promoter, 3×10^{-3} and 5×10^{-4} pmol of specific RNA were synthesized in a 50 μl reaction mixture in 60 min as estimated by Manley *et al.* (1980) and Weil *et al.* (1979a), respectively. In our hands, approximately 0.01 molecule of RNA transcript is made per molecule of DNA in an hour. This is extremely slow

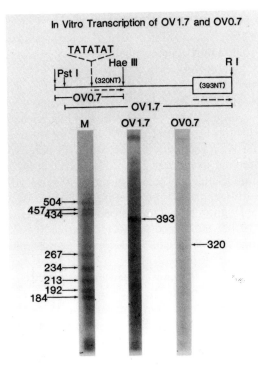

FIG. 10. Analysis of RNA synthesized *in vitro* from OV0.7 and OV1.7. DNA fragment (OV0.7) was isolated from pOV1.7 by Hae III and Eco RI restriction enzymes digestion as illustrated in the map. ³²P RNAs were synthesized and analyzed according to the procedure described in Fig. 2.

as compared to the *in vivo* rate of transcription, but is not surprising since pertinent regulatory factors are likely to be missing in such heterologous *in vitro* transcription systems. The fact that accurate initiation of transcription can be obtained in these heterologous systems using various eukaryotic genes suggests that a universal signal may be encoded in the 5' flanking sequence to the gene which is important for initiation *in vivo*.

The studies on transcription of 5' deletion mutants indicates that the deletion of DNA sequences between position 61$^-$ and 26$^-$ completely abolishes the initiation of synthesis of the specific product. The ovalbumin gene, like many other eukaryotic genes (ad2, globin, SV 40, histones, conalbumin) contains a conserved DNA sequence, TATATAT (Hogness box) 26–32 nucleotides upstream from the cap site. Our results imply that a limited region of 5'-flanking DNA, which includes the Hogness box and 29 nucleotides upstream from it, is required for *in vitro* initiation for the ovalbumin gene. Taken together, these results implicate the conserved DNA sequences, the Hogness box, as the presumptive promoter which is necessary for *in vitro* initiation of polymerase II gene products. However, in view of the low efficiency of the *in vitro* transcription system, these results do not exclude the possibility that other DNA regions might also be important for specific interaction with regulatory proteins.

It should be noted that in Fig. 7 the deletion of different lengths of the 5'-flanking sequence upstream from the Hogness box has a *polar* (gradient) effect on the rate of initiation of the specific 395 nucleotides product. Deletion of all sequences upstream from 61$^-$ gives rise to the highest level of transcription of the 395 nucleotides product while the parental OV1.7 which has the longest-flanking ovalbumin DNA sequence has the lowest level of specific transcription. Similar to our observation, Grosschedl and Birnstiel (1980) have recently reported that deletion of the conserved sequences upstream from the Hogness box of the histone (H2A) gene also enhances the rate of synthesis of the H2A gene product. Whether these sequences can function as binding sites for specific inhibitors of transcription can only be subjected to speculation at present.

The observation that 5'-flanking sequences are potential promoters for initiation of transcription of the ovalbumin and other eukaryotic genes indicates that the recognition sites for eukaryotic polymerase II and polymerase III might be very different at the level of the DNA. Polymerase III appears to require internal DNA sequences and a protein factor which interacts with 5S genes in the internal gene region between position 41$^+$ and 85$^+$ (Sakonju *et al.*, 1980; Bogenhagen *et al.*, 1980; Engelke *et al.*, 1980).

Recently, we have also been able to obtain accurate initiation of transcription of ovomucoid gene fragments *in vitro*. We transcribed an ovomucoid DNA fragment containing the cap site, 154 nucleotides of the gene sequence and only 10 nucleotides of the 5'-flanking sequences. When the Hogness box was deleted, no specific run-off RNA product was detected. Again, this observation is consistent with the hypothesis that the Hogness box is necessary for specific initiation of transcription *in vitro* by RNA polymerase II.

Finally, it should be noted that other Hogness box-type sequences at internal gene positions or 5'-flanking positions (e.g., 1090^- upstream) of the OV1.7 DNA do not serve as a signal for initiation of a specific RNA product *in vitro*. Thus, it is likely that other genomic sequences, in addition to the Hogness box, are essential for specific initiation of gene transcription.

SUMMARY

An *in vitro* system was used for studying initiation of transcription of the ovalbumin gene. The DNA template was a cloned ovalbumin gene fragment that contained 5'-flanking sequences, and part of the 5' end of the ovalbumin gene. An HeLa-cell crude extract was used as the source of RNA polymerase and initiation factors. Correct initiation was judged by the sizes of transcription products generated from ovalbumin templates that were truncated at various positions prior to the 3' end of the gene and by S1 mapping of the RNA products. Transcription of the specific product was carried out by RNA polymerase II as judged from α-amanitin sensitivity. A series of deletion mutants were constructed by trimming 5'-flanking sequences of the ovalbumin DNA template using exonuclease III. The DNAs generated were then recloned in pBR322 and used as templates to determine which sequences were necessary for initiation of transcription. Specific initiation of the ovalbumin gene was unaffected by deletion of all but 61 nucleotides of the 5'-flanking sequence, but was completely abolished by deletion of all but 26 nucleotides of the 5'-flanking sequence. Thus a region between 61 and 26 nucleotides upstream from the cap site, which includes the Hogness box (TATATAT) at position 32^- to 26^- is essential for the correct initiation of the ovalbumin gene. A DNA clone containing 133 nucleotides of 5'-flanking and 43 nucleotides of gene sequence can also serve as template to initiate RNA synthesis. This result suggests that sequences within the gene (downstream from nucleotide position 43) are not important for correct initiation of RNA synthesis. Nevertheless, natural DNA fragments contain false Hogness boxes in addition to the one normally located in the immediate 5'-flanking region of an authentic gene. Most of these Hogness boxes did not serve as promoters for initiation of transcription *in vitro*. These results suggest that the Hogness box is essential, but not sufficient for specific initiation of RNA synthesis.

ACKNOWLEDGEMENTS

We would like to thank Dr. Charles Lawrence for providing HeLa cells and Drs A. Dugaiczyk and T. Schulz for their valuable advice in the construction of the 5' deletion mutants. We would also like to thank Mr Warren Zimmer and Misses Iris Fund and Cindy Herzog for their expert technical assistance. This work was supported by NIH grant HD-08188 and an NIH Center Grant for Population Research and Reproductive Biology, HD-07495-08.

REFERENCES

Birkenmeier, E. H., Brown, D. D. and Jordan, E. (1978). *Cell* **15**, 1077–1086.
Bogenhagen, D. F., Sakonju, S. and Brown, D. D. (1980). *Cell* **19**, 27–35.
Dugaiczyk, A., Woo, S. L. C., Lai, E. C., Mace, M. L., Jr., McReynolds, L. A. and O'Malley, B. W. (1978). *Nature, Lond.* **274**, 328–333.
Dugaiczyk, A., Woo, S. L. C., Colbert, D. A., Lai, E. C., Mace, M. L., Jr. and O'Malley, B. W. (1979). *Proc. natn. Acad. Sci. U.S.A.* **76**, 2253–2257.
Engelke, D. R., Ng, S. Y., Shastry, B. S. and Roeder, R. G. (1980). *Cell* **19**, 717–728.
Grosschedl, R. and Birnstiel, M. L. (1980) *Proc. natn. Acad. Sci. U.S.A.* **77**, 1432–1436.
Maniatis, T., Jeffrey, A. and Van de Sande, H. (1976). *Biochemistry* **14**, 3787–3794.
Manley, J. L., Fire, A., Cano, A., Sharp, P. A. and Gefter, M. L. (1980). *Proc. natn. Acad. Sci. U.S.A.* **77**, 3855–3859.
Maxam, A. M. and Gilbert, W. (1977). *Proc. natn. Acad. Sci. U.S.A.* **74**, 560–564.
Ng, S. Y., Parker, C. S. and Roeder, R. G. (1979). *Proc. natn. Acad. Sci. U.S.A.* **76**, 136–140.
Roop, D. R., Tsai, M.-J. and O'Malley, B. W. (1980). *Cell* **19**, 63–83.
Sakonju, S., Bogenhagen, D. F. and Brown, D. D. (1980). *Cell* **19**, 13–25.
Schmidt, O., Mao, J. L., Silverman, S., Hovemann, B. and Soll, D. (1978). *Proc. natn. Acad. Sci. U.S.A.* **75**, 4819–4823.
Stein, J. P., Catterall, J. F., Woo, S. L. C., Means, A. R. and O'Malley, B. W. (1978). *Biochemistry* **17**, 5763–5772.
Tsai, S. Y., Roop, D. R., Tsai, M.-J., Stein, J. P., Means, A. R. and O'Malley, B. W. (1978). *Biochemistry* **17**, 5773–5780.
Tsai, S. Y., Roop, D. R., Stumph, W. E., Tsai, M.-J. and O'Malley, B. W. (1980). *Biochemistry* **19**, 1755–1761.
Vogelstein, B. and Gillespie, D. (1979). *Proc. natn. Acad. Sci. U.S.A.* **76**, 615–619.
Weil, P. A., Luse, D. S., Segall, J. and Roeder, R. G. (1979a). *Cell* **18**, 469–484.
Weil, P. A., Segall, J., Harris, B., Ng, S. Y. and Roeder, R. G. (1979b). *J. Biol. Chem.* **254**, 6163–6173.
Woo, S. L. C., Beatie, W. G., Catterall, J. F., Dugaiczyk, A., Staden, R., Brownlee, G. G. and O'Malley, B. W. (1981). *Biochemistry*, submitted.
Wu, G. J. (1978). *Proc. natn. Acad. Sci. U.S.A.* **75**, 2175–2179.

Expression of Cloned Eukaryotic Genes in Microorganisms

O. MERCEREAU-PUIJALON,[1] A. GARAPIN,[1] F. COLBÈRE-GARAPIN[2] and P. KOURILSKY[1]

[1] *Unité de Biologie Moléculaire du Gène,*
E.R. C.N.R.S. 201 et S.C.N. I.N.S.E.R.M. 20,
Institut Pasteur,
28 rue du Docteur Roux,
75724 Paris Cedex 15, France

[2] *Unité de Virologie Médicale,*
Institut Pasteur,
28 rue du Docteur Roux,
75724 Paris Cedex 15, France

INTRODUCTION

Recombinant DNA techniques create new experimental situations in which the expression of cloned genes can be studied in a foreign cellular environment. Many eukaryotic genes have been cloned in microorganisms, particularly *E. coli*. Apart from some genes from the lower eukaryotes, it has generally been found that spontaneous expression into protein does not take place to an appreciable extent in *E. coli*. Signals which specify transcription and translation in eukaryotes may not be accurately recognized by the *E. coli* machinery. In addition, many, if not most of the genes from the higher eukaryotes are split by introns which interfere with their expression in the bacterium (see below). In most cases, therefore, expression into protein is observed only upon proper manipulation of a "continuous" version of the genes, often a cloned double-stranded cDNA sequence derived from the mRNA. Here we shall review work carried out in our laboratory and underline some of the difficulties and prospects raised by such analysis.

PROMOTING TRANSCRIPTION AND TRANSLATION IN *E. COLI*

An *E. coli* promoter, located upstream of the foreign DNA sequence, should promote its transcription by *E. coli* RNA polymerase, unless the sequence contains a termination signal recognized by this enzyme. Even so, there exist genetic tools which would help avoiding incomplete transcription (*E. coli* rho ts mutants, transcription from the early promoters of bacteriophage λ)

295

(Adhya and Gottesman, 1978). At present, no difficulties have been reported in promoting transcription of cDNAs (e.g. Kourilsky *et al.*, 1977) but problems may be encountered with genomic fragments of DNA. It is known that some eukaryotic sequences, presumed to act as termination signals, contain a T-rich sequence and thereby resemble some of their prokaryotic counterparts (Lebowitz and Weissman, 1979). We found, for instance, that a piece of *Xenopus laevis* DNA containing several differently oriented tRNA genes (Müller and Clarkson, 1980) reduced transcription downstream about 5-fold, although transcribed under λp_L control in the presence of λN protein, which is known to alleviate many termination signals in *E. coli* (Mercereau-Puijalon and Kourilsky, 1976, unpublished experiments).

Promoting transcription of a foreign gene does not ensure that translation follows. First, the primary sequence of the mRNA may not contain the information needed for initiation of translation (for instance, a ribosome binding site) (Steitz, 1979). Second, there are other, still poorly understood requirements which are, perhaps, related to mRNA secondary structure (Steitz, 1979). A striking example is provided by the lack of expression of the *E. coli lac* Z gene when transcribed through the *lac* operator promoter region from another promoter such as the *trp* promoter (Reznikoff *et al.*, 1974) or λp_L (Mercereau-Puijalon and Kourilsky, 1976). In the latter case, we showed that the entire *lac* Z gene is actively transcribed, but not translated into β-galactosidase, despite the fact that the transcript contains the sequence required for initiation of translation (Mercereau-Puijalon and Kourilsky, 1976). In this inactive transcript, longer than the usual *lac* mRNA in the 5' region, the Shine and Dalgarno sequence may be masked by hybridization with complementary sequences in the operator promoter region (Mercereau-Puijalon, 1980). Similarly, the λ *cro* gene has been expressed in *lac–cro* fusions where the distance between the *lac* Shine and Dalgarno sequence and the *cro* AUG was varied (Roberts *et al.*, 1979). The level of expression could be correlated with the putative secondary structure of the 5' region of the mRNA (Iserentant and Fiers, 1980).

The elongation process itself can be strongly impaired by the secondary structure of the mRNA, as has been reported for the attenuation mechanism in *E. coli* (see Adhya and Gottesman, 1978, for review). The presence, in the eukaryotic sequence, of codons poorly used in *E. coli* may reduce synthesis because the corresponding isoacceptor tRNAs are present in limiting amounts, but the extent of such interference is still unknown.

Finally, even if the protein is synthesized in *E. coli*, it may be susceptible to bacterial proteases and be unstable (Villa-Komaroff *et al.*, 1978) or kill the host microorganism, which could explain some difficulties and failures in observing expression. We describe below two examples of foreign proteins synthesized in *E. coli* after proper manipulation of the gene sequences.

CHICKEN OVALBUMIN AND *HERPES* VIRUS THYMIDINE KINASE SYNTHESIS IN *E. COLI*

To express a chicken ovalbumin cloned cDNA in *E. coli*, we fused it on to the beginning of the *lac* Z gene in such a way that the seventh amino acid of β-galactosidase was connected with the fifth amino acid of ovalbumin. The *E. coli* strains harboring the appropriate recombinant plasmids made large amounts of an ovalbumin-like protein (OLP) (0.2% to 1% of total protein), as measured in a radioimmunoassay (Mercereau-Puijalon *et al.*, 1978). The protein is quite stable, has the expected structural characteristics and exhibits a functional feature of native ovalbumin with regard to secretion (see below).

More recently, we undertook an analysis of the *Herpes* thymidine kinase (TK) gene (Colbère-Garapin *et al.*, 1979). To promote its expression in *E. coli*, we fused an Eco RI fragment which was found to contain the coding sequence, with the beginning of the *lac* Z gene in each of the three translational phases, using the appropriate vectors (Charnay *et al.*, 1978). When we transferred these plasmids into an *E. coli* TK⁻ host, we observed complementation of the bacterial defect by the expressed viral gene in all three cases (Garapin *et al.*, 1980). DNA sequencing showed that, irrespective of the translational frame dictated by the *lac* region, stop signals were encountered in all three phases in the beginning of the *Herpes* virus DNA fragment, ahead of the presumptive TK coding sequence. This, and the fact that the viral enzyme made in *E. coli* has the same molecular weight in all phases demonstrates that reinitiation of translation takes place within the eukaryotic fragment.

It is known, from the well-studied example of *lac i* nonsense mutants that *E. coli* ribosomes are capable of reinitiating translation within a cistron in the same reading frame (Miller, 1978). In the case of mouse dihydrofolate reductase, (dHFR), it has been shown that reinitiation at the dHFR's AUG occurs in a frame different from that of the plasmid gene in which it is embedded (Chang *et al.*, 1980), without any stop signal upstream. Synthesis of other eukaryotic proteins in *E. coli* seems to involve reinitiation events, such as that of hepatitis Hbc antigen (Pasek *et al.*, 1979), influenza hemagglutinin (Emtage *et al.*, 1980), and human leukocyte interferon (Nagata *et al.*, 1980). These examples show that the mechanisms which promote, or interfere with, the initiation of translation have not all been identified or mastered.

OVALBUMIN SYNTHESIS IN YEAST

By using appropriate composite vectors capable of being propagated both in

E. coli and the yeast *Saccharomyces cerevisiae*, we introduced the *lac*-ovalbumin fusion described above into yeast, where it is expressed (Mercereau-Puijalon *et al.*, 1980). Yeasts make a few thousand molecules of OLP per cell (Fig. 1), and there is some evidence that the *E. coli lac* regulatory region can serve as a promoter in yeast. In the course of these studies, we accidentally identified a region of the 2μ plasmid essential for its maintenance in yeast (Mercereau-Puijalon, 1980; Mercereau-Puijalon *et al.*, 1980). Yeast OLP appears to be much less stable than its *E. coli* counterpart, and there is more segregation of the recombinant plasmid in yeast than in *E. coli*. If yeast is to be used as a productive host, it will probably be necessary to improve the stability of the strains and to resolve possible problems related to protein degradation.

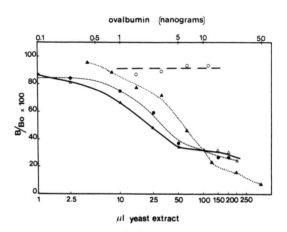

FIG. 1. Assay of OLP in yeast extracts. Yeast cells were transformed with the plasmid vector pFL1, or with two derivatives of it (pOMP 225 and pOMP 915) carrying an in-phase fusion of ovalbumin cDNA with the beginning of the *E. coli lac* Z gene. The presence of OLP in extracts was monitored by radioimmunoassay (see Mercereau-Puijalon *et al.*, 1978). Quantification was done by comparison with the competition curve obtained with pure chicken ovalbumin (▲ ... ▲) pOMP 225 (△——△) and pOMP 915 (●---●) extracts containing competing material, estimated to 20 ng and 5 ng of OLP per mg protein respectively, whereas control pFL1 extracts (○---○) do not. B/B_0 represents the ratio of tracer ^{125}I ovalbumin bound to the antibodies in the presence (B) or in the absence (B_0) of any competing material. For details see Mercereau-Puijalon *et al.* (1980).

INTRONS PREVENT OLP SYNTHESIS IN *E. COLI*

We used the ovalbumin-producing *E. coli* strains to demonstrate that the introduction of introns of the natural ovalbumin gene into the ovalbumin cDNA prevents detectable OLP synthesis (Mercereau-Puijalon and Kourilsky, 1979). There has been no report of prokaryotic split gene to date and we therefore expected to find that *E. coli* is unable to deal with introns. More surprising is the situation in yeast, which has been explored by others. While there exists split chromosomal gene (e.g. actin) (Gallwitz and Sures, 1980), it has been shown that the chromosomal rabbit β-globin gene is not correctly transcribed and processed (Beggs *et al.*, 1980). In contrast, there is correct processing of a variety of mouse, chicken and human genes in a variety of animal cells (Hamer and Leder, 1979; Mantei *et al.*, 1979; Mulligan *et al.*, 1979; Wold *et al.*, 1979; Breatnach *et al.*, 1980; Lai *et al.*, 1980).

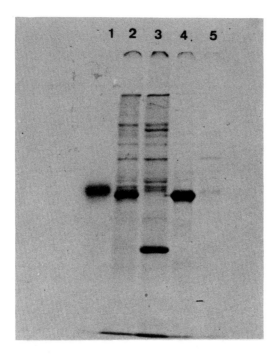

FIG. 2. Size of bacterial OLP synthesized *in vivo*. OLP in bacterial extracts from cells labeled with ^{35}S methionine was immunoprecipitated by specific antibodies. The proteins were separated by electrophoresis on SDS polyacrylamide gels and revealed by autoradiography; lane 1 contains *in vitro* iodinated chicken ovalbumin as a size marker, lane 2 shows intracellular OLP present in sonicated spheroplasts and lane 4 periplasmic OLP secreted into the periplasmic space through the bacterial inner membrane; lane 3 contains intracellular ovalbumin-deleted protein synthesized by pOMP 29. The ovalbumin coding sequence in this plasmid has been deleted from codons 17 to 143. Lane 5 shows the immunoprecipitate obtained with intracellular extracts of control bacteria carrying pBR322. For details see Mercereau-Puijalon, 1980.

FUNCTIONAL ANALYSIS OF FOREIGN GENES EXPRESSED IN *E. COLI*

Expression of a eukaryotic gene in *E. coli* may facilitate a functional analysis pertinent to the eukaryotic situation. Mutants can be obtained in *E. coli* both by the usual genetic methods and by *in vitro* manipulation of the cloned sequence, and provide useful correlations between structural changes and functional alterations.

In general, it is possible to isolate non-expressing mutants from expressing strains, provided that selective or, at the least, screen procedures are available. As summarized above, we have constructed, from an *E. coli* TK⁻ strain, bacteria which become TK⁺ upon introduction of the manipulated *Herpes simplex* virus TK gene. This opens the possibility of isolating HSVI TK⁻ gene mutants, not making active enzyme in *E. coli*, which can then be reintroduced and studied in mouse fibroblasts.

We have been interested in the secretory properties of ovalbumin, first observed by Fraser and Bruce (1978), and have found that, in our ovalbumin-producing *E. coli* strains, 50 % or more of the OLP accumulates in the periplasmic space. We were able to show that there is no cleavage of amino acids at the N terminal of OLP during transport (see Fig. 2). This result indicates that amino acids 1 to 5 (substituted by seven amino acids of *E. coli* β-galactosidase in our construction) are not important for secretion through the bacterial inner membrane. This is in agreement with the proposal that, contrary to most secretory proteins (Blobel *et al.*, 1979) the signal sequence of ovalbumin would be internal to the protein (Palmiter *et al.*, 1978) and not be cleaved (Lingappa *et al.*, 1979). At present, there is no direct proof that the determinants involved in OLP transport in *E. coli* are actually the same as those involved in ovalbumin secretion in chicken or animal cells. This possibility can, however, be tested by the introduction of a mutated gene unable to promote secretion in *E. coli*, in mouse fibroblasts, where some expression is to be expected (Lai *et al.*, 1980). It has recently been shown that the preproinsulin signal sequence is properly cleaved in *E. coli* upon transport into the periplasm (Talmadge *et al.*, 1980). These results raise the interesting possibility that secretion of eukaryotic proteins may, to a certain extent, be studied in *E. coli*.

EXPRESSION AS A TOOL FOR GENE ISOLATION

Such functional analysis of expressed foreign genes are of value in a limited number of situations. Apart from maximizing expression in *E. coli* for applied purposes—which we shall not discuss here—there exists a methodological motivation which has received less emphasis. This is to use expression in the host microorganism as a tool for the isolation of genes. Experimental criteria are here quite different from those set upon maximized expression: with very sensitive detection methods, even limited expression will be sufficient to isolate a producing clone, and thereby the gene or a sequence characteristic of it. Appropriate tools for *in situ* screening (with antibodies) have been set up by us (Sanzey *et al.*, 1976) and others (Skalka and Shapiro, 1976) with increased sensitivity (Broome and Gilbert, 1978; Erlich *et al.*, 1978). These methods have not proved as useful as had been anticipated, because of the lack of a reliable expression system in *E. coli*. If this problem is to be solved, initiation restart mechanisms such as those observed in the case of dHFR (Chang *et al.*, 1980) and TK (Garapin *et al.*, 1981) expression might prove useful in elaborating a general system. Indeed, the use of other microorganisms, yeast and animal cells which have different translation characteristics, should also be considered. Although alternative procedures are available—hybridization of mRNA to cloned DNA followed by *in vitro* protein synthesis (e.g. Nagata *et al.*, 1980), synthetic oligonucleotides (Noyes *et al.*, 1979), etc.—it seems likely that the isolation of genes on the basis of their intracellular expression during cloning would help solve many presently difficult or tedious problems.

ACKNOWLEDGEMENTS

We would like to thank many colleagues for help and discussions, particularly F. Brégégère, A. Royal, B. Cami and O. Le Bail. This work was supported by grants from C.N.R.S. (E.R 201 and A.T.P. 6514246), I.N.S.E.R.M. (S.C.N. 20 and C.R.A.T. 7279104), D.G.R.S.T. (grant Nos 7872931 and 7950500) and F.M.R.F.

REFERENCES

Adhya, S., Gottesman, M. (1978). *A. Rev. Biochem.* **47**, 967–996.
Beggs, J. D., Van den Berg, J., Van Ooyen, A. and Weissman, C. (1980). *Nature, Lond.* **283**, 835–840.
Blobel, G., Walter, P., Chang, C. N., Goldman, B. M., Erickson, A. H. and Lingappa, V. R. (1979). *Symp. Soc. exp. Biol. Great Britain,* **33**, 9–36.
Breathnach, R., Mantei, N. and Chambon, P. (1980). *Proc. natn. Acad. Sci. U.S.A.* **77**, 740–744.
Broome, S. and Gilbert, W. (1978). *Proc. natn. Acad. Sci. U.S.A.* **75**, 2746–2750.

302   O. MERCEREAU-PUIJALON *et al.*

Chang, A. C. Y., Erlich, H. A., Gunsalus, R. P., Nunberg, J. H., Kaufman, R. J., Schimke, R. T. and Cohen, S. N. (1980). *Proc. natn. Acad. Sci. U.S.A.* **77**, 1441–1446.

Charnay, P., Perricaudet, M., Galibert, F. and Tiollais, P. (1978). *Nucl. Acids Res.* **5**, 4479–4494.

Colbère-Garapin, F., Chousterman, S., Horodniceanu, F., Kourilsky, P. and Garapin, A. C. (1979). *Proc. natn. Acad. Sci. U.S.A.* **76**, 3755–3759.

Emtage, J. S., Tacon, W. C. A., Catlin, G. H., Jenkins, B., Porter, A. G. and Carey, N. H. (1980). *Nature, Lond.* **283**, 171–174.

Erlich, H. A., Cohen, S. N. and McDevitt, H. O. (1978). *Cell* **13**, 681–689.

Fraser, T. H. and Bruce, B. J. (1978). *Proc. natn. Acad. Sci. U.S.A.* **75**, 5936–5940.

Gallwitz, D. and Sures, I. (1980). *Proc. natn. Acad. Sci. U.S.A.* **77**, 2546–2550.

Garapin, A., Colbère-Garapin, F., Cohen-Solal, M., Horodniceanu, F. and Kourilsky, P. (1981). *Proc. natn. Acad. Sci. U.S.A.* **78**, 815–819.

Hamer, D. H. and Leder, P. (1979). *Nature, Lond.* **281**, 35–40.

Iserentant, D. and Fiers, W. (1980). *Gene* **9**, 1–12.

Kourilsky, P., Gros, D., Rougeon, F. and Mach, B. (1977). *Nature, Lond.* **267**, 637.

Lai, E. C., Woo, S. L., Bordelon-Riser, M. E., Fraser, T. H. and O'Malley, B. W. (1980). *Proc. natn. Acad. Sci. U.S.A.* **77**, 244–248.

Lebowitz, P. and Weissman, S. (1979). *Curr. Top. Microbiol. Immunol.* **87**, 43–172.

Lingappa, V. R., Lingappa, J. R. and Blobel, G. (1979). *Nature, Lond.* **281**, 117–121.

Mantei, N., Boll. W. and Weissan, C. (1979). *Nature, Lond.* **281**, 40–46.

Mercereau-Puijalon, O. (1980). Doctoral Thesis, University of Paris VII.

Mercereau-Puijalon, O. and Kourilsky, P. (1976). *J. molec. Biol.* **108**, 733–751.

Mercereau-Puijalon, O. and Kourilsky, P. (1979). *Nature, Lond.* **279**, 647–649.

Mercereau-Puijalon, O., Royal, A., Cami, B., Garapin, A., Krust, A., Gannon, F. and Kourilsky, P. (1978). *Nature, Lond.* **275**, 505–510.

Mercereau-Puijalon, O., Lacroute, F. and Kourilsky, P. (1980). *Gene* **11**, 163–167.

Miller, J. H. (1978). *In* "The Operon" (J. H. Miller and W. S. Reznikoff, eds), pp. 31–88. Cold Spring Harbor Laboratory, New York.

Müller, F. and Clarkson, S. G. (1980). *Cell* **19**, 345–353.

Mulligan, R. C., Howard, B. H. and Berg, P. (1979). *Nature, Lond.* **277**, 108–114.

Nagata, S., Taira, H., Hall, A., Johnsrud, L., Streuli, M., Ecsödi, J., Boll, W., Cantell, K. and Weissmann, C. (1980). *Nature, Lond.* **284**, 316–320.

Noyes, B. E., Mevarech, M., Stein, R. and Agarwal, K. L. (1979). *Proc. natn. Acad. Sci. U.S.A.* **76**, 1770–1774.

Palmiter, R. D., Gagnon, J., Walsh, K. A. (1978). *Proc. natn. Acad. Sci. U.S.A.* **75**, 94–98.

Pasek, M., Goto, T., Gilbert, W., Zink, B., Schaller, H., MacKay, P., Leadbetter, G. and Murray, K. (1979). *Nature, Lond.* **282**, 575–579.

Reznikoff, W. S., Michels, C. A., Cooper, T. A., Silverstone, A. E. and Magasanik, B. (1974). *J. Bacteriol.* **117**, 1231–1239.

Roberts, J., Kacich, R. and Ptashne, M. (1979). *Proc. natn. Acad. Sci. U.S.A.* **76**, 760–764.

Sanzey, B., Mercereau-Puijalon, O., Ternynck, T. and Kourilsky, P. (1976). *Proc. natn. Acad. Sci. U.S.A.* **73**, 3394–3397.

Skalka, A. and Shapiro, L. (1976). *Gene* **1**, 65–79.

Steitz, J. A. (1979). *In* "Biological Regulation and Development—I. Gene Expression" (R. F. Goldberg, ed.), pp. 349–389. Plenum Press, New York.

Talmadge, K., Stahl, S. and Gilbert, W. (1980). *Proc. natn. Acad. Sci. U.S.A.* **77**, 3988–3992.

Villa-Komaroff, L., Efstratiadis, A., Broome, S., Lomedico, P., Tizard, R., Naber, S. P., Chick, W. L. and Gilbert, W. (1978). *Proc. natn. Acad. Sci. U.S.A.* **75**, 3727–3731.

Wold, B., Wigler, M., Lacy, E., Maniatis, T., Silverstein, S. and Axel, R. (1979). *Proc. natn. Acad. Sci. U.S.A.* **76**, 5684–5688.

Organization of Human Genomic DNA-encoding Pro-opiomelanocortin

ANNIE C. Y. CHANG, MADELEINE COCHET
and STANLEY N. COHEN

*Departments of Genetics and Medicine, Stanford University School of Medicine,
Stanford, California 94305, U.S.A.*

INTRODUCTION

The cloning in *E. coli* of complementary DNA (cDNA) that corresponds to
the bovine messenger RNA (mRNA) for the ACTH-β-LPH precursor
protein has recently been reported (Nakanishi *et al.*, 1979). This pituitary
protein is known to include several smaller peptides having biological
activities: α-melanotropin (α-MSH) and corticotropin-like intermediate lobe
peptide (CLIP) are contained within corticotropin (ACTH); γ-LPH and β-
endorphin are peptide components of β-lipotropin (β-LPH). β-MSH is part
of γ-LPH (Chrétien *et al.*, 1977; Li and Chung, 1976; Ling *et al.*, 1976;

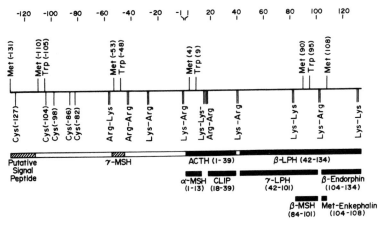

FIG. 1. Schematic representation of the component peptides contained within the pro-
opiomelanocortin (Nakanishi *et al.*, 1979). The positions of relevant amino acid residues are
shown relative to the first amino acid of ACTH (position $+1$). The locations of the component
peptides (black bars) are shown in parentheses. Details are given in Nakanishi *et al.* (1979).

305

Roberts and Herbert, 1977; Scott *et al.*, 1973). A new hormone, named γ-MSH because of its structural similarity to α- and β-MSH, was discovered in the previously "cryptic" pre-ACTH part of the precursor peptide (Nakanishi *et al.*, 1979). The 29 500-dalton protein precursor encoded by this mRNA has been called pro-opiomelanocortin (POMC) (Gossard *et al.*, 1980). From the nucleotide sequence of a 1091 base-pair cDNA insert, the amino acid sequence of the pre-ACTH part of the peptide was predicted and the structural organization of the biologically active component peptides of the POMC molecule was shown (Fig. 1) (Nakanishi *et al.*, 1979).

In order to elucidate the organization of the chromosomal DNA segment encoding POMC and to investigate inter-species variation in this genetic region, we have used a bovine cDNA clone as a probe to isolate and characterize the corresponding genomic DNA segment from a human fetal DNA library. The procedures used and the results obtained have been reported in detail elsewhere (Chang *et al.*, 1980).

ISOLATION OF HUMAN POMC GENOMIC DNA SEQUENCE

Using the *in situ* hybridization technique described as in Benton and Davis (1977) we screened a total of 1.5×10^5 plaques of λ Charon 4A bacteriophage containing DNA fragments generated by partial digestion of human fetal DNA with the Hae III and Alu I endonucleases. The pSNAC20, plasmid which contains a cDNA insert encoding bovine POMC, was used as a probe under hybridization conditions that would allow annealing to occur in the presence of possible base-pair mismatch between the bovine cDNA sequence and complementary human genomic DNA sequences.

Phage from the positive-reacting plaques was tested with four separate probes synthesized by nick-translation of bovine POMC cDNA fragments generated using combinations of Pst I, Sst II and Eco RI enzymes as shown in Fig. 2, panel (a). One plaque (λ8A) reacted strongly with all four probes.

DNA from phage λ8A was shown by Southern blotting procedures (Southern, 1975) to contain a 2.7 kb Eco RI-generated DNA fragment that includes sequences common to all four probes, suggesting that this fragment spans most or all of the pSNAC20 insert. The fragment was isolated from λ8A-phage DNA and introduced into the pBR322 cloning vector to yield the pACYC401 plasmid.

The extent of homology between the bovine POMC cDNA and sequences contained in the 2.7 kb Eco RI fragment was determined by heteroduplex analysis (Fig. 3). Such duplexes showed a region that extends for 850 ± 25 base-pairs without interruption and that is bracketed by single-stranded DNA arms of 862 ± 45 (segment a) and 935 ± 37 (segment b) nucleotides.

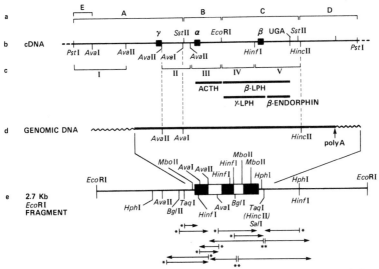

FIG. 2. Analysis of the structural organization of a 2.7 kb human genomic DNA fragment encoding POMC. Panel (b) shows certain restriction endonuclease cleavage sites on cDNA corresponding to bovine POMC mRNA. The boxes show the positions of α-, β- and γ-MSH sequences. Broken lines flanking the Pst I site represent plasmid vector sequences. Panels (a) and (c) show the extent of bovine cDNA fragments that were nick-translated into probes for localization of human genomic POMC sequences on λ8A DNA and on the 2.7 kb fragment. Thickened lines show positions of the sequences encoding ACTH, β-LPH, γ-LPH and β-endorphin on the bovine cDNA. In panel (d), the straight line represents human genomic sequences that correspond to the peptide-coding sequence of bovine cDNA; vertical lines represent endonuclease cleavage sites common to both species. The arrow indicates the site at which polyA is added in bovine mRNA. Panel (e) indicates an endonuclease cleavage map of the 2.7 kb Eco RI fragment containing the human POMC sequences. The block shows the extent of the POMC structural gene. Darkened segments indicate the conserved regions, while white boxes indicate the variable regions of the peptide-coding segment. The arrows below panel (e) show the DNA fragments sequenced and the direction of sequencing; asterisks indicate the 5′ labeled ends. Experimental details are given in Chang *et al.* (1980).

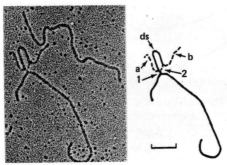

FIG. 3. Heteroduplex analysis of 2.7 kb fragment with plasmid DNA of pBR322 and pSNAC 20. The dashed and solid lines represent single-stranded DNA and double-stranded DNA, respectively. Experimental details are given in Chang *et al.* (1980). Scale bar = 0.5 kb.

These findings suggest that the majority of the bovine POMC cDNA sequences contained in the 2.7 kb Eco RI human genomic fragment exist as an uninterrupted segment lacking an intervening sequence.

RESTRICTION ANALYSIS OF THE HUMAN GENOMIC DNA FRAGMENT ENCODING POMC

A restriction endonuclease cleavage map was constructed for the pACYC401 plasmid by cleaving the DNA with combinations of the restriction endonucleases shown in Fig. 2, panel (e); the component segments of POMC were assigned to individual DNA fragments by filter-paper blotting techniques, again using fragments of the nick-translated bovine POMC cDNA as probes. This analysis showed that most of the human genomic DNA sequence corresponding to the bovine POMC DNA is located between the Ava II and Hinc II restriction sites in the 2.7 kb Eco RI fragment (Fig. 2, panel d). The primary DNA sequence was determined for this region using the strategy shown in Fig. 2, panel (e). The nucleotide sequence determined for the human genomic DNA, the differences found for the corresponding segment of bovine POMC cDNA, and the predicted amino acid sequences for bovine and human POMC are shown in Fig. 4.

ANALYSIS OF THE ORGANIZATION OF HUMAN GENOMIC POMC SEQUENCE

Comparison of the genomic DNA and cDNA sequences indicates continuity from a site corresponding to the polyA addition locus in the 3′ non-coding region of the bovine cDNA to a position corresponding to the *cys* residue 92 amino acids before the beginning of ACTH; there the two sequences diverge sharply. The di-nucleotide AG, which is characteristic of the 3′ termini of intervening sequences (Breathnach *et al.*, 1978), occurs at this point and a pyrimidine-rich region is seen just upstream from the putative intron–exon junction.

Despite the potential for evolutionary drift in codon usage in the two species without alteration of the amino acid sequence, the nucleotide sequences of three regions are seen to be surprisingly well conserved within the human and bovine POMC genes: (1) the segment encoding the ACTH; (2) the region extending from β-MSH to the end of β-LPH and including the sequences for Met-enkephalin and β-endorphin; and (3) the segment extending from γ-MSH towards the amino terminus of the POMC protein. In these regions, all amino acids are identical except at seven locations where one-base substitution results in amino acid differences. Two regions are divergent in the two species: the segments encoding the amino acids immediately preceding the *lys-arg* pair that separates ACTH from the preceding segment of POMC, and the segment immediately following ACTH (i.e. the first 120 base-pairs of β-LPH).

A 24 bp nucleotide segment and a *leu* residue at position 72 in the human

```
5'.. CTGTTCACAAAAGCTAGGGGTGGCTAGATGGCTAGACAAACCATGGAATGGGAAGGGAAGTGTGTTGCAGTTGCAGGCAGAAGCATGAAGGGGATGGGACA

    AAAGAGGCGGTGGCAAGATCTTAGATGCCCACGAGTGCCAAGAAAGCAGGTGGGCAGACCTGCCTGTAGGGAGGCCTCGACGCTTGACACGCCCGACACTG

    TGCCCTGTGTCCTCGGCACGTGGCGAGGGCGGCCAGGGCCAGGCGCAGTGACGGGCGCGGCAGCCGGGCCGGGGTGCGGGGCACGGGCTGCCCTCATGCCC

                                   -90                                        -80
HUMAN  ┌A.A.                Cys Ile Gly Ala Cys Lys Pro Asp Leu Ser Ala Glu Thr Pro Met Phe Pro Gly Asn Gly
       └GEN. DNA  TCGCGTCTTCCCCCAAGAGTGT ATC GGC GCC TGC AAG CCC GAC CTC TCG GCC GAG ACT CCC ATG TTC CCG GGA AAT GGC
BOVINE ┌cDNA                          C       CGG                    C           G G G         C C  C
       └A.A.                                  Arg                                            Val

      -70                                         -60                                     -50
Asp Glu Gln Pro Leu Thr Glu Asn Pro Arg Lys Tyr Val Met Gly His Phe Arg Trp Asp Arg Phe Gly│Arg Arg│Asn     Ser Ser
GAC GAG CAG CCT CTG ACC GAG AAC CCC CGG AAG TAC GTC ATG GGC CAC TTC CGC TGG GAC CGA TTC GGC│CGC CGC│AAC --- AGC AGC
     T       G       T                                       T               C              │    T G│T GGT
                                                                                            │       │    Gly

      -40                                  -30                                    -20
Ser Ser Gly Ser Ser Gly Ala Gly Leu Gln Lys Arg Glu Asp     Val Ser Ala Gly Val Asp Cys Gly Pro Leu Pro Glu Gly Gly Pro
AGC AGC GGC AGC AGC GGC GCA GGG CAG AAG CGC GAG GAC --- GTC TCA GCG GAG GAC GAA GAC TGC GGC CCG CTC CCT GAG GGC GGC CCC
         A GTT G G       G CC            G GAA  G G G  T          G CC       G  C G       --- --- --- --- ---
         Val Gly        Ala              Glu Glu   Ala Val        Gly Pro        Arg

      -10                         -1  +1                                  +10
Glu Pro Arg Ser Asp Gly Ala Lys Pro Gly Pro Arg Glu Gly│Lys Arg│Ser Tyr Ser Met Glu His Phe Arg Trp Gly Lys Pro Val
GAG CCC CGC AGC GAT GGT GCC AAG CCG GGC CCG CGC GAG GGC│AAG CGC│TCC TAC TCC ATG GAG CAC TTC CGC TGG GGC AAG CCG GTG
--- --- --- G       AC    G A  T                      A │       │A           T                           A
          Gly      Asp   Glu Thr                     Asp

      +20                              +30                              +40
Gly Lys Lys Arg Arg Pro Val Lys Val Tyr Pro Asn Gly Ala Glu Asp Glu Ser Ala Glu Ala Phe Pro Leu Glu Phe│Lys Arg│Glu
GGC AAG AAG CGG CGC CCA GTG AAG GTG TAC CT AAC GGC GCC GAG GAC GAG TCG GCG GAG GCC TTC CCC CTG GAG TTC│AAG AGG│GAG
         G                        C                                        C       T        C A        │       │
                                                                          Gln

                       +50                            +60                            +70
Leu Thr Gly Gln Arg Leu Arg Glu Gly Asp Gly Pro Asp Gly Pro Ala Ala Thr Ala Gln Glu Pro Gly Asp Leu Glu His Ser Leu
CTG ACT GGG CAG CGA CTC CGG GAG GGA GAT GGC CCC GAC GGC CCC GCC GCG ACG CAG GGG CCA GGC GAC CTG GAG CAC AGC CTG
    C    G  A  G         GA  C   C  AG       G  C  AG  T  A T         GCC C  AG  CT  G              T T G
                        Glu         Glu Gln Ala Arg        Glu Ala Gln       Glu Ser       Ala Ala Arg Ala Glu          Thy Gly

                                          +80                             +90
Leu Val                           Ala Ala Glu│Lys Lys│Asp Glu Gly Pro Tyr Arg Met Glu His Phe Arg Trp Gly Ser Pro Pro
CTG GTG --- --- --- --- --- --- GCG GCC GAG│AAG AAG│GAC GAG GGC CCC TAC AGG ATG GAG CAC TTC CGC TGG GGC AGC CCG CCC
---           GCG GAG GCG GAG GCT GAG     │       │     TC  G      T  A     A
              Ala Glu Ala Glu Ala Glu     │       │     Ser             Lys

                          +100                            +110                           +120
Lys Asp Lys Arg Tyr Gly Gly Phe Met Thr Ser Glu Lys Ser Gln Thr Pro Leu Val Thr Leu Phe Lys Asn Ala Ile Ile Lys Asn
AAG GAC AAG CGC TAC GGC GGT TTC ATG ACC TCC GAG AAG AGC CAG ACG CCC CTA GTG ACG CTG TTC AAA AAC GCC ATC ATC AAG AAC
             G                                                       A         T  C

      +129
Ala Tyr Lys Lys Gly Glu
GCC TAC AAG AAG GGC GAG TGA GGGCACAGCGGGCCCCAGGG--CTCACCTCCCCACAGGAGGTCGACCCCAAAGTCCCCTTGGCTCTCCCCTGCCCTGCTGCCGCCTCCC
  C       C       G         ---       GC T-  G GG-A T-T    TG  G -- CTCT  G  C -- A ---            G
  His              Gln

AGC-TGGGGG--G--TCG----TGGCAGATAAT--------------CAGCCTCTTAAAGCTTGCCTGTAGTTAGGAAATAAAACCTTTCAAATTTCACATCCACCTCTGACTTTGAAT
  C    T AG AT   CCCA    - G  GGCGCCAGGTATC C A      - T      A                          G      G↑
                                                                                              polyA
```

FIG. 4. Nucleotide sequence of human genomic DNA strand corresponding to POMC mRNA. Above the nucleotide sequence is the predicted amino acid sequence; nucleotide and amino acid differences in the human v. bovine sequences are indicated. Broken lines represent nucleotides absent in the cDNA or genomic sequences. The pairs of basic amino acids (*lys* and *arg*) that separate conserved and non-conserved regions of the POMC coding sequence are indicated by boxes. The site of polyA addition is indicated in the 3' non-coding region of the sequence. Amino acids are not assigned for the region that preceeds the *cys* at amino acid position −92. Amino acids are numbered with plus numbers in the direction of the carboxy terminus of POMC from the first amino acid of ACTH, and with minus numbers in the direction of the amino terminus of the peptide for the pre-ACTH segment, as in Nakanishi *et al.* (1979).

gene are absent in the bovine sequence; however, a stretch of 18 bp and two triplets present in the bovine sequence are lacking in the human genomic DNA. Thus, the overall lengths for the bovine and human peptides predicted from the nucleotide sequences differ by only one amino acid. It is also worth noting that the 3′ non-coding region of the bovine cDNA corresponds closely to the human genomic DNA segment that follows the translational stop codon terminating the POMC protein, indicating that the region is highly conserved despite its non-coding nature.

DISCUSSION

The nucleotide sequences of human genomic DNA and bovine cDNA encoding POMC show general continuity from a position 276 base-pairs before the beginning of ACTH to a position corresponding to the 3′ terminus of the POMC mRNA. Immediately 5′ to this site of major divergence are structural features that make it highly likely that the divergence represents an intron–exon junction rather than a site of inter-species dissimilarity. The recently obtained amino acid sequence for this segment of the POMC peptide (Seidah *et al.*, 1980) is consistent with this interpretation.

The regions of interspecies similarity and divergence in POMC are largely delineated by pairs of basic amino acid residues that separate the structural and functional domains of the POMC protein, suggesting that the separate domains have undergone evolutionary change at different rates. The two segments of greatest DNA sequence dissimilarity are located immediately preceding and immediately following the nucleotide sequence encoding corticotropin, which is highly conserved.

The availability of a cloned segment of human genomic DNA encoding POMC enables the analysis of adjacent DNA segments and the assignment of the POMC gene to a particular chromosomal location by *in situ* hybridization procedures. Furthermore, the absence of non-coding intervening sequences in component peptide hormones of human POMC will allow the functional expression of the genomic DNA sequences in bacterial cells.

ACKNOWLEDGEMENTS

These studies were supported by grants GM 27241 and AI 08619 from the NIH and in part by a grant from the American Cancer Society to S.N.C. M.C. is a postdoctoral research fellow of the CNRS.

REFERENCES

Benton, W. and Davis, R. (1977). *Science, N.Y.* **196**, 180–182.

Breathnach, R., Benoist, C., O'Hare, K., Gannon, F. and Chambon, P. (1978). *Proc. natn. Acad. Sci. U.S.A.* **75**, 4853–4857.

Chang, A. C. Y., Cochet, M. and Cohen, S. N. (1980). *Proc. natn. Acad. Sci. U.S.A.* **77**, 4890–4894.

Chrétien, M., Seidah, N., Benjannet, S., Dragon, N., Routhier, R., Motomatsu, T., Crine, P. and Lis, M. (1977). *Ann. N.Y. Acad. Sci.* **297**, 84–107.

Gossard, F., Seidah, N., Crine, P., Routhier, R. and Chrétien, M. (1980). *Biochem. biophys. Res. Commun.* **92**, 1042–1051.

Li, C. and Chung, D. (1976). *Proc. natn. Acad. Sci. U.S.A.* **73**, 1145–1148.

Ling, N., Burgus, R. and Guillemin, R. (1976). *Proc. natn. Acad. Sci. U.S.A.* **73**, 3942–3946.

Nakanishi, S., Inoue, A., Kita, T., Nakamura, M., Chang, A. C. Y., Cohen, S. N. and Numa, S. (1979). *Nature, Lond.* **278**, 423–427.

Roberts, J. and Herbert, E. (1977). *Proc. natn. Acad. Sci. U.S.A.* **74**, 5300–5304.

Scott, A. P., Ratcliff, J. G., Rees, L. H., Landon, J., Bennett, H. P. J., Lowry, P. J. and McMartin, C. (1973). *Nature, Lond. New Biol.* **244**, 65–67.

Seidah, N., Benjannet, S., Routhier, R., Lis, M. and Chrétien, M. (1980). *Biochem. biophys. Res. Commun.* **95**, 1417–1424.

Southern, E. (1975). *J. molec. Biol.* **98**, 503–517.

Construction of Functional Yeast Minichromosomes Containing Centromeric DNA

JOHN CARBON and LOUISE CLARKE

Department of Biological Sciences, University of California,
Santa Barbara, California 93106, U.S.A.

INTRODUCTION

We are interested in the structure and function of yeast chromosomes; in particular in the mechanism by which chromosomes are replicated and transported during mitotic and meiotic cell divisions. Of central importance in this process is the centromeric region of the chromosome, which serves as an attachment point for spindle fibres or for membrane sites involved in chromosome segregation during cell division (in the case of yeast, see Peterson and Ris, 1976).

The molecular biology of yeast has advanced to a stage where the isolation and characterization of a functional centromere, as cloned centromeric DNA, has become a practical possibility. Yeast is an ideal organism for this effort, because (1) extensive genetic mapping studies have identified several genes that are very closely linked to centromeres; (2) yeast chromosomes do not contain highly repetitive DNA sequences in the centromere regions; and (3) autonomously replicating plasmid vectors are available containing yeast chromosomal replicators (Stinchcomb *et al.*, 1979; Kingsman *et al.*, 1979); these plasmids serve as excellent vehicles to assay for centromere function of cloned DNA sequences *in vivo*. In this chapter we shall describe the isolation and preliminary characterization of functional centromeric DNA from yeast chromosome III. Plasmids containing a functional centromere along with a chromosomal replicator behave in yeast essentially as chromosomes, segregating properly through mitosis and the two meiotic divisions, and thus are excellent model systems for the study of chromosome function.

MOLECULAR CLONING OF CENTROMERE-LINKED GENES

Genetic mapping studies have localized many genes to the centromere regions of the yeast chromosomes (Mortimer and Hawthorne, 1975). On chromosome III, leu2 mutations (β-isopropylmalate dehydrogenase) map approximately 8 centiMorgans (cM) to the left of the centromere as the chromosome is usually drawn (Mortimer and Hawthorne, 1975); pgk (3-phosphoglycerate kinase) mutations occur about 2 cM to the right of the centromere (Lam and Marmur, 1977; Mortimer, R. K., personal communication); and cdc10 mutations are very tightly linked to centromere (no recombinations between cdc10 and centromere in analysis of over 200 tetrads (Culbertson et al., 1977; Mortimer, R. K., personal communication)). Yeast DNA segments containing these genes have been obtained by a variety of molecular cloning techniques.

The LEU2 gene was originally isolated by selecting for complementation of E. coli leuB (β-isopropylmalate dehydrogenase) mutations with a yeast DNA segment cloned in plasmid Col E1 (Ratzkin and Carbon, 1977). The PGK gene was isolated on a hybrid plasmid by screening an appropriate E. coli recombinant "yeast library" with ^{125}I-labeled antibody directed against yeast 3-phosphoglycerate kinase (Hitzeman et al., 1979). Recently, an 8 kilobasepair (kb) segment of yeast DNA containing CDC10 gene was obtained in a recombinant plasmid (pYe(CDC10)1) by transformation of a suitable trp1 cdc10 yeast mutant with a hybrid plasmid DNA pool, selecting for complementation of these mutations by the transforming plasmid DNA (Clarke and Carbon, 1980). In this case, the plasmid pool contained random fragments of yeast DNA sealed with poly(dA:dT) connectors into the shuttle vector pLC544, which contains the yeast TRP1 gene and the yeast chromosomal replicator ars1 in plasmid vector pBR313 (Stinchcomb et al., 1979; Kingsman et al., 1979).

It is possible to isolate relatively long contiguous regions of chromosomal DNA by the technique of overlap hybridization or chromosome "walking" (Chinault and Carbon, 1979). This procedure involves the screening of clone libraries, constructed from randomly sheared yeast DNA segments, for hybridization with specific ^{32}P-labeled DNA probes, thereby detecting other cloned DNA segments that may contain common or overlapping sequences. Using this technique, a total of approximately 70 kb of DNA from the centromere region of chromosome III has been isolated and mapped (Chinault and Carbon, 1979; Chinault, Clarke, Kingsman, Gimlich and Carbon, unpublished results). Of most interest, however, is the contiguous stretch of about 25 kb of DNA that extends from the LEU2 gene to the CDC10 gene (Chinault and Carbon, 1979; Clarke and Carbon, 1980), since this region can be shown to contain a functional centromere (see Fig. 1).

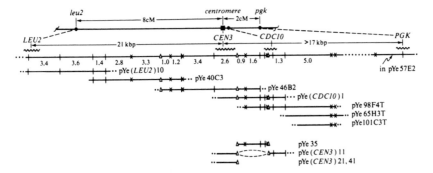

FIG. 1. Genetic and physical maps of the centromere region of *S. cerevisiae* chromosome III. The restriction map gives the location of Eco RI (—+—), Hind III (—×—), and Bam HI (—△—) sites in the DNA. Numbers refer to kilobasepairs. The order of inserts in various overlapping plasmids is indicated below the restriction map. Small dots at the ends of inserts pertain to those sheared segments of yeast DNA that are joined to their respective plasmid vectors by poly (dA·dT) connectors. The isolation and characterization of plasmids pYe(*LEU2*)10, pYe40C3, pYe46B2, pYe(*CDC10*)1 and pYe57E2, have been described (Chinault and Carbon, 1979; Clarke and Carbon, 1980; Hitzeman *et al.*, 1979). Plasmids pYe98F4T, pYe65H3T and pYe101C3T were obtained by overlap hybridization and characterized as described by Chinault and Carbon (1979), with the modification that a single restriction fragment (from pYe(*CDC10*)1) purified by agarose gel electrophoresis was used as probe in the hybridizations. Plasmids pYe(*CDC10*)1, pYe98F4T, pYe65H3T, pYe1013T, pYe35 and pYe(*CEN3*)11 all contain the *TRP1 ars1* vector pLC544 (Kingsman *et al.*, 1979). The vector portion of the remaining plasmids is ColE1.

IDENTIFICATION OF A FUNCTIONAL YEAST CENTROMERE ON PLASMID pYe(*CDC10*)1

Plasmids containing a yeast chromosomal replicator, such as *ars1* (Stinchcomb *et al.*, 1979) or *ars2* (Hsiao and Carbon, 1979), can replicate autonomously in yeast cells but are mitotically quite unstable, presumably due to lack of an attachment point (centromere) for the mitotic spindle apparatus (see Table I). Because of high segregation rates, it is impossible to carry out meaningful tetrad analyses on strains carrying these plasmids, since the plasmids rarely survive through the meiotic divisions, sporulation and germination. Plasmid pYe(*CDC10*)1 behaves quite differently, however, because in all yeast transformants carrying this element the plasmid is maintained quite stably through both mitotic (Table I) and meiotic cell divisions (Table II). Furthermore, this plasmid can be shown by genetic and molecular techniques to be present in controlled copy number in an unintegrated state.

The mitotic segregation rates in yeast of various plasmids containing cloned DNA segments from the *CDC10* region and an *ars1* replicator are

TABLE I

Mitotic stability in yeast of various plasmids and minichromosomes containing DNA from the centromere region of chromosome III

Plasmid	Number of individual transformants tested		Average percentage of Trp^+ or Leu^+ transformants remaining after non-selective growth	
	Strain XSB52-23C	Strain RH218	Strain XSB52-23C	Strain RH218
pYe(CDC10)1	5	ND	97%	ND
pYe(CEN3)11	5	3	95%	83%
pYe(CEN3)41	2	ND	91%	ND
pYe35	5	5	<1%	7%
pYe98F4T	5	ND	<1%	ND
pYe65H3T	2	ND	<1%	ND
pYe101C3T	2	ND	<1%	ND
pYe(CEN3)21	4	ND	<1%	ND
pYe(ars1-ars2)1	ND	8	ND	5%

Yeast strains XSB52-23C(cdc10 gal leu2-3 leu2-112 trp1) and RH218a (CUP1 gal2 mal SUC trp1) were transformed with the above plasmid DNAs and in each case several individual transformants were isolated and grown non-selectively overnight on YPD (rich) medium. Each culture was subsequently streaked for single colonies on YPD agar, allowed to grow 2 days at 23°C (XSB52-23C) or 32°C (RH218) and replica-plated on to minimal agar medium with or without tryptophan or leucine, depending on the marker carried by the plasmid. From 40 to 100 individual clones of each original transformant were scored for plasmid loss after growth at 23°C (XSB52-23C) or 32°C (RH218). Preparations of media and transformation of yeast were carried out as described by Hsiao and Carbon (1979). (ND = not determined.)

shown in Table I. Note that only those plasmids containing DNA from a region approximately 3 kb to the left of the *CDC10* gene (towards *LEU2*) are mitotically stable in yeast. Specifically, the stabilizing element appears to lie to the *LEU2* side of the leftward Bam H1 site in the pYe(*CDC10*)1 insert (Fig. 1), since recloning experiments show that pYe(*CEN3*)11, without the central 3.5 kbp Bam HI fragment, contains the stabilizing element, but the recloned 3.5 kbp Bam HI fragment (in pYe35) or DNA sequences to the right of this fragment do not (see Fig. 1 and Table I).

The location of the stabilizing element was more exactly determined by recloning a 2.0 kb Bam HI-Hind III DNA fragment from pYe(*CDC10*)1 into the plasmid vector pGT12 (Tschumper and Carbon, 1980). The recloned fragment consists of about 1.6 kb of yeast DNA extending from the Bam HI site at the left end of the insert in pYe(*CDC10*)1 to a Hind III site that is in the original pLC544 vector (the 2.0 kb fragment cloned

TABLE II
Meiotic segregation of the minichromosomes

Genetic cross number	Minichromosome in cross	Minichromosome marker scored	Distribution in tetrads of genetic marker on minichromosome (%)					Test for centromere linkage of marker on minichromosome			Reference centromere marker (chromosome)
			4+:0−	3+:1−	2+:2−	1+:3−	0+:4−	PD	NPD	T	
1	pYe(CDC10)1	TRP1	1(6%)	0	10(63%)	0	5(31%)	2	8	0	met14(XI)
2	pYe(CDC10)1	CDC10	1(8%)	0	11(92%)	0	0	2	8	1	ade1(I)
3	pYe(CDC10)1	TRP1	1(7%)	0	11(79%)	0	2(14%)	4	7	0	met14(XI)
4	pYe(CDC10)1	TRP1	4(21%)	0	11(58%)	0	4(21%)	ND	ND	ND	—
5	pYe(CEN3)11	TRP1	4(27%)	0	9(60%)	0	2(13%)	4	5	0	ade1(I)
								5	4	0	cdc10(III)
6	pYe(CEN3)41	LEU2	3(14%)	3(14%)	13(62%)	1(5%)	1(5%)	7	6	0	trp1(IV)
								7	5	0	cdc10(III)
7	pYe(CDC10)1 × pYe(CDC10)1	TRP1	6(35%)	0	9(53%)	1(6%)	1(6%)	ND	ND	ND	—
8	pYe(CDC10)1	TRP1	10(24%)	1(2%)	31(74%)	0	0	15	16	0	met14(XI)
	× pYe(CEN3)41	LEU2	8(19%)	2(5%)	24(57%)	0	8(19%)	10	14	0	met14(XI)
								17	2	0	TRP1(mini)

In all the above crosses the marker used to follow the minichromosome was wild-type on the minichromosome and mutant in both parents. The crosses were: (1) XSB52-23Cα/pYe(CDC10)1(cdc10 leu2-112 trp1 gal/CDC10 TRP1) with X2928-3D-Aa(ade1 leu1 met14 trp1 ura3); (2) SB17Aa/pYe(CDC10)1(adel cdc10 leu2-3 leu2-112 met14 trp1/CDC10 TRP1) with 6204-18Aα(cdc10 thr4 leu2-3 leu2-112); (3) SB1Cα/pYe(CDC10)1(his2 leu1 met14 trp1/CDC10 TRP1) with XSB52-23Cα(cdc10 leu2-3 leu2-112 trp1 gal); (4) SB17Bα/pYe(CDC10)1(ade1 cdc10 leu2-3 leu2-112 met14 trp1/CDC10 TRP1) with X2928-3D-1Aa; (5) XSB52-23Cα/pYe(CEN3)11 with X2928-3D-1Aa; (6) XSB52-23Cα/pYe(CEN3)41 leu2-3 leu2-112 met14 trp1/CDC10 TRP1) with X3144-1Da(ade2 arg9 can1 his2 his6 leu2 pet8 trp1); (7) SB14Ba/pYe(CDC10)1 (ade1 his2 trp1/CDC10 TRP1) with SB17Bα/pYe(CDC10)1(ade1 cdc10 leu2-3 leu2-112 met14 trp1/CDC10 TRP1); (8)SXB52-23Cα/pYe(CEN3)41 with SB17Aa/pYe(CDC10)1.

Transformation of yeast in the presence of polyethylene glycol results in a high proportion of diploid transformants (a/a or α/α). Diploids are easily recognized by the extremely low spore viability obtained when they are crossed with a haploid, and in this way were eliminated from the group of haploid transformants used in the above crosses. (PD = parental ditype; NPD = non-parental ditype; T = tetratype; ND = not determined.)

includes a small 0.4 kb fragment of the pLC544 vector). As shown in Fig. 2, the resulting plasmid pYe(*CEN3*)41 contains a yeast *LEU2* gene, the *ars1* chromosomal replicator, and the 2.0 kbp Bam HI-Hind III fragment in a pBR322-derived vector. Yeast *leu2* mutants are transformed to *LEU2*⁺ with high frequency by pYe(*CEN3*)41 DNA and the transformants are again quite stable with regard to mitotic divisions (Table I).

Since the stabilizing element found on pYe(*CDC10*)1 possesses all of the functional properties usually associated with centromere function (see below), we have termed this element *CEN3* (centromere from chromosome III).

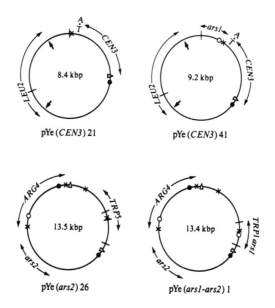

FIG. 2. Physical maps of various plasmid DNAs. The maps show the location of Eco RI (—+—), Hind III (—×—), Bam HI (—△—), Pst I (⊼), Sal I (—●—) and Bgl II (—O—) sites in the DNAs and indicate approximate locations of pertinent replicators, centromeres and genes. The use of these plasmids is described in the text.

PLASMIDS CONTAINING *CEN3* SEGREGATE AS CHROMOSOMES IN MEIOSIS

The meiotic segregation of plasmids containing *CEN3* was studied first by crossing haploid *cdc10 trp1* strains containing pYe (*CDC10*)1 with haploids containing either a *cdc10* or *trp1* mutation and other centromere-linked markers. After sporulation of the diploids, the progeny from the tetrad spores were screened for the presence of the minichromosome by scoring for the $CDC10^+$ and $TRP1^+$ markers carried by pYe(CDC10)1. The results are summarized in Table II.

In crosses 1–4, only one parent contained the minichromosome and both parents were mutant for the marker used to score the presence of pYe(CDC10)1. Note that in the majority of the tetrads, the minichromosome appears in two of the four spores; no $3+:1-$ or $1-:3+$ distributions were observed among more than 60 tetrads. Complete loss of the minichromosome ($0+:4-$ segregation) occurred in about one out of four tetrads scored in these crosses, although this meiotic instability can vary markedly in different crosses (Table II). Apparent diploidization or non-Mendelian segregation ($4+:0-$) also occurs, but at relatively low frequency.

The tetrads were scored for the distribution of centromere-linked markers on other chromosomes to order the spores with regard to the first and second meiotic divisions. In those tetrads where the minichromosome segregated $2+:2-$, a comparison of the distribution among the four spores of a genetic marker on pYe(CDC10)1 and centromere-linked markers (met14 and ade1) on other chromosomes indicated that the minichromosome markers (CDC10 and TRP1) were behaving as true centromere-linked genes (all parental ditypes and non-parental ditypes, few or no tetratypes; see Table II). Thus, the minichromosome appears in the two sister spores produced by the second meiotic division, and migrates randomly to either pole in the first meiotic division. No linkage of minichromosome markers to genes on other chromosomes (including chromosomes III and IV) has ever been observed, indicating that the minichromosome is behaving as an independent genetic element and is not integrated into any of the parental chromosomes. This was also confirmed by Southern hybridization experiments (see below).

Plasmids containing a recloned DNA segment bearing CEN3 (e.g., pYe(CEN3)11 and pYe(CEN3)41) also segregate as chromosomes in meiosis (see crosses 5 and 6, Table II), distributing predominantly $2+:2-$ in the tetrads. Again, no linkage to centromere markers on other chromosomes, including III and IV, was observed; however the minichromosome genes do behave as centromere-linked elements.

Crosses of two haploids, each containing a minichromosome, indicate that diploid minichromosomes do not always pair in meiosis, but they do tend to migrate to opposite poles in the first division (see crosses 7 and 8, Table II). The minichromosomes containing CEN3 do not appear to pair with any of the normal chromosomes, since minichromosome:chromosome pairing would be expected to generate a significant number of tetrads containing only two viable spores (Strathern et al., 1979). However, in most of the crosses shown in Table II very few bilethals were observed, and in those cases where only two viable spores were obtained they were usually not second-division sister spores.

PLASMIDS CONTAINING *CEN3* DO NOT INTEGRATE INTO THE YEAST GENOME

Although the genetic data shown in Table II indicate that the *CEN3* minichromosomes are not integrated into chromosomes I, III, IV or XI, it could still be possible that integration has occurred on some other chromosome at a site very near the centromere. This possibility was eliminated by isolation of total DNA from several of the progeny from cross 1 and 6, and assaying for the presence of unintegrated and integrated plasmid DNAs by Southern hybridizations (Southern, 1975), using ^{32}P-labeled vector sequences (pBR322) as probe. For example, pYe(*CEN3*)41 DNA contains single restriction sites for restriction endonucleases Bam HI, Bgl II and Sal I (see Fig. 2). The Bam HI site is at the junction of the vector and the yeast DNA segment that includes the centromere (from III), the Bgl II site is in the DNA segment from chromosome IV that includes *ars1*, and the Sal I site is in the pBR322 vector. Total yeast DNAs isolated from various *LEU2*$^+$ or *leu2* progeny from cross 6 were digested separately with each of the above restriction enzymes, subjected to agarose gel electrophoresis, and denatured DNA fragments blotted through on to nitrocellulose by the method of Southern (1975). Regardless of which of the three restriction enzymes was used, hybridization with ^{32}P-labeled pBR322 DNA gave a single labeled band in the case of the *LEU2*$^+$ DNAs, a band of identical size to unit-length linear pYe(*CEN3*)41 DNA standards (Fig. 3). As expected, yeast DNA from *leu2*$^-$ progeny do not contain DNA sequences complementary to the labeled pBR322 probe. Identical results have been obtained using total DNA prepared from the progeny of the pYe (*CDC10*)1 cross 1 digested with Sal I.

The only reasonable interpretation of the Southern hybridization experiments combined with the genetic data is that minichromosomes containing the *CEN3* element exist autonomously in an unintegrated state in yeast cells.

BOTH A FUNCTIONAL CENTROMERE AND A CHROMOSOMAL REPLICATOR ARE REQUIRED FOR MITOTIC STABILIZATION OF THE MINICHROMOSOMES

The stable maintenance of plasmids through both mitotic and meiotic cell divisions requires both a chromosomal replicator (in this case *ars1*) and the centromere (*CEN3*). For example, the plasmid pYe(*CEN3*)21 (Fig. 2), which contains the 2.0 kb Bam HI-Hind III fragment including *CEN3* but no *ars1* element, is maintained very poorly in yeast transformants and has

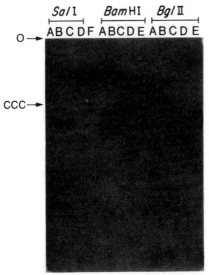

FIG. 3. Southern blot hybridization autoradiogram of total yeast DNAs from pYe(*CEN3*)41-containing strains with ^{32}P pBR322 DNA as probe. Crude yeast DNA was prepared as described by Stinchcomb *et al.* (1979) from one *Leu⁻* progeny (lanes A) and two *Leu⁺* progeny (lanes B and C) of a single 2+:2− ascus in cross 6 and from a *Leu⁻* progeny of a 3+:1− ascus in the same cross (lanes D). Approximately 2 mg of each DNA preparation were digested with Bam HI, Sal I, or Bgl II and electrophoresed in 0.75% agarose. Other lanes on the gel contained 2 ng of pYe(*CEN3*)41 DNA mixed with 2 mg of crude *Leu⁻* yeast DNA, either undigested (lane F) or digested (lanes E) with restriction enzyme. Blot hybridization was carried out according to Southern (1975). The probe was pBR322 DNA labeled with ^{32}P dTTP *in vitro* by nick translation. The origin (O) and position of covalently closed circular pYe(*CEN3*)41 DNA (CCC) are indicated on the autoradiogram.

obvious difficulties with DNA replication (see Table I). Centromere-containing plasmids without a chromosomal replicator can replicate poorly in yeast; *leu2* host cells containing pYe(*CEN3*)21 grow with a generation time of 12 hours in selective media containing no leucine, but segregation rates are quite high (Table I).

The presence of *CEN3* is essential for mitotic and meiotic stabilization of the minichromosomes; two chromosomal replicators are insufficient. For example, plasmid pYe(*ars1-ars2*)1 (see Fig. 2) contains two replicators, *ars1* and a yeast chromosomal replicator (*ars2*) from the *ARG4* region (Hsiao and Carbon, 1979), but no *CEN3*. This plasmid is extremely unstable mitotically (Table I), much too unstable for meiotic tetrad analysis to be carried out. In addition, Stinchcomb *et al.* (1979) have shown that the presence of multiple copies of *ars1* on the same plasmid does not lead to increased mitotic stability. Recent studies (Hsiao, Clarke and Carbon, unpublished results) have shown that plasmids containing *CEN3* plus the *ars2* replicator are mitotically stable in yeast, indicating that the requirement for *ars1* can be met by other chromosomal replicators.

We conclude that stable chromosome maintenance through mitotic and meiotic cell divisions requires at least one chromosomal replicator in addition to a functional centromere. Minichromosomes containing both *CEN3* and *ars1* are not completely stabilized, however, since the minichromosome is lost completely in a significant proportion of the tetrads examined (Table II). The frequency of loss of the minichromosome may not be very different from the rate of loss of other aneuploid chromosomes, which are sometimes quite unstable in yeast (Parry and Cox, 1971).

CEN3 SHOWS NO STRONG CROSS-HOMOLOGY WITH OTHER YEAST DNA SEQUENCES

Since yeast contains at least 17 chromosomes, it is of interest to ask if the various centromeres contain DNA sequences in common. Yeast genomic DNA was digested with Eco RI, fractionated on agarose gels, and hybridized by the method of Southern (1975) to high-specific-activity ^{32}P-labeled pYe(*CDC10*)1 DNA (Clarke and Carbon, 1980). Under conditions where strong homology of at least 100 bp would have been easily detected, no background of cross-hybridization of probe to genomic Eco RI fragments other than to those from the *CDC10* region could be detected. Thus, the centromere sequence is either unique or the homology with other centromeric DNAs is too short or imperfect to be detected by the standard Southern filter hybridization procedures. These studies are being pursued using hybridization conditions capable of detecting weak homologies. At this point, however, it seems possible that there is not a common centromere sequence occurring on all the yeast chromosomes.*

DISCUSSION

It is quite likely that *CEN3* is the centromeric DNA from chromosome III. This DNA element occurs between the *LEU2* and *CDC10* genes, about 13 kb from *LEU2* and 3–4 kb from *CDC10*. This places *CDC10* to the right of the centromere, rather than immediately to the left (*LEU2* side), as

---

* *Note added in proof.* Recent nucleotide sequence comparisons of functional centromere DNAs from yeast chromosomes III and IX (*CEN3* and *CEN11*) indicate that, although some sequence homologies are apparent, no regions of continuous homology longer than 14 bp occur (M. Fitzgerald-Hayes and L. Clarke, unpublished results).

it is usually drawn (Mortimer and Schild, 1980). The DNA is continuous through the centromere, as has also been shown by the isolation of a circular form of chromosome III as covalently closed circular DNA (Strathern et al., 1979).

Minichromosomes containing CEN3 plus a chromosomal replicator (ars1 or ars2) are mitotically stable, presumably because the CEN3 locus provides an attachment point for the spindle apparatus. Thus, DNA segments containing CEN3 will serve as excellent probes for the isolation of spindle proteins or other centromere-associated proteins from yeast nuclei. Similarly, the function of the centromere and associated proteins during the meiotic divisions can now be studied with the aid of the minichromosomes containing CEN3.

It is significant that small circular double-stranded DNA molecules containing only a chromosomal replicator, the CEN3 sequence, and a yeast gene (for identification) can segregate properly through both mitotic and the first and second meiotic divisions. The occasional appearance of tetrads containing no minichromosome $(0+:4-$ segregation) could indicate that association of CEN3 with the proper nuclear binding sites is imperfect, although aneuploid chromosomes in yeast usually do show some instability. If proper meiotic chromosome pairing is necessary for optimal stabilization of chromosomes through the first meiotic division, the apparent lack of significant pairing in the case of the minichromosomes could be an important factor leading to the observed meiotic instability. The presence of a significant number of $4+:0-$ segregations of the minichromosomes is also interesting, since diploidization of aneuploid yeast chromosomes does not always occur at this high frequency. Alternatively, it is possible that in a small percentage of the sporulating cells the minichromosome escapes from the attachment site and is replicated in the absence of normal controls to produce several copies that distribute randomly, appearing in all four spores.

Finally, plasmid vectors containing CEN3 plus a suitable replicator element and driver genes should be excellent vehicles for the stable maintenance of cloned DNA segments in yeast. It will also be of considerable interest to determine if the yeast centromere will function when introduced into other cell types on the appropriate self-replicating DNA vectors.

ACKNOWLEDGEMENTS

The authors are grateful to Robert Gimlich for excellent technical assistance. We owe a great deal to Drs Terry Cooper and Robert Mortimer, who have helped us learn the intricacies of yeast genetics, and who have clarified our thinking with many profitable discussions. Many of the DNA segments from the *LEU2, CDC10* and *PGK* areas were isolated by Drs Barry Ratzkin, Craig Chinault, Alan Kingsman and Ronald Hitzeman. This work was supported by a research grant (CA-11034) from the National Cancer Institute, National Institutes of Health.

REFERENCES

Chinault, A. C. and Carbon, J. (1979). *Gene* **5**, 111–126.
Clarke, L. and Carbon, J. (1980). *Proc. natn. Acad. Sci. U.S.A.* **77**, 2173–2177.
Culbertson, M. R., Charnas, L., Johnson, M. T. and Fink, G. R. (1977). *Genetics* **96**, 745–764.
Hitzeman, R. A., Chinault, A. C., Kingsman, A. J. and Carbon, J. (1979). *In* "Eucaryotic Gene Regulation", (T. Maniatis and C. F. Fox, eds), Vol. 14, pp. 57–68. Academic Press, New York and London.
Hsiao, C. and Carbon, J. (1979). *Proc. natn. Acad. Sci. U.S.A.* **76**, 3829–3833.
Kingsman, A. J., Clarke, L., Mortimer, R. K. and Carbon, J. (1979). *Gene* **7**, 141–153.
Lam, K. and Marmur, J. (1977). *J. Bact.* **130**, 746–749.
Mortimer, R. K. and Hawthorne, D. C. (1975). *In* "Methods in Cell Biology" (D. M. Prescott, ed.), Vol. XI, pp. 221–233. Academic Press, New York and London.
Mortimer, R. K. and Schild, D. (1980). *Microbiol. Rev.* **44**, 519–571.
Parry, E. M. and Cox, B. S. (1971). *Genet. Res., Camb.* **16**, 333–340.
Peterson, J. B. and Ris, H. (1976). *J. Cell Sci.* **22**, 219–242.
Ratzkin, B. and Carbon, J. (1977). *Proc. natn. Acad. Sci. U.S.A.* **74**, 487–491.
Southern, E. M. (1975). *J. molec. Biol.* **98**, 503–517.
Stinchcomb, D. T., Struhl, K. and Davis, R. W. (1979). *Nature, Lond.* **282**, 39–43.
Strathern, J. N., Newlon, C. S. Herskowitz, I. and Hicks, J. B. (1979). *Cell* **18**, 309–319.
Tschumper, G. and Carbon, J. (1980). *Gene* **10**, 157–166.